Diagnosis of Mycobacterium

Amit Singh • Divakar Sharma
Editors

Diagnosis of Mycobacterium

Springer

Editors
Amit Singh
Department of Gastroenterology & HNU
All India Institute of Medical Sciences
New Delhi, India

Divakar Sharma
Department of Microbiology
Maulana Azad Medical College
New Delhi, India

ISBN 978-981-99-5623-4 ISBN 978-981-99-5624-1 (eBook)
https://doi.org/10.1007/978-981-99-5624-1

This Springer imprint is published by the registered company Springer Nature Singapore Pte Ltd.
The registered company address is: 152 Beach Road, #21-01/04 Gateway East, Singapore 189721, Singapore

Dedicated to
My teacher Prof. Sarman Singh for his guidance, wisdom, and inspiration
My mother (late Smt. Somvati Singh), wife (Dr. Arti) and daughter (Anaya) for their love, support, and patience

Foreword

Dr. Amit Singh and Dr. Divakar Sharma deliver a highly useful work in the form of a book entitled ***Diagnosis of Mycobacterium*.** This book contains a collection of chapters on recent developments in the diagnosis of Mycobacterium from various types of samples. The editors have recruited a vast array of tuberculosis specialists to write different chapters that constitute this book. The chapters are structured in such a way that the reader may easily find pearls about tuberculosis diagnostics and its applications to timely manage mycobacterium-related current issues, especially drug resistance. I am sure researchers who are working in the fields of diagnosis of mycobacterium and related research will find this compilation useful and enjoyable.

Laboratory of Molecular Epidemiology
and Evolutionary Genetics
St. Petersburg Pasteur Institute,
St. Petersburg, Russia

Igor Mokrousov

Preface

Recent advancements in the field of diagnostic biology have dramatically increased the complexity of this field. The diagnosis of mycobacterium is widely available and plays an increasing role in the diagnosis and management of tuberculosis. However, proper knowledge for recommending tests for the diagnosis of different forms of tuberculosis is crucial for correct and timely diagnosis. Inappropriate knowledge about the tests leads to delays in diagnosis, misinterpretation of results, and hampers the management of mycobacterial infection, which can contribute to the development of MDR and XDR tuberculosis. This book is the result of 15 years of working experience in the field of tuberculosis research, diagnosis, and challenges. It stands out among other books as it specially focuses on the diagnosis of mycobacterium in humans, pediatrics cases, zoonotic TB, and NTMs. This book includes the most historic diagnosis, i.e., smear microscopy, clinical diagnosis, radiological diagnosis to more advanced microbiological and molecular diagnosis of TB infection. The diagnosis of latent TB infection is also addressed, which is crucial for TB control. Since TB diagnostics are continuously evolving, the book also includes future diagnostic techniques such as nano-biosensor-based diagnosis. The mycobacterium diagnostic methods are still evolving and this book will be helpful to upgrade the knowledge about the relevant diagnostic methods used for the detection of mycobacterium from various clinical samples. Tremendous advances have been made in the field of diagnostics in the last few years, particularly resulting from the introduction of molecular biology tools.

I have had the privilege of working and establishing a lab for the routine microbiological and molecular diagnosis of mycobacterium at the All India Institute of Medical Sciences, New Delhi, and All India Institute of Medical Sciences, Bhopal, under the guidance of eminent clinical microbiologist Prof. Sarman Singh who significantly contributed to the field of mycobacteriology. I was privileged to work under his guidance. We believe that primarily in countries with high TB endemicity, a quick and handy book on the diagnosis of mycobacterium is useful for microbiologists, clinicians, teachers, researchers, and PG and UG students of medicine and microbiology.

I wish to acknowledge the outstanding assistance from all the authors, contributors, colleagues, and physicians in the preparation of this book. Without them, this book may not have been possible. I would also like to thank all the patients and collaborators for their support from AIIMS, New Delhi, ICGEB, New Delhi, AIIMS, Bhopal, and Delhi University for improving and developing newer diagnostics for tuberculosis. Furthermore, I wish to thank all the helpful and kind people from the publishers involved in the production of this book. We hope that our initiative will inspire the interest of our microbiologists, physicians, students, and researchers not only in the solution of TB clinical and laboratory diagnostics dilemmas but also prompt them to present innovative queries on routine and basic diagnosis, nurturing a continuous bidirectional exchange among the realms of health care and those of clinical and basic research.

This book divided into several chapters. The first chapter covers the basics of various diagnostics of mycobacterium and then the clinical, radiological, pathological, biochemical, and hematological-based diagnostics. Moreover, it covers the entire field of mycobacterium diagnosis from microscopy to molecular tests, providing the latest content. However, the reader should consider that this field is still changing rapidly and day-to-day improvements and newer diagnostics will get included. Suggestions for further improvements to be considered in future editions are highly appreciated. Please contact the editor at amit4us@gmail.com, amit.singh@cup.edu.in, divakarsharma88@gmail.com

Enjoy reading this new book!

New Delhi, India

Amit Singh
Divakar Sharma

Contents

Editors and Contributors

About the Editors

Amit Singh is an Assistant Professor of Microbiology at Central University of Punjab, India. He has completed his MS degree from Jiwaji University Gwalior, India, and Doctoral Degree in Clinical Microbiology and Molecular Medicine from AIIMS, New Delhi, India. He did postdoctoral research in the field of virology and mycobacteriology at AIIMS, New Delhi. After completion of postdoctoral training, he joined as Senior Resident in the Department of Microbiology at AIIMS, Bhopal. After SRship he again joined AIIMS, New Delhi, as a Senior Scientist where he was actively involved in molecular diagnosis, providing training and quality control of COVID-19 throughout pandemics. He has been working in the field of diagnosis of infectious diseases for the last 14 years and is an active clinical and molecular microbiologist, mycobacteriologist. His research focuses on the exploration of drug resistance mechanism of *M. tuberculosis* and ESKAPE pathogens through omics approaches. He is also actively involved in the development of point-of-care and molecular diagnostics for infectious diseases. His contribution to this field includes more than 70 peer-reviewed publications and book chapters of international repute. He is an academic editor, guest editor, editorial board member, and reviewer of various international and national journals. Also he is a member of various national and international societies.

Divakar Sharma has completed his MS degree in Biotechnology from Chatrapati Sahu Ji Maharaj University Kanpur, India, and his PhD degree in Biotechnology from the Jiwaji University, Gwalior, India, as well as ICMR-National JALMA Institute for Leprosy and Other Mycobacterial Diseases, Agra, India. He has completed two Post-Doctoral Fellowships research at ICMR-National JALMA Institute for Leprosy and Other Mycobacterial Diseases, Agra, India (2015–2017), and Aligarh Muslim University, Aligarh (2017–2019). After completion of the Post-Doctoral Fellowships, he worked in the Indian Institute of Technology, Delhi, India, as a Principal Project Scientist (2019–2020). From 2021 to 2022, he worked at Maulana Azad Medical Collage, Delhi, as a Research Officer on a project collaborated with John Hopkins University, USA. Presently, Dr. Sharma is a Scientist at the Lady Hardinge Medical College, Delhi, India.

He has been an active researcher in Tuberculosis Research and Antimicrobial Resistance (AMR) on opportunistic infections for the last 15 years. His research focuses on drug resistance exploration of *M. tuberculosis* and opportunistic pathogens through proteomics, genomics, and bioinformatics-based approaches. His contribution to this field includes more than 60 peer-reviewed articles, 16 book chapters, and 1 book of international repute. He is the editorial board member for various visible journals of international repute affiliated to well-known publishers like BMC, Frontiers, Hindawi, and Bentham Science. He is also an expert's panelist member of several international journals. He has delivered several invited lectures and presented various papers in recognized platforms. Apart from that, he is a life member of various societies in India like Proteomics Society of India, Indian Society for Mass Spectrometry, and Association of Microbiologist of India.

Contributors

Rekha Chaubey Department of Pathology, All India Institute of Medical Sciences, New Delhi, India

Shivendra K. Chaurasiya Department of Biological Science and Engineering, Maulana Azad National Institute of Technology, Bhopal, India

Rima Dada Department of Anatomy, All India Institute of Medical Sciences, New Delhi, India

Sudip Kumar Datta Department of Laboratory Medicine, All India Institute of Medical Sciences, New Delhi, India

Bhavna Dhingra Department of Pediatrics, All India Institute of Medical Sciences, Bhopal, India

Chandan Kumar Dubey 3B Blackbio Biotech India Ltd, Bhopal, India

School of Sciences, SAM Global University, Raisen, India

Rekha Dwivedi Department of Neurology, All India Institute of Medical Sciences, New Delhi, India

Praveen Gautam 3B Blackbio Biotech India Ltd, Bhopal, India

Aishwarya Govindaswamy Department of Microbiology, Apollo Proton Cancer Centre, Chennai, India

Anil Kumar Gupta IRCH, All India Institute of Medical Sciences, New Delhi, India

Sudheer Gupta 3B Blackbio Biotech India Ltd, Bhopal, India

Gayathri S. Kamath Department of Zoology, Advanced Centre for Regenerative Medicine and Stem cell in Cutaneous Research (AcREM-Stem), University of Kerala, Thiruvananthapuram, Kerala, India

Alkesh Khurana Department of Pulmonary Medicine, All India Institute of Medical Sciences, Bhopal, India

Amber Kumar Department of Pediatrics, All India Institute of Medical Sciences, Bhopal, India

Bipul Kumar CSIR-Institute of Genomics and Integrative Biology, New Delhi, India

Department of Gastroenterology & HNU, All India Institute of Medical Sciences, New Delhi, India

Fidha Latheef Department of Zoology, Advanced Centre for Regenerative Medicine and Stem cell in Cutaneous Research (AcREM-Stem), University of Kerala, Thiruvananthapuram, Kerala, India

Anvita Gupta Malhotra Department of Microbiology, All India Institute of Medical Sciences, Bhopal, India

Bijina J. Mathew Department of Biological Science and Engineering, Maulana Azad National Institute of Technology, Bhopal, India

Sandeep R. Mathur Department of Pathology, All India Institute of Medical Sciences, New Delhi, India

Anand Kumar Maurya Department of Microbiology, All India Institute of Medical Sciences, Bhopal, Madhya Pradesh, India

Thangarasu Mohanraj Department of Natural Products Chemistry, School of Chemistry, Madurai Kamaraj University, Madurai, Tamil Nadu, India

Shunmuga Nainar Shunmuga Nathan Department of Natural Products Chemistry, School of Chemistry, Madurai Kamaraj University, Madurai, Tamil Nadu, India

Sanjay Singh Negi Department of Microbiology, All India Institute of Medical Sciences (AIIMS), Raipur, Chhattisgarh, India

Sreejith Parameswara Panicker Department of Zoology, Advanced Centre for Regenerative Medicine and Stem cell in Cutaneous Research (AcREM-Stem), University of Kerala, Thiruvananthapuram, Kerala, India

Rupal Rai Department of Biological Science and Engineering, Maulana Azad National Institute of Technology, Bhopal, India

Daisy Rani Rasaily Department of Pathology, Sikkim Manipal University of Medical Science, Gangtok, Sikkim, India

Greeshma Sasikumar Department of Zoology, Advanced Centre for Regenerative Medicine and Stem cell in Cutaneous Research (AcREM-Stem), University of Kerala, Thiruvananthapuram, Kerala, India

Guru Datta Satyarthee Department of Neurosurgery, All India Institute of Medical Sciences, New Delhi, India

Tushar Sehgal Department of Laboratory Medicine, All India Institute of Medical Sciences, New Delhi, India

Piraman Shakkthivel Department of Nanoscience and Technology, Alagappa University, Karaikudi, Tamil Nadu, India

Divakar Sharma Department of Microbiology, Maulana Azad Medical College, New Delhi, India

Kuldeep Sharma Department of Microbiology, All India Institute of Medical Sciences (AIIMS), Raipur, Chhattisgarh, India

Pawan Sharma International Centre for Genetic Engineering and Biotechnology (ICGEB), New Delhi, India

Shadma Siddiqui School of Sciences, SAM Global University, Raisen, India

Amit Singh Department of Gastroenterology & HNU, All India Institute of Medical Sciences, New Delhi, India

Department of Microbiology, Central University of Punjab, Bathinda, India

Anirudh K. Singh School of Sciences, SAM Global University, Raisen, India

Parul Singh Department of Microbiology, All India Institute of Medical Sciences, Gorakhpur, India

Pushpendra Singh Department of Microbiology, All India Institute of Medical Sciences (AIIMS), Raipur, Chhattisgarh, India

Saumya Srivastava Department of Ocular Pharmacology, All India Institute of Medical Sciences (AIIMS), New Delhi, India

Anjana Suresh Department of Zoology, Advanced Centre for Regenerative Medicine and Stem cell in Cutaneous Research (AcREM-Stem), University of Kerala, Thiruvananthapuram, Kerala, India

Prabhakar Tiwari Department of Anatomy, All India Institute of Medical Sciences, New Delhi, India

Shipra Tomar Department of Microbiology, All India Institute of Medical Sciences, Bhopal, Madhya Pradesh, India

Manjari Tripathi Department of Neurology, All India Institute of Medical Sciences, New Delhi, India

Vairathevar Sivasamy Vasantha Department of Natural Products Chemistry, School of Chemistry, Madurai Kamaraj University, Madurai, Tamil Nadu, India

Anjali Yadav Department of Anatomy, All India Institute of Medical Sciences, New Delhi, India

Abbreviations

°C	Degree centigrade
2D	Two dimensional
3D	Three dimensional
ABPA	Allergic bronchopulmonary aspergillosis
ACF	Active case finding
ADA	Adenosine deaminase
AFB	Acid-fast bacilli
Ag85	Antigenic protein-85
ahpC	Alkylhydroxy peroxidase C gene
AI	Artificial intelligence
AIDS	Acquired immunodeficiency syndrome
AIIMS	All India Institute of Medical Sciences
ALIS	Amikacin liposome inhalation solution
AST	Antimicrobial susceptibility test
ATS	American Thoracic Society
ATT	Anti-tubercular treatments
BAL	Bronchoalveolar lavage
BALF	Bronchoalveolar lavage fluid
BCG	Bacille Calmette-Guérin
BDQ	Bedaquiline
BSA	Bovine serum albumin
BSL-3	Bio-safety level-3
BSL-III	Bio-safety level-III
BTB	Bovine tuberculosis
BTS	British Thoracic Society
CA	Chronoamperometry
CAMHB	Cation-adjusted Mueller Hinton broth
CBC	Complete blood count
CBNAAT	Cartridge-based nucleic acid amplification tests
CCT	Chest computed tomography
CCT	Comparative cervical test

CD4 & CD8T	Cluster of differentiation 4 & 8T
CDC	Centre for Disease Control
CECT	Contrast-enhanced computerized tomographic scan
CFP-10	Culture filtrate protein-10
CFT	Caudal fold test
CFU	Colony forming unit
CHIT	Chitosan
CIT	Cervical intradermal test
CLSI	The Clinical and Laboratory Standards Institute
cm	Centimeter
CMI	Cell-mediated immune response
CNPA	Chronic necrotizing pulmonary aspergillosis
CNS	Central nervous system
CO_2	Carbon dioxide
COPD	Chronic obstructive pulmonary disease
COVID-19	Coronavirus disease
CPA	Chronic pulmonary aspergillosis
CRISPER	Clustered regularly interspaced short palindromic repeats
CSF	Cerebrospinal fluid
CT scan	Computerized tomography
CUP	Central University of Punjab
CV	Cyclic voltammetry
CXR	Chest X-ray
Cy5	Cyanine-5
DC-SIGN	Dendritic cell-specific intercellular adhesion molecule-3-grabbing non-integrin
DIC	Disseminated intravascular coagulation
DIG	Digoxigenin
DNA	Deoxyribonucleic acid
DOTS	Directly observed treatment short-course
DPV	Differential pulse voltammetry
DR	Direct repeats
DST	Drug susceptibility testing
DTH	T-cell mediated delayed-type hypersensitivity response
EG	Ethylene glycol
EIS	Electrochemical impedance spectroscopy
ELISA	Enzyme-linked immunosorbent assays
ELISPOT	Enzyme-linked immunospot
EMB	Ethambutol
EPTB	Extrapulmonary tuberculosis
ERS	European Respiratory Society
ESAT-6	Early secretory antigenic target-6
ESCMID	European Society of Clinical Microbiology and Infectious Diseases
ESR	Erythrocyte sedimentation rate

Fc	Ferrocene
FDA	Food and Drug Administration
FDG	Fluorodeoxyglucose
FET	Field effect transistor
FIND	Foundation for Innovative New Diagnostics
FLQ	Fluoroquinolone
FNA	Fine needle aspiration
FNAC	Fine needle aspirate cytology
FRET	Förster resonance energy transfer
GER	Gastro-esophageal reflux
GOPS- 3	Glycidoxypropyltrimethoxysilane
GPL	Glycopeptidolipids
H	Isoniazid
H/O	History of
HIV	Human immunodeficiency virus
hr.	Hour
HRCT	High-resolution computed tomography
HRP	Horseradish peroxidase
Hsp	Heat shock protein
IAP	Indian Association of Pediatrics
ICT	Immunochromatographic test
IDSA	Infectious Disease Society of America
IFN- γ	Interferon-gamma
IgG	Immunoglobulin
IGRA	Interferon gamma release assay
IHC	Immuno-histochemistry
IL-1	Interleukin 1
INH	Isoniazid
inhA	Isonicotinhydratase A
IS*6110*	Insertion sequence element 6110
kasA	Beta-ketoacyl synthases gene
katG	Catalase/peroxidase gene
LAM	Lipoarabinomannan
LAMP	Loop-mediated isothermal amplification
LCR	Ligase chain reaction
LED-FM	LED fluorescence microscopy
LJ	Löwenstein–Jensen medium
LMICs	Low- and middle-income countries
LN	Lymph node
LOC	Lab-on-chip
LoD	Limit of detection
LoQ	Limit of quantification
LPA	Line probe assay
LSCM	Laser scanning confocal microscopy
LSG-NF	Laser-scribed graphene nanofiber

LTBI	Latent tuberculosis infection
MABC	*M. abscessus* complex
MAC	*Mycobacterium avium complex*
MAC-PD	Mycobacterium avium complex pulmonary disease
MALDI-TOF-MS	Matrix-assisted laser desorption/ionization-time of flight mass spectrometry
Man-LAM	Mannose-capped lipoarabinomannan
MB	Methylene blue
MB-7H10	Middlebrook 1H10
MB-7H9	Middlebrook 7H9
MBs	Magnetic beads
MCP-1	Monocyte chemoattractant protein-1
MDR	Multi-drug resistance
MGIT	Mycobacterium Growth Indicator Tube
MIC	Minimum inhibitory concentration
MIRU	Mycobacterial interspersed repetitive units
MIRU-VNTR	Mycobacterium interspersed repetitive unit—variable number of tandem repeats
ML	Machine learning
ml	Milliliter
mm	Millimeter
MMS	Magnetic microspheres
MNPs	Magnetic nanoparticles
MOTT	Mycobacteria other than tubercle bacilli
mPCR	Multiplex PCR
MRI	Magnetic resonance imaging
Mtb/MTB	*Mycobacterium tuberculosis*
MTBC	*Mycobacterium tuberculosis* complex
MTC	*Mycobacterium tuberculosis* complex
NAA	Nucleic acid amplification
NAAT	Nucleic acid amplification test
NALC-NaOH	N-acetyl-l-cysteine sodium hydroxide solution
NaOH	Sodium hydroxide
NCC	Nanocrystalline cellulose
NCCT	Non-contrast CT scan
ndh	NADH dehydrogenase gene
NGS	Next generation sequencer
NK	Natural killer cells
NSCLC	Non-small cell lung cancer
NTEP	National Tuberculosis Elimination Program
NTM	Nontuberculous mycobacteria
NTM-LD	Nontuberculous mycobacterial lung disease
OADC	Oleic acid, albumin, dextrose, catalase
OMPA	Oligomethoxyphenylacetonitrile
PAMAM	Polyamidoamine

PANTA	Polymyxin B, amphotericin B, nalidixic acid, trimethoprim, and azlocillin
PCR	Polymerase chain reaction
PDITC	1,4-Phenylene diisothiocyanate
PEGMA	Poly (ethylene glycol) methacrylate
PET	Positron emission tomography
PHAS	Public Health Action Support
PLWHA	People living with HIV/AIDS
PMSSQ	Polymethylsilsesquioxane
POC	Point-of-care
POCT	Point-of-care test
PPD	Purified protein derivative
pre-XDR	Pre-extensively drug-resistant
PTB	Pulmonary tuberculosis
PZA	Pyrazinamide
QDs	Quantum dots
QFT-G	QuantiFERON-TB Gold
R	Rifampicin
RBCs	Red blood cells
RCTs	Randomized clinical trials
RD	Region of difference
RFLP	Restriction fragment length polymorphism
RGM	Rapid-growing mycobacteria
RIF	Rifampicin
RNA	Ribonucleic acid
RNTCP	Revised National Tuberculosis Control Programme
rpoB	β subunit of bacterial RNA polymerase
RRDR	RIF resistance-determining region
rRNA	Ribosomal RNA
RR-TB	Rifampicin-resistant tuberculosis
SA	Streptavidin
SAMs	Self-assembled monolayers
SARS-CoV-2	Severe acute respiratory syndrome coronavirus 2
SDG	Sustainable Development Goals
SLID	Second-line injectable drugs
SNPs	Single nucleotide polymorphisms
STS	Senior treatment supervisors
SWV	Square wave voltammetry
TB	Tuberculosis
TBL	Lymph node tuberculosis
TBM	Tubercular meningitis
TDR-TB	Total drug resistance tuberculosis
temp	Temperature
THMS	Triple-helix molecular switch
TMB	Tetramethyl benzidine

TNF	Tumor necrosis factor
TNF-α	Tumor necrosis factor-alpha
TST	Tuberculin skin test
TTD	Time to detection
TU	Tuberculin unit
UCNPs	Upconversion Nanoparticles
U-DST	Upfront Universal-Drug Sensitivity Testing strategy
UNION	International Union Against Tuberculosis and Lung Disease
USG	Ultrasonography
UV	Ultraviolet
VNTR	Variable number of tandem repeats
VOBa	Volatile organic biomarkers
WBCs	White blood cells
WGS	Whole genome sequencing
WHO	World Health Organization
XDR	Extensively drug resistance
ZN	Ziehl-Neelsen
zTB	Zoonotic tuberculosis

Chapter 1
Introduction to the Diagnosis of Mycobacterium

Amit Singh, Guru Dutta Satyarthee, and Divakar Sharma

Abstract *Mycobacterium tuberculosis (Mtb)* is a causative agent of tuberculosis. It is slightly curved, rod shaped, and arranged in singles or in groups. It manifests as pulmonary, extrapulmonary, and sometimes miliary disease. It grows very slowly and takes up to 21–42 days to become a visible colony in the Löwenstein-Jensen (LJ) growth media. Cases of tuberculosis are increasing, and timely management is an essential component for the control of the disease, in which laboratory diagnosis plays an important role. Various diagnostics tools have been developed and used for the diagnosis of tuberculosis, drug-resistant tuberculosis, nontuberculosis mycobacterium (NTM), and zoonotic tuberculosis. The available techniques involved more ancient like microscopy, culture to the most advanced molecular tests, matrix-assisted laser desorption/ionization-time of flight-mass spectrometry (MALDI-TOF-MS), and point-of-care (POC) devices (e.g., LAM). These available tools are currently used in the management of tuberculosis. Researchers are also exploring newer, advanced, fast, and cheaper diagnostics for tuberculosis to overcome the current limitations. In the introductory chapter, we outline and summarize the different laboratory tests used for the diagnosis of tuberculosis, which are further elaborated in the dedicated chapter of the book.

Keywords Tuberculosis · LJ culture · Microscopy · GeneXpert · NAAT · Automated culture · Zoonotic tuberculosis · NTM

A. Singh (✉)
Department of Gastroenterology & HNU, All India Institute of Medical Sciences, New Delhi, India

Department of Microbiology, Central University of Punjab, Bathinda, India

G. D. Satyarthee
Department of Neurosurgery, All India Institute of Medical Sciences, New Delhi, India

D. Sharma
Department of Microbiology, Maulana Azad Medical College, New Delhi, India

A. Singh, D. Sharma (eds.), *Diagnosis of Mycobacterium*,
https://doi.org/10.1007/978-981-99-5624-1_1

1.1 Introduction

In 2022, tuberculosis (TB) caused an estimated 1.6 million deaths, including 187,000 deaths among people positive with HIV/AIDS and 1.4 million among HIV negatives, and gave rise to 10.6 million new cases globally, becoming the leading infectious cause of mortality in the world [1]. The number of new cases of rifampicin resistant was reported to be 450,000 in 2021. Approximately three to five million cases are not reported or diagnosed, which might have contributed to the increase in tuberculosis cases globally [1]. The most common manifestation of tuberculosis is known as pulmonary tuberculosis (PTB), which mainly infects the lungs. The bacterium may spread to other organs through a hematogenous route and may cause infection in other organs; this is called extrapulmonary tuberculosis (EPTB) [2]. Depending on the site of infection, various clinical samples were collected. In PTB cases, sputum, induced sputum, bronchoalveolar lavage (BAL), etc. were collected, and for EPTB cases, pleural fluid, pericardial fluid, blood, bone marrow, cerebrospinal fluid (CSF), tissue, pus, swabs, urine, endometrium aspirate, lymph nodes, fine-needle aspiration cytology (FNAC) samples, bronchial secretions, and gastric lavage [3, 4]. Depending on the samples, different diagnostic techniques were used to diagnose tuberculosis.

The diagnosis of TB includes conventional (culture, Ziehl-Neelsen (ZN), and fluorescent stained microscopy) [5], radiological (X-ray, CT scan, MRI) [6], and molecular diagnostics (GeneXpert, polymerase chain reaction (PCR), reverse transcription polymerase chain reaction (RT-PCR), line probe assay (LPA)) [7–11]. Microscopy is a rapid test and may be completed within half an hour to detect acid-fast bacilli (AFB) in smear examination [12]. Since its limit of detection requires a minimum bacterial load of 10,000/ml of clinical sample, it lacks sensitivity and specificity [10, 13, 14]. Further, it does not differentiate between infectious and environmental mycobacteria. The culture method is considered a gold standard for tuberculosis diagnosis as it provides a specific growth identification feature to characterize up to species level and can further be used for drug susceptibility tests [15, 16]. The culture technique is more sensitive and more specific than other conventional diagnostics methods but is time-consuming. *Mycobacterium tuberculosis* grows very slowly, and Löwenstein-Jensen (LJ) culture takes 4–8 weeks, whereas automated liquid culture methods take 2–6 weeks [17]. The molecular test is rapid, sensitive but requires laboratory infrastructure, is expensive, and needs trained manpower to perform it [13]. Further research needs to identify novel biomarkers for the development of rapid, inexpensive, and sensitive point-of-care tests for the effective management of tuberculosis and to achieve World Health Organization's (WHO's) sustainable development goal (SDG) to end TB [1, 18]. The major challenges for the diagnosis of zoonotic tuberculosis [19], TB in pediatric cases and vulnerable populations [9, 20].

1.2 Laboratory Diagnosis of Mycobacterium

Laboratory diagnosis starts with sample collection. The identification of the appropriate site of infection and collecting samples for the diagnosis of mycobacterium are important tasks. The procedure and quality of the sample determine the correct laboratory diagnosis of the disease. The most commonly used sample for *Mtb* culture is sputum from patients with pulmonary TB [4, 17]. For EPTB, various clinical samples were collected depending on the site of infection on sterile container [3, 17]. The samples include tissue, pleural fluid, pericardial fluid, blood, bone marrow, CSF, tissue, pus, swabs, urine, endometrium aspirate, lymph nodes, FNAC samples, bronchial secretions, and gastric lavage. After receiving the samples, proper storage conditions and procedures should be followed to proceed to diagnostics. An appropriately collected sample is the key to the diagnosis of TB [17]. The collected samples were further processed for mycobacterial diagnosis, depending on the facility available. The diagnosis of tuberculosis is divided into conventional, radiological, and molecular methods; immunological diagnosis; and newer diagnosis (Figs. 1.1 and 1.2).

1.3 Conventional Methods

1.3.1 Microscopy

Smear microscopy is the most rapid and a cost-effective method for the detection of tuberculous bacilli [17]. The smears are prepared from processed or direct samples and further stained with ZN stain or fluorescent stained and observed under standard light microscopy and a fluorescent microscope, depending on the staining techniques used. The ZN-stained tubercle bacilli under a light microscope appear as red pink, long, or slightly curved bacilli [17]. The rhodamine-auramine stained bacilli, when observed under a fluorescent microscope, appears yellow or orange under ultraviolet light and bright orange bacilli in pale green background when stained by acridine orange [17]. Some histopathological methods (e.g., immunohistochemistry (IHC)) were also used for the diagnosis of tuberculosis.

1.3.2 Culture

The culture method is more sensitive and to date is considered a gold standard method for the diagnosis of tuberculosis. Culture is also essential for performing drug-susceptibility testing (DST) and genotyping through mycobacterial interspersed repetitive unit-variable number tandem repeat (MIRU-VNTR), restriction fragment length polymorphism (RFLP), or spoligotyping methods [12, 21, 22].

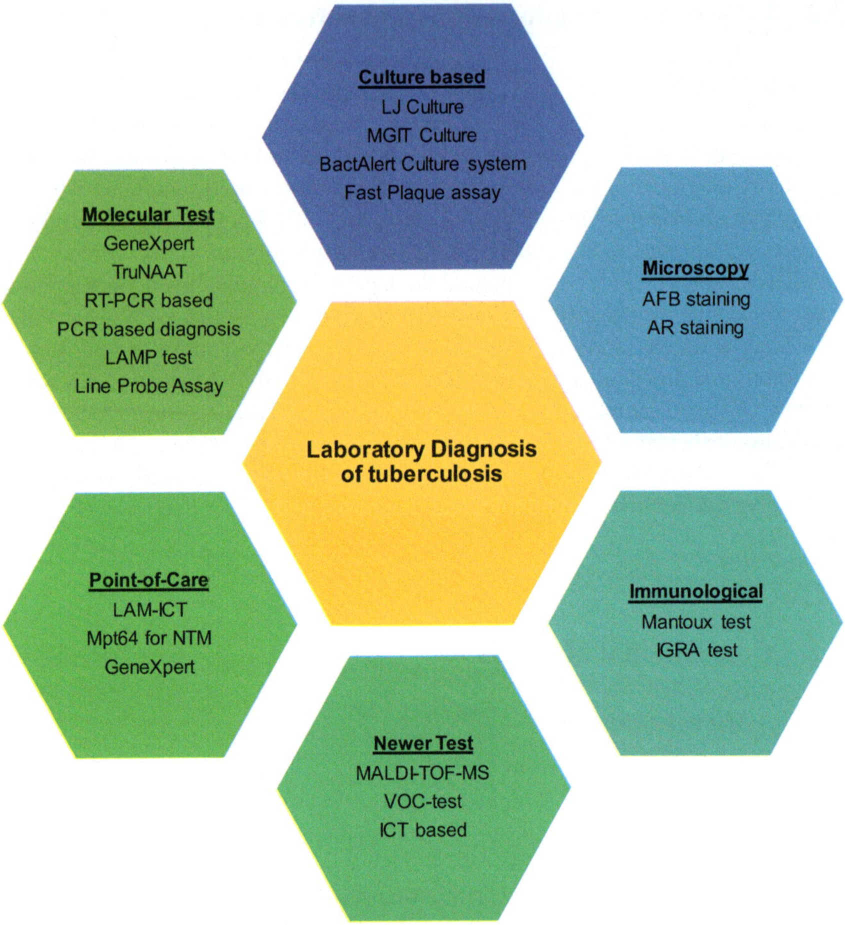

Fig. 1.1 Laboratory diagnosis of mycobacterium

Sterile samples were directly inoculated on culture media, while sputum or nonsterile samples were processed through Patroff's modified methods and then incubated at 37 °C for up to 8 weeks [17]. *Mtb* shows rough, tuff, buff, and creamy colonies in the LJ medium [5, 17]. In the liquid culture medium, the bacilli grow on surface and form cords and wrinkle pellicle. They grow more rapidly in liquid culture media as compared to LJ or other solid media. Automated culture systems (Mycobacteria Growth Indicator Tube (MGIT), BactAlert, VersaTREK, etc.) are more rapid, and growth is detected through the production of fluorescence or a change in color due to the depletion of carbon dioxide (CO_2) in the medium [15, 17, 23]. Further *Mtb* identification in culture is performed through biochemical tests (nitrate reductase, niacin test, arylsulfatase, urease, and tween 80 hydrolysis), molecular tests, or immunochromatographic test (ICT)-based MPT64 strip-based tests (Fig. 1.1) [17].

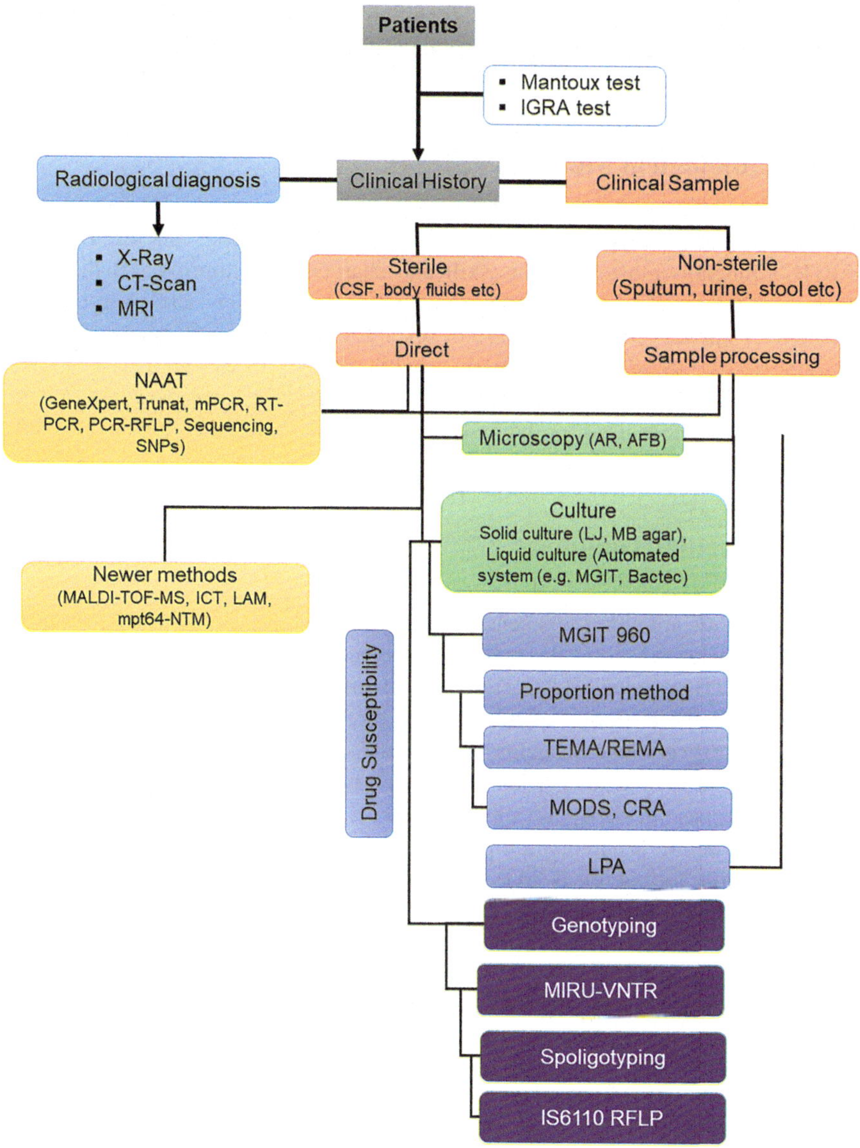

Fig. 1.2 Flow chart of various dianostics methods for the diagnosis of mycobacterium

The bacteriophage-based test was also developed to detect TB in sputum samples [17]. Nonpathogenic mycobacteria (i.e., sensor cells) were used as control bacteria in the test. The phage replicates, infects, and lyses the sensor cells, leaving zones of clearing (holes) in the agar media. The zone of clearing indicates that the patient's sputum contains viable TB bacilli. The test is cheaper, sensitive (only requires 100 bacilli/ml sputum), and fast with a 2-day turnaround time. Its main limitations are it is technically demanding and can be applied for sputum specimens only.

1.4 Radiological Methods

Radiological diagnosis is an important tool for the diagnosis of PTB and EPTB. Clinicians must learn this skill and sharpen it to improve TB diagnosis. Errors in radiograph interpretation can lead to both under- and overdiagnosis. Remember impact of rotation, phase of respiration, exposure, motion, artifact and cofounders like thymic shadow. Radiographic findings in tuberculosis have been well described [24]. Chest X-ray (CXR), CT scan, and MRI are most commonly the radiological methods used for the presumptive diagnosis of tuberculosis (Fig. 1.2).

1.5 Molecular Diagnosis

Earlier methods are either less sensitive, time-consuming, and have a slow growth rate of *Mtb*, whereas molecular methods have become the first choice for the rapid detection of TB. Various Nucleic Acid Amplification Test (NAAT)-based tests were developed for the detection of tuberculosis directly from patient samples. These tests can identify etiological agents and also provide information on drug resistance. The cartridge-based GeneXpert and TruNAAT are the current choice for molecular diagnosis because of their ease of performance, but these are very expensive and need infrastructure. PCR or multiplex polymerase chain reaction (mPCR) is the most common format of NAAT; other amplification techniques include loop-mediated isothermal amplification (LAMP), ligase chain reaction (LCR), strain displacement amplification, CRISPER-Cas, and real-time-based tests. Some molecular tests, like Xpert MTB/RIF, Xpert MTB/XDR, LPA, and real-time-based tests are capable of providing information on drug resistance. Whole-genome sequencing or targeted sequencing methods are also in use for the identification and detection of TB either from samples or culture [10, 11, 23, 25, 26].

1.6 Immunological Methods

Immunological methods used for the immunodiagnostics of TB can provide indirect evidence on the current or past infections/exposure to tubercle bacilli. Tuberculin skin test (TST) has limited application due to cross-reactivity and poor sensitivity [8, 13]. Also, WHO bans all existing serological tests for use in the diagnosis of tuberculosis due to their poor sensitivity and specificity and at the same time encourages identifying novel biomarkers to develop newer immunodiagnostics for the detection of tuberculosis [14, 27].

1.6.1 Antigen Detection

The detection of tuberculosis antigens in patient samples provides direct evidence of TB. Lipoarabinomannan (LAM) is a promising antigen for antigen detection for TB and is used as a point-of-care test for the rapid detection of TB in urine and other samples. LAM is a promising target for antigen detection for TB diagnosis due to its temperature stability and its possibility to be detected in urine [28]. LAM-based assays are included in the WHO-TB diagnosis retooling program.

1.6.2 Antibody Detection

Antibody detection against TB is rapid, cost-effective, simple, and easy to perform, but due to its poor sensitivity and specificity, WHO bans the use of serological tests for tuberculosis diagnosis and instructed researchers to find novel biomarkers for the development of antibody-based detection methods [13].

1.6.3 Tuberculin Skin Test or PPD Test

The PPD (purified protein derivative) test determines previous exposure to TB bacilli or whether the patient received bacillus Calmette-Guérin (BCG) vaccination. The standard recommended Mantoux test, in which 0.1 mL of liquid containing 5 TU (tuberculin units) PPD is injected into the top layers of skin of the forearm and observed for presence or absence and the amount of induration (localized swelling) within 48–72 h after the injection [9].

1.6.4 Interferon-Gamma Release Assays (IGRAs)

IGRA overcomes the problems confronting tuberculin skin test (TST) interpretation. There are two available formats for IGRA: (1) Quantiferon-TB Gold and (2) T Spot-TB test. IGRA assay is based on the ability of TB antigens, such as ESAT-6 and CFP-10, to stimulate the host for the production of interferon-gamma (IFN-γ). The amount of interferon-γ released from the cells is estimated. Results are interpreted both qualitatively (positive, negative, or indeterminate) and quantitatively.

Immunization with BCG does not affect the test result of the IGRA assay because the antigens used are specific to *Mtb* and absent in BCG [29].

1.7 Newer TB Diagnosis

Rapid technological development in the laboratory diagnosis of tuberculosis, especially molecular diagnosis, has reduced the time required for identification and susceptibility testing. Continuous effort is made for the increase of reproducibility, improvement of performance, and cost-effectiveness. Many rapid, portable, and sensitive biosensors, electro-sensors with instant "on-the-spot" interpretation, have been developed for tuberculosis detection based on different biological elements, like DNA, RNA, or protein recognition systems and basic signal transducer principles. The combination of biosensing technology and nanotechnology is very promising. Many point-of-care devices are in development that will further help in enhancing TB detection [13].

Currently, nanoparticles have shown a potential for the diagnosis of infectious diseases. The engineering of biomaterials and nanocarriers represents an exciting opportunity in the field of diagnostics (30). Various innovative and efficient nanodiagnostics have been developed by researchers for infectious diseases, including TB. Laksanasopin et al. [31] developed a smartphone-based POC to diagnose infectious diseases by connecting traditional immunoassay into a smartphone via accessories such as dongles [13, 31, 32].

1.8 Conclusions

This is now exciting era for development of newer TB diagnostics. The development of various advanced biosensors and nanotechnology enhance the possibility to developing rapid, sensitive point-of-care tests. Till now, rapid, high-specific molecular assays for TB detection and drug susceptibility tests cannot replace the standard diagnostic methods (such as culture, microscopy, clinical and radiological assessments, and conventional DST for active TB in PTB and EPTB samples). In this book, we are covering all the diagnostics of *Mycobacterium*, from microscopy to the more advanced point-of-care tests for the detection of tuberculosis.

References

1. WHO. Global tuberculosis report 2022 [Internet]. 2022 [cited 2023 May 5]. https://www.who.int/teams/global-tuberculosis-programme/tb-reports/global-tuberculosis-report-2022.

2. Purohit M, Mustafa T. Laboratory diagnosis of extra-pulmonary tuberculosis (EPTB) in resource-constrained setting: state of the art, challenges and the need. J Clin Diagn Res. 2015;9(4):EE01–6.
3. Sharma SK, Mohan A. Extrapulmonary tuberculosis. Indian J Med Res. 2004;120(4):316–53.
4. Ling DI, Flores LL, Riley LW, Pai M. Commercial nucleic-acid amplification tests for diagnosis of pulmonary tuberculosis in respiratory specimens: meta-analysis and meta-regression. PLoS One. 2008;3(2):e1536.
5. Kubica GP, Dye WE, Cohn ML, Middlebrook G. Sputum digestion and decontamination with N-acetyl-L-cysteine-sodium hydroxide for culture of mycobacteria. Am Rev Respir Dis. 1963;87:775–9.
6. Andreu J, Cáceres J, Pallisa E, Martinez-Rodriguez M. Radiological manifestations of pulmonary tuberculosis. Eur J Radiol. 2004;51(2):139–49.
7. Rufai SB, Singh A, Kumar P, Singh J, Singh S. Performance of Xpert MTB/RIF assay in diagnosis of pleural tuberculosis by use of pleural fluid samples. J Clin Microbiol. 2015;53(11):3636–8.
8. Rufai SB, Singh S, Singh A, Kumar P, Singh J, Vishal A. Performance of Xpert MTB/RIF on Ascitic fluid samples for detection of abdominal tuberculosis. J Lab Physicians. 2017;9(1):47–52.
9. Singh S, Singh A, Prajapati S, Kabra SK, Lodha R, Mukherjee A, et al. Xpert MTB/RIF assay can be used on archived gastric aspirate and induced sputum samples for sensitive diagnosis of paediatric tuberculosis. BMC Microbiol. 2015;15:191.
10. Steingart KR, Schiller I, Horne DJ, Pai M, Boehme CC, Dendukuri N. Xpert® MTB/RIF assay for pulmonary tuberculosis and rifampicin resistance in adults. In: The Cochrane Collaboration, Steingart KR, editors. Cochrane Database of Systematic Reviews [Internet]. Chichester: Wiley; 2014 [cited 2014 Mar 9]. http://doi.wiley.com/10.1002/14651858.CD009593.pub3.
11. Gopinath K, Singh S. Multiplex PCR assay for simultaneous detection and differentiation of Mycobacterium tuberculosis, Mycobacterium avium complexes and other Mycobacterial species directly from clinical specimens. J Appl Microbiol. 2009;107(2):425–35.
12. Singh A, Gopinath K, Singh N, Singh S. Deciphering the sequential events during in vivo acquisition of drug resistance in Mycobacterium tuberculosis. Int J Mycobacteriol. 2014;3(1):36–40.
13. Singh A, Kumar Gupta A, Gopinath K, Sharma P, Singh S. Evaluation of 5 novel protein biomarkers for the rapid diagnosis of pulmonary and extra-pulmonary tuberculosis: preliminary results. Sci Rep. 2017;7:44121.
14. Singh S, Katoch VM. Commercial serological tests for the diagnosis of active tuberculosis in India: time for introspection. Indian J Med Res. 2011;134(5):583.
15. Bardarov S, Dou H, Eisenach K, Banaiee N, Ya S u, Chan J, et al. Detection and drug-susceptibility testing of M. tuberculosis from sputum samples using luciferase reporter phage: comparison with the Mycobacteria Growth Indicator Tube (MGIT) system. Diagn Microbiol Infect Dis. 2003;45(12573551):53–61.
16. Katoch VM. New generation methods for drug susceptibility testing for tuberculosis. Indian J Tuberc. 2008;55(2):61–3.
17. Kent PT, Kubica GP. Public health mycobacteriology: a guide for the level III laboratory. Washington, DC: U.S. Department of Health and Human Services; 1985. 207p.
18. WHO-STAG TB. Strategic and technical advisory group for tuberculosis (STAG-TB) [Internet]. Geneva: World Health Organization (WHO); 2011. p. 1–40. https://cdn.who.int/media/docs/default-source/documents/tuberculosis/stag-tb-report-11.pdf?sfvrsn=7eccaf2_2&download=true.
19. Rahman MT, Sobur MA, Islam MS, Ievy S, Hossain MJ, El Zowalaty ME, et al. Zoonotic diseases: etiology, impact, and control. Microorganisms. 2020;8(9):1405.
20. Chauhan LS, Arora VK, Central TB Division, Directorate General of Health Services, Ministry of Health and Family Welfare, Indian Academy of Pediatrics. Management of pediatric tuberculosis under the revised National Tuberculosis Control Program (RNTCP). Indian Pediatr. 2004;41(9):901–5.

21. Brudey K, Driscoll JR, Rigouts L, Prodinger WM, Gori A, Al-Hajoj SA, et al. Mycobacterium tuberculosis complex genetic diversity: mining the fourth international spoligotyping database (SpolDB4) for classification, population genetics and epidemiology. BMC Microbiol. 2006;6:23.
22. Singh A, Gopinath K, Sharma P, Bisht D, Sharma P, Singh N, et al. Comparative proteomic analysis of sequential isolates of Mycobacterium tuberculosis from a patient with pulmonary tuberculosis turning from drug sensitive to multidrug resistant. Indian J Med Res. 2015;141(1):27–45.
23. Rufai SB, Kumar P, Singh A, Prajapati S, Balooni V, Singh S. Comparison of Xpert MTB/RIF with line probe assay for detection of rifampin-monoresistant Mycobacterium tuberculosis. J Clin Microbiol. 2014;52(6):1846–52.
24. Bomanji JB, Gupta N, Gulati P, Das CJ. Imaging in tuberculosis. Cold Spring Harb Perspect Med. 2015;5(6):a017814.
25. Kumar P, Pandya D, Singh N, Behera D, Aggarwal P, Singh S. Loop-mediated isothermal amplification assay for rapid and sensitive diagnosis of tuberculosis. J Infect. 2014;69(6):607–15.
26. Sankar MM, Kumar P, Munawwar A, Kumar M, Singh J, Singh A, et al. Usefulness of multiplex PCR in the diagnosis of genital tuberculosis in females with infertility. Eur J Clin Microbiol Infect Dis. 2013;32(3):399–405.
27. Singh S, Singh J, Kumar S, Gopinath K, Balooni V, Singh N, et al. Poor performance of serological tests in the diagnosis of pulmonary tuberculosis: evidence from a contact tracing field study. PLoS One. 2012;7(7):e40213.
28. Patel VB, Singh R, Connolly C, Kasprowicz V, Zumla A, Ndungu T, et al. Comparison of a clinical prediction rule and a LAM antigen-detection assay for the rapid diagnosis of TBM in a high HIV prevalence setting. PLoS One. 2010;5(12):e15664.
29. Schepers K, Mouchet F, Dirix V, De Schutter I, Jotzo K, Verscheure V, et al. Long-incubation-time gamma interferon release assays in response to purified protein derivative, ESAT-6, and/or CFP-10 for the diagnosis of Mycobacterium tuberculosis infection in children. Clin Vaccine Immunol. 2013;21(2):111–8.
30. Gupta AK, Singh A, Singh S. Diagnosis of tuberculosis: nanodiagnostics approaches. In: NanoBioMedicine [Internet]. Springer; 2019 [cited 2023 May 5]. p. 261–83. https://www.ncbi.nlm.nih.gov/pmc/articles/PMC7122355/
31. Laksanasopin T, Guo TW, Nayak S, Sridhara AA, Xie S, Olowookere OO, et al. A smartphone dongle for diagnosis of infectious diseases at the point of care. Sci Transl Med. 2015;7(273):273re1.
32. Gupta AK, Srivastava S, Singh A, Singh S. Development of rapid immuno-diagnostic test for the early detection of tuberculosis. Int J Mycobacteriol. 2016;5(Suppl 1):S114–5.

Chapter 2
Clinical and Radiological Diagnosis of Tuberculosis

Amber Kumar and Bhavna Dhingra

Abstract Accurate and timely diagnosis of tuberculosis (TB), including drug sensitivity patterns, is essential for the patient's health and the population in contact with them. In the diagnosis of tuberculosis, strong clinical suspicion required in populations having constitutional symptoms is defined as:

- Persistent unexplained fever documented ≥2 weeks and/or
- Unremitting cough ≥2 weeks and/or
- Unexplained documented loss of weight >5% in the last 3 months despite adequate nutrition

The diagnosis of TB is based on the isolation of pathogens. Paucibacillary cases and inaccessible sites of infection are the limitations in isolating mycobacteria. Although tuberculosis can affect any organ, the lung remains the most commonly involved organ. Recently, the World Health Organization (WHO) has replaced the term smear-negative TB with clinically diagnosed TB. Often, tuberculosis is first suggested by an imaging study, especially in inaccessible sites. An upfront chest X-ray (CXR) is recommended in all suspected cases of pulmonary tuberculosis. In the era of universal drug sensitivity testing, all suspected cases of tuberculosis based on clinical or radiological findings must be submitted for microbiological confirmation, at least for rifampicin and isoniazid sensitivity. In cases of nonspecific shadows on chest X-rays, antibiotics should be given to look for the resolution of shadows. Contrast-enhanced computerised tomographic scan (CECT) of the chest and flexible bronchoscopy may be required for confirming the diagnosis. In diagnosed tuberculosis cases, imaging is often required to assess disease extent, evaluate therapeutic response, or look for residual disease for treatment extension. Imaging is also crucial for guiding aspiration, biopsies and therapeutic drainage of fluid collections.

Keywords Tuberculosis · Microbiological · Diagnosis · Clinical · Radiological · Intrathoracic · Lymph node CNS · Abdominal · Osteoarticular

---

A. Kumar · B. Dhingra (✉)
Department of Pediatrics, AIIMS Bhopal, Bhopal, India
e-mail: bhavna.pediatrics@aiimsbhopal.edu.in

A. Singh, D. Sharma (eds.), *Diagnosis of Mycobacterium*,
https://doi.org/10.1007/978-981-99-5624-1_2

2.1 Introduction

The lifetime risk of developing tuberculosis is about 5–10% in people infected with *Mycobacterium tuberculosis*, primarily within 5 years of infection [1]. Microbiological diagnosis of tuberculosis is essential for tuberculosis control but is often ignored and neglected. Acid-fast smear of an expectorated sputum specimen was the first investigation to confirm microbiologically discovered over a century ago. We are still relying on it with a few modifications. Unfortunately, around 50% of cases are smear-negative, resulting in the under-diagnosis of a substantial percentage of the world's tuberculosis cases. The late diagnosis of these cases in an advanced stage of the disease leads to significant transmission in the community by that time. The risk of tuberculosis is increased in any patient with the following demographic characteristics: residence in a high-TB prevalence state; contact with an active pulmonary case of tuberculosis; residence in a setting with community exposure, such as a nursing home, guest house, community shelter, prison in which occult exposure to tuberculosis might have occurred; or underlying immunosuppression due to HIV infection, other illnesses, or the administration of immunosuppressive drugs, which reduce immunity, particularly biologicals [2–4]. Failure to consider tuberculosis in the differential diagnosis of at-risk patients leads to delays in the diagnosis and administration of antitubercular drugs [5, 6]. Continued education and community awareness about tuberculosis are essential in this situation.

2.2 Available Diagnostic Tests for TB [7, 8]

- Smear: it is commonly performed on sputum and gastric aspirates. Use singularly and if only rapid nucleic acid amplification test (NAAT) is not available. It is positive in only 10–15% of children.
- Culture: liquid culture takes 2–6 weeks time in isolation, while in vitro drug sensitivity testing takes more time. The time involved in sensitivity testing decreases when used with line probe assay (LPA).
- Nucleic acid amplification test (NAAT): NAAT-like cartridge-based (CBNAAT) or line probe assay (LPA) significantly increases the yield. As we have moved towards an upfront universal drug sensitivity testing (U-DST) strategy by culture and/or molecular methods for all TB cases (new as well as retreatment cases), microbiological diagnosis must be attempted in all presumptive cases, especially in drug-resistant (DR) TB suspects. Collect good-quality samples for testing—smear, NAAT, LPA and culture. Always try to send for all investigations simultaneously. Where sputum is not available for examination, use alternative specimens, like gastric aspirate (GA), induced sputum (IS), bronchoalveolar lavage (BAL) and appropriate extrapulmonary TB (EPTB) specimens. Recent advances in technology has made a complete shift in the paradigm of writing

simple diagnosis like Mycobacterium tuberculosis (MTB) to MTB with its sensitivity pattern. The approved rapid NAAT tests have made it possible to simultaneously detect the presence of MTB and resistance to rifampicin, even in smear-negative samples directly. The rapid and direct detection of resistance to isonicotinic acid hydrazide (INH) is also possible but less universally (for smear-positive specimens). Hence, the current strategy focuses more on microbiological diagnosis at all ages and for all types of system involvement.

- Chest radiography: nonspecific findings and variable interpretation are its limitations, but it is very useful in sputum-negative probable TB cases, especially in children where TB is paucibacillary, and getting adequate good-quality specimens is challenging.
- TB skin test and interferon-gamma release assays (IGRAs): IGRAs and skin tests neither diagnose active TB disease nor predict the risk of developing active TB disease in infected individuals. IGRAs are technically complex and costly investigations. Blood samples must be processed within 8–30 h (preferably within 6 h) of collection while the white blood cells (WBC) remain viable. Errors during the collection and transport of blood specimens or while performing and interpreting the assay can decrease the accuracy of IGRAs. C-Tb is the next-generation skin test for the detection of tuberculosis, which is an easy-to-use point-of-care specific test (based on ESAT-6 and CFP-10 antigens of MTB and unaffected by bacille Calmette-Guérin (BCG) vaccination status. It is an easy replacement for the purified protein derivative (PPD) test (tuberculin skin test (TST)), as administered via intradermal injection of 0.1 mL using the Mantoux technique. An induration of ≥5 mm indicates an infection with *Mycobacterium tuberculosis*. The safety profile is comparable to TST (local skin reactions).

This chapter will review approaches to the clinical and radiological diagnosis of pulmonary and extrapulmonary tuberculosis.

2.3 Clinical Suspicion of Tuberculosis

Detailed history and examination are essential for the symptom characterisation of tuberculosis, followed by focused investigations. The proper characterisation of constitutional symptoms increases the positive predictive value of these symptom complexes in diagnosing tuberculosis [7, 8]:

- Fever for 2 weeks or more: persistent (>2 weeks) and unexplained fever (>38 °C). reported by a person/guardian or objectively recorded at least once.
- Persistent (>2 weeks), non-remitting cough.
- A definite weight loss (≥5% loss in the past 3 months) or failure to gain weight in the past 3 months despite adequate nutrition.
- Persistent peripheral lymphadenopathy (>2 cm in size).
- Insidious onset meningitis.
- Gibbus of the spine.

- Exposure to an infectious TB case (patient with lung or airway involvement): the risk is highest among contacts of infectious sputum-positive cases. In symptomatic patients, the exposure may be from sputum-negative index cases with unrecognised coexisting pulmonary involvement. Contact with a person with active tuberculosis within the last 2 years (more significant in the last 3 months) may be considered contact positive.

2.4 Intrathoracic TB

In all suspected cases of intrathoracic TB based on constitutional symptoms, chest X-ray and sputum examination should be done at the first visit. However, in children under 5 years, getting the sputum specimen is challenging, so a chest X-ray should be done first, followed by gastric aspirate or induced sputum, depending on the chest X-ray findings (Fig. 2.1 *Algorithm*). In adults, both the X-ray and sputum results should be interpreted simultaneously in the presence of constitutional symptoms of TB.

2.5 Chest Radiology

A chest X-ray is an essential tool for diagnosing pulmonary TB and osteoarticular TB. Clinicians must know the TB-suggestive shadows on chest X-rays. Mistakes in interpretation can lead to both under- and over-diagnosis. Poor-quality images, like significantly rotated, expiratory, and under- and over-exposed films, should not be reported for diagnosing tuberculosis. Remember the cofounders, like a thymic shadow. Radiographic findings in tuberculosis have been well described [6, 14–18].

2.5.1 Radiological Manifestations of TB

- Highly suggestive imaging findings for pulmonary TB.
  - Hilar/mediastinal lymphadenopathy.
  - Chronic fibro cavitary disease.
  - Miliary pattern.
- Non-specific radiology for pulmonary TB.
  - Consolidations.
  - Non-homogenous opacity.
  - Ground glassing.
  - Thin-walled cavities.

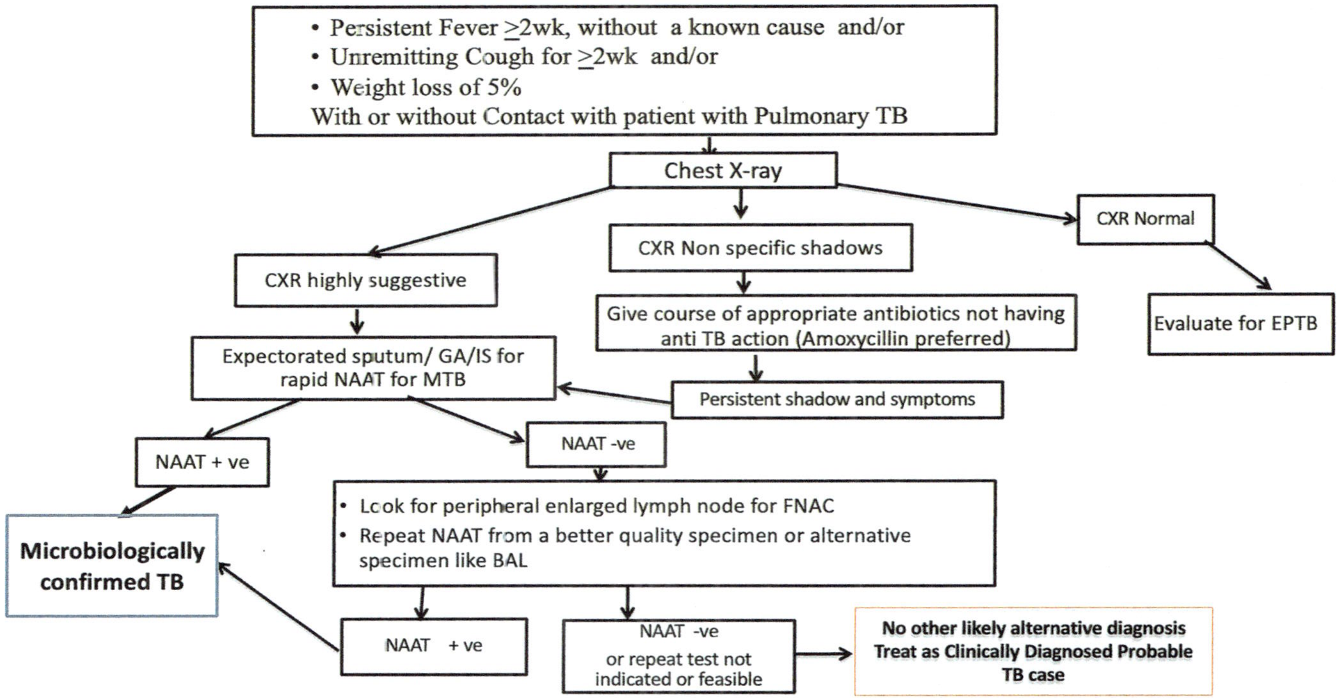

Fig. 2.1 Algorithm for an approach to paediatric pulmonary tuberculosis [7, 8]. (Algorithm adapted from IAP NTEP Pediatric TB guidelines 2019 & 2021, central TB division MOHFW New Delhi India)

- Lymphadenopathy on X-ray.
- Direct signs:
  - Sharply demarcated and lobulated density obliterating the hilar point.
  - Other structures making up the mediastinum and hilum can obscure it.

Indirect signs:

- The contour and calibre of the trachea.
- Compression of the tracheobronchial tree or splaying of the carina.

Remember in the current era of universal DST, all cases with radiological findings must have tests for the microbiological confirmation of MTB and its resistance to rifampicin (and INH). Antibiotics should always have been given in non-specific shadows, and bacteriological tests are pursued only in cases with persistent pneumonia. In case the need is felt, a specialist opinion may be taken. Investigations like CT chest and flexible bronchoscopy may be needed to establish the diagnosis (Figs. 2.2 and 2.3).

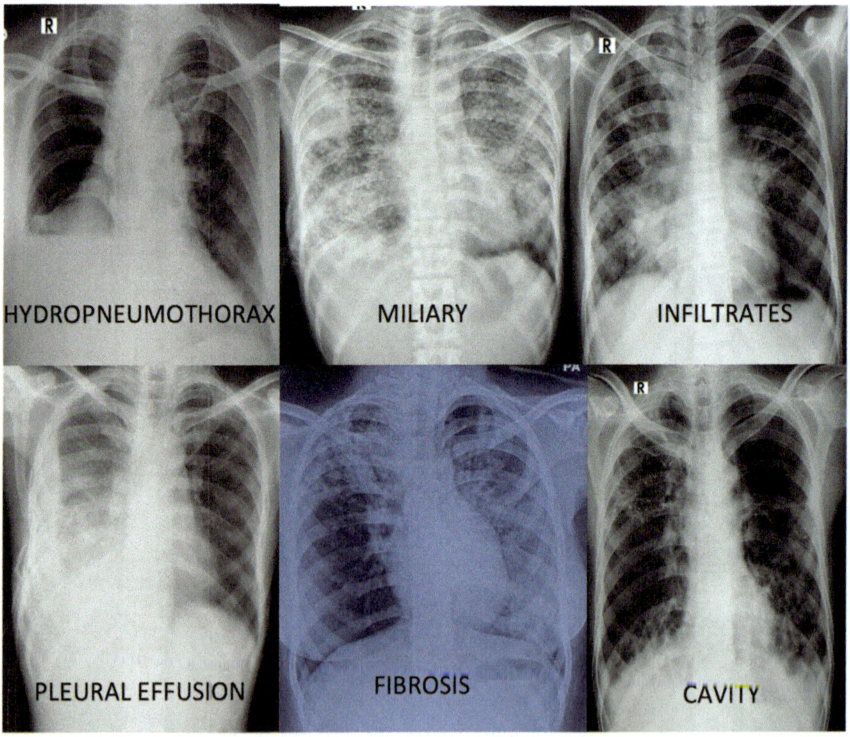

Fig. 2.2 Chest X-ray of intrathoracic tuberculosis showing varied presentations: hydropneumothorax (**a**), miliary pattern both lung fields (**b**), infiltrate right lung and left upper and lingular lobes (**c**), pleural effusion (**d**), fibrosis right upper and left lung (**e**), cavitary lesion (**f**)

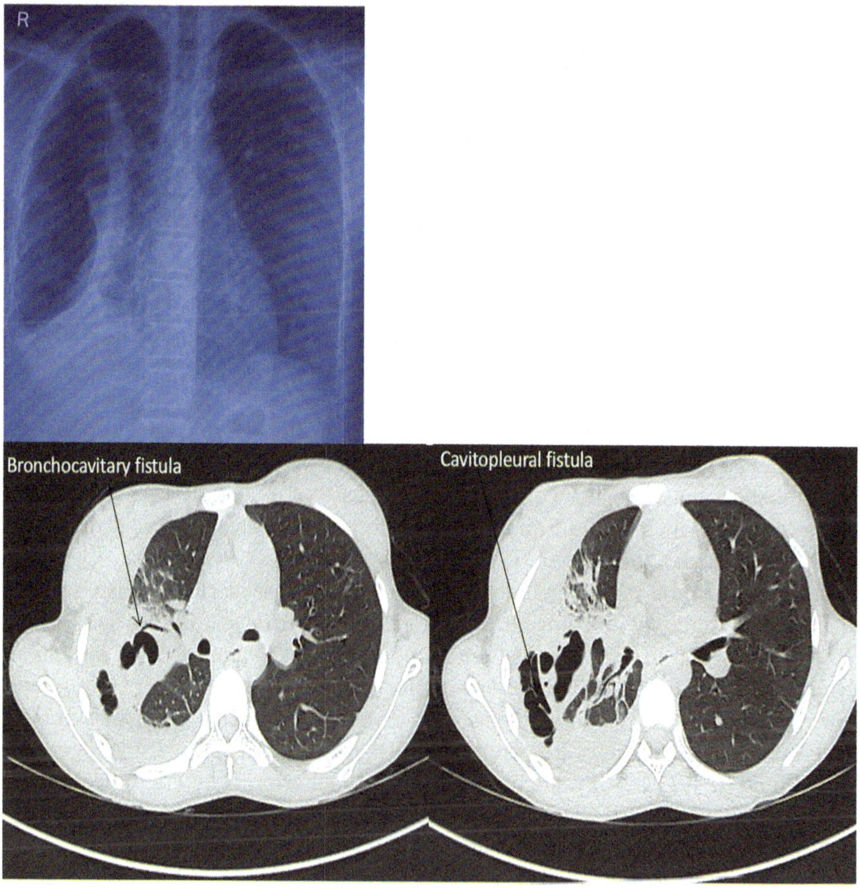

Fig. 2.3 Chest X-ray showing cavitary consolidation of the right lung in a 12-year-old girl. NCCT chest of the same patient showed cavitary consolidation in the right lung with broncho-cavitary and cavitopleural fistula

2.6 Lymph Node Tuberculosis (TBL)

Progressive enlargement of lymph nodes (cervical/axillary/inguinal regions), size >2 cm, matted lymph node, chronic sinus with constitutional symptoms and/or history of TB contact is suggestive of TB lymphadenitis. Non-response to oral antibiotics is also indicative of investigation for TB lymphadenitis. TBL should not be suspected with a small (<2 cm) lymph node persisting >1 year without systemic symptoms and a history of scalp lesions/skin lesions. Waxing and waning lymphadenitis because of recurrent cough and cold are against the diagnosis of TB lymphadenitis. Essential investigations are fine-needle aspiration cytology (FNAC), acid-fast bacilli (AFB) smear, cartridge-based nucleic acid amplification test (CBNAAT (Xpert MTB/RIF) and chest X-ray to look for mediastinal lymph nodes with ultrasonography (USG) abdomen to look for abdominal LN. Optional investigations are lymph node biopsy, if diagnosis is uncertain and CECT neck and chest if mediastinal widening in chest X-ray and diagnosis uncertain on aspiration [7, 8].

2.7 CNS Tuberculosis

CNS tuberculosis includes tubercular meningitis and/or tuberculomas. Both are associated with high morbidity and mortality if treatment is delayed [7, 8].

CNS tuberculosis should be suspected in a patient having a combination of fever with one or more of the following: headache, vomiting, irritability/lethargy/drowsiness, malaise, anorexia, weight loss, stagnation/regression of developmental milestones in children, seizures, and weakness of the arms/legs. Symptoms usually have more than 5–7 day duration with insidious onset. Fever may be absent in those only with tuberculomas. A history of contact with a case of infectious TB is essential to suspect TBM. The history of treatment for primary complex or other forms of TB must be checked.

The examination should include an assessment of the sensorium (Glasgow Coma Scale) and signs of meningeal irritation (neck stiffness, Kernig's sign and Brudzinski's sign). CNS examination including fundus examination of eye should be done for papilledema, optic atropy and/or choroid tubercles. Peripheral lymphadenopathy and chest examination are also essential to look for TB elsewhere.

Essential investigations are complete blood count (CBC), blood sugar, cerebrospinal fluid (CSF) cytology and biochemistry, Mycobacteria Growth Indicator Tube (MGIT) culture, NAAT (CBNAAT/Xpert MTB-Rif/TruNAAT/UltraNAAT), aerobic culture, HIV, CT head with contrast (MRI brain preferred), CXR, and gastric lavage/induced sputum in patients where CXR is abnormal. Desirable investigations are USG abdomen and CECT chest/abdomen to look for extra-cranial sites of infection.

2.8 Abdominal Tuberculosis

Abdominal tuberculosis may present in one or more of the following forms [7–9]:

- Intestinal tuberculosis: small bowel, large bowel.
- Nodal tuberculosis: mesenteric, retroperitoneal.
- Peritoneal tuberculosis: ascitic/wet type and dry type.
- Visceral tuberculosis: liver, spleen, pancreas, or as a part of disseminated tuberculosis.

Abdominal tuberculosis may present in isolation or in association with other forms of tuberculosis. Abdominal tuberculosis should be suspected in a patient having the presence of two or more of the following symptoms for >2 weeks: fever, recurrent or chronic abdominal pain, abdominal distension, abdominal mass, chronic diarrhoea, anorexia, unexplained weight loss, features of subacute intestinal obstruction. Other less common symptoms are upper or lower GI bleeding, jaundice, gastric outlet obstruction, constipation, chronically discharging perianal fistula, obstructive jaundice, dysphagia or odynophagia. The patient may or may not have systemic features of fever, weight loss or anorexia. Isolated recurrent or chronic pain without other symptoms is usually not due to TB. History should be asked for the involvement of other sites, as well as similar illnesses in the past, H/O ATT intake in the past, H/O contact for TB, and H/O ATT intake in the family (current or in the last 2 years), including HIV status.

- Intrathoracic: cough with or without sputum, shortness of breath, noisy breathing.
- Central nervous system (CNS): altered behaviour, altered sensorium, seizures, vomiting, headache, neck pain.
- Tuberculous lymphadenitis (LN TB): cervical swelling or swelling anywhere else in the body.
- Osteoarticular TB: back pain, stiffness in the back, chronic joint swelling.
- Non-healing or chronically discharging sinus.

Examination findings

- Anthropometry: malnutrition.
- General examination: pallor, icterus, lymphadenopathy, pedal oedema.
- Per abdomen: abdominal distension, ascites, doughy feel of the abdomen.
- Hepatomegaly/splenomegaly/hepatosplenomegaly.
- Palpable lump (matted mesenteric lymph nodes).
- Abdominal distension with a hyper-resonant note on percussion with increased or diminished bowel sounds (gaseous due to obstruction).
- Examination of other systems according to symptoms.
- Red flags for the diagnosis of abdominal TB with chronic abdominal pain.

The following USG findings are suggestive of abdominal TB:

- Abdominal LN with central necrosis, conglomerate lymph nodes, peripheral enhancement.
- Loculated ascites, mesenteric thickening, Caked omentum, multiple SOLs in the liver or spleen.
- Abdominal lymphadenopathy (SAD >15 mm) with other abnormal findings.

Diagnostic workup of abdominal TB is a multimodality evaluation, including a combination of clinical examination, laboratory, radiology, endoscopy, microbiology and histopathology. Do not investigate for TB or start anti-tuberculosis therapy (ATT) in case of recurrent/chronic abdominal pain without red flag signs and recurrent/chronic abdominal pain with sub-centimetric mesenteric lymphadenopathy.

2.9 Osteoarticular TB

Depending on the site involved, it can be divided into three types [7, 8].

2.9.1 *Pott's Spine*

This is the commonest among osteoarticular TB, almost 50% of osteoarticular TB cases. Presenting features are insidious onset back pain (thoracic is the commonest, followed by lumbar/cervical) for more than 6 weeks and/or localised/referred root pain. Constitutional symptoms of TB (fever/weight loss) may be associated with it. Examination may be done to look for local tenderness or deformity. In advanced disease, neurological complications like paraparesis (in 20–50 cases), cauda equina syndrome or paraspinal muscle wasting may develop. X-ray spine is the first line of investigation that may show end plate erosions, joint space narrowing or collapse, decreased vertebra height, and paravertebral soft tissue shadow. In the early stage, the X-ray spine may be normal. The next imaging is MRI spine (preferable) and CT scan (if MRI is not feasible). Imaging findings may include marrow oedema, bony destruction (intervertebral disc/adjacent vertebral bodies/opposing end plates),or abscess in the prevertebral, paravertebral and/or epidural area. In cases of abnormal chest X-ray, sputum or gastric aspirate (GA) samples is to be sent for NAAT. FNAC for smear and NAAT are indicated in enlarged peripheral lymph nodes.

2.9.2 *TB Dactylitis (Short Bones)*

The swelling of the short tubular bones of the hands and feet is known as dactylitis. The proximal phalanx or metacarpals of the index/middle/ring fingers are commonly affected. Multiple or consecutive bone involvement is common in children, compared to a single bone in adults. It may present without pyrexia or signs of inflammation. X-ray of involved bones may show a diaphyseal expansile lesion. Periosteal reaction is uncommon in contrast to pyogenic. Healing is by gradual sclerosis. X-ray film of the chest is also indicated to look for pulmonary focus. Submit sputum or GA for NAAT if CXR is abnormal. FNAC for smear and NAAT are indicated in enlarged peripheral lymph nodes.

2.9.3 TB Arthritis

It commonly affects large joints. The hip/knee is the commonest weight-bearing joint. Insidious onset joint pain and swelling is the presenting feature. TB arthritis is characteristically mono-articular. Associated pulmonary or lymph node TB should be looked for. Classical X-ray picture is a triad of peri-articular osteoporosis, peripherally located osseous erosion and gradual joint space narrowing, also known as Phemister's triad. In the early stages of synovitis and arthritis, radiographic features may include joint space widening due to joint effusion. Bony ankylosis is uncommon in tubercular arthritis. USG/MRI of the joint is the next line of investigation. X-ray film of the chest is also indicated to look for pulmonary focus. In cases of abnormal chest X-ray, sputum or GA for NAAT is to be sent. FNAC for smear and NAAT are indicated in enlarged peripheral lymph nodes.

2.10 Summary of Imaging Findings

See Table 2.1.

2.11 Learning Points

1. Tuberculosis infection is not a tubercular disease. One-third of the world's population is infected, but only a few develop active diseases, depending on host immunity and the virulence of the organism.
2. Transmission of tuberculosis is by droplet infection. Coughers are the most infectious, especially patients with parenchymal cavitary disease. People with close contact and immunodeficient patients are at higher risk of infection.
3. Diagnosis of tuberculosis is by identification of the organism. Radiology with constitutional symptoms is supportive and very useful in diagnosing paucibacillary cases and in children where obtaining sputum is challenging.
4. Proper characterisation of constitutional symptoms results in an increased positive predictive value of this symptom complex in diagnosing tuberculosis.
5. Identifying tuberculosis-suggestive X-ray findings is very important in diagnosing smear-negative cases.
6. We are moving towards the upfront universal drug sensitivity testing strategy (U-DST) to diagnose drug resistance at the beginning of treatment to prevent the transmission of drug-resistant strains in the community and amplify drug resistance in index cases.
7. Pulmonary tuberculosis without treatment progresses to cause lung destruction and death, but effective timely diagnosis and appropriate treatment can save most of these cases with good outcomes.

Table 2.1 Summary of imaging findings of all forms of tuberculosis [10–17]

Thoracic	
Primary TB	
Lymphadenopathy	Usually unilateral; commonly paratracheal, hilar or subcarinal; usually >2 cm in diameter; can feature "rim-sign": Enhancing the rim of granulomatous tissue and a low-attenuation necrotic Centre
Parenchymal involvement	Ghon focus: Dense consolidation, preferentially in lower and middle lobes, may heal with residual calcified granuloma, Ghon focus + hilar lymphadenopathy = Ghon's complex
Miliary TB	Pulmonary hematogenous dissemination: 1–4 mm granulomata, randomly distributed throughout the lungs
Pleural effusion	Commonly on the side of the primary pulmonary focus, may be complicated by broncho-pleural fistula or empyema necessitans, can result in pleural thickening and calcification, may lead to fibrothorax
Post-primary pulmonary TB	
Parenchymal involvement	Poorly defined consolidation; predominant in apical and posterior segments of upper lobes and superior segments of lower lobes; cavitations are common; endobronchial spread results in "tree-in-bud" appearance; cavities can progress to TB empyema, broncho-pleural fistula, empyema necessitans and pulmonary arterial pseudoaneurysm
Airway involvement	Airway narrowing with irregular wall thickening; luminal obstruction resulting in lobar collapse, hyperinflation, and/or tree-in-bud configuration
Pleural effusion	Less common than in primary TB; usually small, associated with parenchymal disease
Cardiac	*Pericardial effusion and pericardial constriction*
Abdominal	
Lymphadenopathy	Abdominal LN with central necrosis, conglomerate lymph nodes, peripheral enhancement; abdominal lymphadenopathy (SAD >15 mm) with other abnormal findings
Peritoneal	Loculated ascites, mesenteric thickening, caked omentum, multiple spaces occupying lesions in the liver or spleen; peritoneal enhancement common; fibrotic type: Omental and mesenteric masses
Gastrointestinal	Commonest at the ileo-cecal junction, may extend to terminal ileum and cecum, concentric thickening of the bowel in CECT
Renal	Parenchymal nodules or abscesses, urinary epithelial thickening or enhancement, urinary strictures, hydronephrosis, may lead to renal cortical thinning, scarring and calcifications
Musculoskeletal	
Tuberculous spondylitis (Pott's disease)	Most common in thoracic spine > lumbar spine; anterior vertebral body adjacent to the end plate is the usual starting point, followed by an intervertebral disc; associated paravertebral abscesses are common
Tuberculous arthritis	Monoarticular, affecting large weight-bearing joints; imaging findings: Osteopenia, synovitis, soft-tissue swellings, marginal erosions, cartilage destruction

(continued)

Table 2.1 (continued)

Tuberculous dactylitis	The proximal phalanx or metacarpals of the index/middle/ring fingers are commonly affected. Multiple or consecutive bone involvement is common in children, compared to a single bone in adults. It may present without pyrexia or signs of inflammation. X-ray of involved bones may show a diaphyseal expansile lesion. Periosteal reaction is uncommon in contrast to pyogenic. Healing is by gradual sclerosis
Central nervous system	
Tuberculous meningitis	Abnormal meningeal enhancement exudates predominantly in basal cisterns, which may lead to deep infarcts, hydrocephalus, cranial nerve involvement; spinal meningitis: Subarachnoid space obliteration, matting of nerve roots, nodular and linear intradural enhancement
CNS Parenchymal	Tuberculomas: May be solitary, multiple or military; CT and MRI: Round or lobulated masses with peripheral oedema, ring-like or homogeneous enhancement, mid-line shift, lipid (cell wall) and lactate (anaerobic glycolysis) pick in MRS

References

1. Comstock GW, Cauthen GM. In: Reichman LB, Hershfield ES, editors. Tuberculosis. A comprehensive international approach, vol. 66. New York: Marcel Dekker; 1993. p. 23–48.
2. Mohan AK, Cote TR, Block JA, et al. Tuberculosis following the use of etanercept, a tumour necrosis factor inhibitor. Clin Infect Dis. 2004;39:295–9.
3. Centres for Disease Control and Prevention. Tuberculosis associated with blocking agents against tumor necrosis factor- alpha – California, 2002–2003. MMWR Morb Mortal Wkly Rep. 2004;53:683–6.
4. Keane J, Gershon S, Wise RP, et al. Tuberculosis associated with infliximab, a tumor necrosis factor alpha-neutralizing agent. N Engl J Med. 2001;345:1098–104.
5. Okur E, Yilmaz A, Saygi A, et al. Patterns of delays in diagnosis amongst patients with smear-positive pulmonary tuberculosis at a teaching hospital in Turkey. Clin Microbiol Infect. 2006;12:90–2.
6. Ellis SM. The spectrum of tuberculosis and non-tuberculous mycobacterial infection. Eur Radiol. 2004;14(Suppl. 3):E34–42.
7. Draft updated IAP RNTCP Ped TB guidelines 2019, Central TB division MOHFW, New Delhi; 21.
8. Sharma SK, Ryan H, Khaparde S, Sachdeva KS, Singh AD, Mohan A, et al. Index-TB guidelines: guidelines on extrapulmonary tuberculosis for India. Indian J Med Res. 2017;145(4):448–63.
9. World Gastroenterology Organisation (WGO). Practice guideline - digestive tract tuberculosis; 2021.
10. Geng E, Kreiswirth B, Burzynski J, et al. Clinical and radiographic correlates of primary and reactivation tuberculosis: a molecular epidemiology study. J Am Med Assoc. 2005;293:2740–5.
11. Skoura E, Zumla A, Bomanji J. Imaging in tuberculosis. Int J Infect Dis. 2015;32:87–93.
12. Burrill J, Williams CJ, Bain G, Conder G, Hine AL, Misra RR. Tuberculosis: a radiologic review. RadioGraphics. 2007;27:1255–73.
13. Bhalla AS, Goyal A, Guleria R, Gupta AK. Chest tuberculosis: radiological review and imaging recommendations. Indian J Radiol Imaging. 2015;25:213–25.
14. Bomanji JB, Gupta N, Gulati P, Das CJ. Imaging in tuberculosis. Cold Spring Harb Perspect Med. 2015;5:a017814.
15. Van Dyck P, Vanhoenacker FM, Van den Brande P, De Schepper AM. Imaging of pulmonary tuberculosis. Eur Radiol. 2003;13:1771–85.

16. Nachiappan AC, Rahbar K, Shi X, Guy ES, Barbosa EJM Jr, Shroff GS, et al. Pulmonary tuberculosis: role of radiology in diagnosis and management. Radiographics. 2017;37:52–72.
17. Jeong YJ, Lee KS. Pulmonary tuberculosis: up-to-date imaging and management. Am J Roentgenol. 2008;191:834–44.

Chapter 3
Pathology-Based Diagnosis of Tuberculosis

Prabhakar Tiwari, Anjali Yadav, Rekha Dwivedi, Rekha Chaubey, Daisy Rani Rasaily, Manjari Tripathi, Sandeep R. Mathur, and Rima Dada

Abstract *Mycobacterium tuberculosis (Mtb)* is a life-threatening infectious pathogen that causes tuberculosis (TB). The MTBC complex comprises eight species, including *Mtb*, *M. cannettii*, *M. bovis*, *M. africanum*, *Bacillus Calmette-Guerin (BCG)*, *M. caprae*, *M. microti*, and *M. pinnipedi*. The methods used for TB laboratory diagnosis are constantly changing to produce more rapid, sensitive, less expensive, and accurate results. Currently, acid-fast staining remains a fundamental component of the pathological diagnostic algorithm, enabling morphological diagnosis. Respiratory specimens like sputum, bronchial aspirates, bronchoalveolar lavage fluid (BALF), and biopsy samples are stained and examined under a microscope based on the clinical manifestation of tuberculosis. Mycobacteria staining dyes include crystal violet, arylmethane dyes, carbolfuchsin dye, and Ziehl-Neelsen staining. Also, these staining methods generally exhibit high sensitivity and specificity. Immunohistochemistry (IHC), which recognizes released mycobacterial antigens like MPT64 on formalin-fixed tissue biopsies, is a technique for making an etiological diagnosis of TB in a histological specimen. These antigens have been detected primarily in *Mtb* and are not typically found in nontuberous mycobacteria. TB can be differentiated by considering various factors, including endoscopy, clinical manifestations, and pathological examinations. In this chapter, we provide a brief overview of the numerous diagnostic techniques used to evaluate pathological and histological differential diagnoses based on pathology-driven examinations. These techniques utilize a variety of specimens that have been extensively explored for TB diagnosis thus far. Furthermore, we discuss into the future the prospects and

P. Tiwari (✉) · A. Yadav · R. Dada
Department of Anatomy, All India Institute of Medical Sciences, New Delhi, India

R. Dwivedi · M. Tripathi
Department of Neurology, All India Institute of Medical Sciences, New Delhi, India

R. Chaubey · S. R. Mathur
Department of Pathology, All India Institute of Medical Sciences, New Delhi, India

D. R. Rasaily
Department of Pathology, Sikkim Manipal University of Medical Science, Gangtok, Sikkim, India

A. Singh, D. Sharma (eds.), *Diagnosis of Mycobacterium*,
https://doi.org/10.1007/978-981-99-5624-1_3

significance of these histopathological techniques, which hold the potential to diagnose TB with increased accuracy and sensitivity.

Keywords Tuberculosis · Pathology · Diagnosis · *Mycobacterium tuberculosis (Mtb)* · Microscopy · Staining

3.1 Introduction

Mycobacterium tuberculosis (Mtb) is estimated to infect approximately one-third of the global population. The histological diagnosis of tuberculosis (TB) has been an important area in the field of anatomical pathology for an extended period. Currently, approximately 10–15% of *Mtb* infections are classified as extrapulmonary TB (EPTB). The rates are increasing in parallel with the rise in HIV incidence. Globally, It has been estimated that there were around 4,50,000 incident cases of multidrug-resistant tuberculosis (MDR-TB) or rifampicin-resistant TB (RR-TB) in 2021, representing a 3.1% increase from the 4,37,000 cases reported in 2020 [1]. Due to the limitations of TB diagnosis using staining methods or other approaches, the sensitivity and specificity of different diagnosis techniques vary. Different lab methods used for the diagnosis of Tuberculosis included Ziel-Neelson staining for mycobacterial assessment, mycobacterial culture, and molecular and serological techniques, histomorphological analysis emerges as the most feasible and viable tool for the diagnosis of tuberculosis (TB) [2–4]. Histopathological investigation appears to be an efficient approach to diagnosing TB for certain individuals in specific cases. Granulomatous and nongranulomatous reactions like foamy macrophages or mycobacterial spindle cell pseudotumors can be observed in mycobacterial infections, including TB. The presence of TB bacilli has been confirmed in mycobacterial culture using molecular and serological techniques [5]. ZN staining can be employed to establish the relationship between tissue reaction and mycobacterial infection as beaded bacilli are frequently detected in the necrotic zone of tissue specimens. The ZN stain, with a sensitivity ranging from 0 to 44% for this infection, exhibits a relatively low sensitivity for detecting *Mtb*. In addition, a diagnosis usually required the presence of at least ~104 bacilli per slide [6, 7]. Furthermore, this stain has limited diagnostic relevance and is inadequate for the pathological evaluation of *Mtb* antigens in the tissue due to its mechanism and relatively low sensitivity and specificity, as mentioned earlier. Therefore, current approaches such as immunohistochemistry (IHC) are well-established methods for detecting mycobacterial antigens and have found particular use in research initiatives. This method relies on the generation of various polyclonal and monoclonal antibodies in response to tissue antigens. It has been shown that this approach can detect mycobacterial antigens with a sensitivity range of 64–100%. The positive tissue staining technique for *Mtb* has been examined in only a limited number of research studies [8]. While recent textbooks do not offer a standardized approach for morphological evaluation concerning the IHC staining or ZN staining of granulomatous reactions, pathology reference books only provide brief discussions on these procedures. Recent studies on the variable morphology of tuberculosis bacteria emphasize the significance of differentiating various morphological attributions associated with TB antigens in tissue using different staining techniques. Considering the limitations and available tools, in this chapter, we discuss the details of pathological tissue staining methods and their microscopic examinations, immunohistochemistry (IHC), and microbial culturing approaches. Also, we discuss the clinical manifestations of tuberculosis in the pulmonary and extrapulmonary regions. Subsequently, we delve into the differential diagnosis of tuberculosis and the scope of these approaches for future investigations.

3.2 Histopathological Techniques for the Diagnosis of TB

Mtb has the ability to mimic numerous infectious diseases. It is imperative to identify and use the proper diagnostic tools. Also, patients who have latent tuberculosis, HIV, and alcoholism comprise a large pool, *which* significantly hinders efforts to manage the disease. The common histological techniques utilized for the pathological diagnosis of TB are depicted in Fig. 3.1. However, detailed descriptions of each method are provided below.

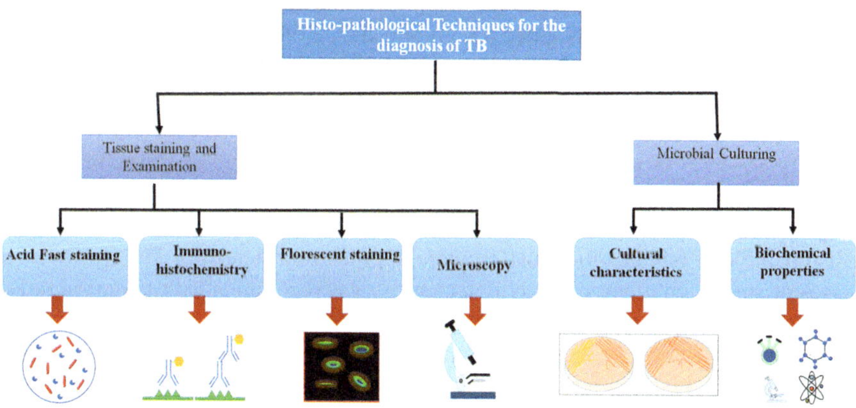

Fig. 3.1 Methods utilizing histopathology for the detection of active TB: this figure depicts the various pathological techniques for the diagnosis of TB using different staining and detection methods

3.2.1 Tissue Staining Methods

In staining procedures, ZN staining is the most commonly used technique for the detection of acid-fast bacteria, particularly mycobacteria. It plays an important role in enhancing the accuracy of TB diagnosis. This approach is considered easy, rapid, and cost-effective for the identification of mycobacteria. However, it is not entirely specific as it can also bind and/or detect other nontuberculous mycobacteria. ZN staining is commonly used in histological sections, contributing to the enhancement of diagnostic accuracy. By utilizing ZN staining to recognize beaded bacilli in tissue samples, it is possible to establish a connection between the tissue reaction and mycobacterial infection. These beaded bacilli are more frequently found in the necrotic zone. The sensitivity range of ZN staining for this infection is reported to be between 0 and 44%, indicating its limited effectiveness for detecting *Mtb* for diagnostic purposes in terms of both sensitivity and specificity [7].

3.2.2 Microscopic Examination

Typically, both ZN stain and mycobacterium culture are used under the microscope for the clinical detection of *Mtb*. Histological examinations of specimens have revealed the characteristic feature of TB lymphadenitis. The histological identification of TB has been a serious issue in anatomical pathology for a very long time. Histological examinations of specimens have revealed the characteristic feature of TB lymphadenitis which included a high prevalence of giant cells (88%), the presence of caseation (84%), an abundance of epithelioid cells (80%), and the occurrence of granulomas with caseation (68%), lymphocytes (31%), and a smaller population of histiocytes (4%). Extrapulmonary Mtb infections account for approximately 10–15% of overall cases, around 15–20% in immunocompetent patients, and approximately 50% in HIV patients [9–11]. Conventional ZN and fluorescence techniques are limited in their ability to diagnose TB due to their low sensitivity. Histopathology remains the most accurate approach for detecting TB. However, in cases of suspected tuberculous lymphadenitis, it is recommended to confirm negative histopathology results with a more sensitive and targeted method, such as polymerase chain reaction (PCR) or immunohistochemistry (IHC), before reporting them [12, 13].

3.2.3 Immunohistochemistry (IHC)

Various staining techniques have limited diagnostic utility and a poor ability to determine the pathophysiological nature of *Mtb* antigens in the tissue. This inadequacy is primarily attributed to their mechanism, as well as their comparatively low sensitivity and specificity [14]. IHC is a method that has gained widespread acceptance for its use in research projects looking for mycobacterial antigens. This method is based on the staining of mycobacteria in tissue using various polyclonal and monoclonal antibodies. According to numerous studies, its sensitivity ranges from ~64 to 100% for the identification of mycobacterial antigen by IHC, proving both the involvement of granulomatous tissue and the immunolocalization of *Mtb* antigen. Although these techniques are more expensive and require more time, they produce more accurate results than acid-fast staining [8]. In addition, polyclonal antibodies can produce results that are not specific because of cross-reactions with other nontuberculous mycobacteria. Also, due to their potential for cross-reaction activity with other nontuberculous mycobacteria, polyclonal antibodies may provide less or nonspecific results [13].

3.2.4 *Mycobacterial Culturing*

This method is considered a gold standard for confirming the presence of an active TB infection. Also, it can effectively distinguish between different subspecies of mycobacteria and assess their susceptibility to various antibiotics. In approximately 60–70% of respiratory specimens with positive cultures, acid-fast bacilli (AFBs) can frequently be detected by microscopy. The sensitivity and specificity of TB culture for pulmonary TB (PTB) forms are approximately 80–85% and around 98.5%, respectively [15]. However, culturing needs biosafety level-3 (BSL-3) laboratories and is costly and time-consuming. Detecting pathogen excretion and determining the likelihood of transmission require the use of viable samples, which also allows for ongoing research. It is essential to assess the infection rate both before and after implementing additional precautions, like vaccination. It is also important to evaluate the efficacy of vaccines in preventing TB infection. After culture isolation, the microbe can be identified using either colony traits, biochemical testing, or nucleic acid recognition techniques [16].

3.3 Pathological Manifestations of TB

The pathophysiology of TB is characterized by the formation of a lesion consisting of inflammatory cells, such as lymphocytes, macrophages, epithelioid cells, lymphocytes, and, in some cases, Langhans multinucleated giant cells. Necrosis in caseating tubercles is also a hallmark feature of Tuberculosis. There may also be the presence of nonnecrotizing granulomatous encapsulation [17]. According to international guidelines for TB management, individuals who have an unexplained cough persisting for 2 weeks or longer or unexplained abnormalities on chest radiographs are recommended to undergo a diagnostic evaluation of TB [1]. In addition, the patient's underlying medical conditions should be determined, especially diabetes and HIV infection, which raise the risk of TB. The advancement in the medical history or genetic history of TB or latent tuberculosis infection (LTBI) must be recorded throughout the examination [18]. Patients who do not take their prescribed medication as directed or do not receive adequate treatment for their TB condition are more likely to relapse or develop drug resistance [19]. All patients' first isolates of *Mtb* should undergo drug susceptibility testing (DST) to check for resistance to the first-line anti-TB drugs isoniazid (INH), ethambutol (EMB), rifampin (RIF), and pyrazinamide (PZA). A patient is diagnosed with MDR-TB disease if the organisms are resistant to at least the two most effective first-line anti-TB drugs, isoniazid and rifampicin [20]. A patient is diagnosed with extensively drug-resistant tuberculosis (XDR-TB) if their TB isolate is resistant to INH, RIF, any fluoroquinolone, and at least one of three injectable second-line medications (i.e., amikacin, kanamycin, or capreomycin) [21]. Based on these medical assessments, PTB and EPTB are the two main forms of clinical presentation of TB.

3.3.1 Pulmonary TB (PTB)

Globally, PTB continues to be a serious issue for public health. PTB, also known as TB infection, predominantly affects the lungs. Patients with relevant clinical symptoms, as defined by the World Health Organization (WHO) consolidated guidelines, are facing challenges due to population aging and the increasing prevalence of MDR-TB. It also highlighted the urgency of current research to identify novel approaches for diagnosing TB and MDR-TB, as well as to advance TB control. A combination of techniques, including nucleic acid amplification, sputum microscopy, culture in solid and liquid media, and chest X-ray (CXR), should be used to achieve a prompt and precise/accurate diagnosis of PTB. Chest computed tomography (CCT) or histological analysis of biopsy samples as well as novel molecular diagnostic techniques are recommended for patients with smear-negative PTB, clinically diagnosed TB, and/or MDR-TB [22].

3.3.2 Extrapulmonary TB (EPTB)

EPTB is developing in any organ or tissue outside of the lungs. It can affect various anatomical sites, including the lymph nodes, pleura, abdomen, genitourinary system, skin, bones, joints, and meninges. In the case of EPTB, various clinical specimens, including urine, cerebrospinal fluid (CSF), pleural fluid, pus, and or biopsy specimens, may be submitted for analysis instead of sputum. The diagnosis of EPTB poses significant challenges due to the diverse clinical samples obtained from different organs. This variability in sample types makes access difficult and can decrease the sensitivity of the diagnostic tests [23]. It has been shown that both PTB and EPTB can show symptoms that overlap with other diseases. In cases where

EPTB is suspected, the gold standard for laboratory confirmation of TB disease is the culture examination of the specimen. The growth of bacteria is necessary for conducting DST and genotyping to establish appropriate treatment strategies [19]. Positive cultures for *Mtb* in extrapulmonary specimens confirm the diagnosis of tuberculosis [24]. Generally, tissue biopsy produces more positive culture findings than fluid aspiration. When the outcomes of the polymerase chain reaction (PCR) tests, culture, and biopsy histology are coupled, diagnostic accuracy might rise even further [23]. In cases where a positive culture is not available, the diagnosis of TB disease can be performed based on clinical signs and symptoms.

The detailed diagnoses for pathological evaluations of PTB and EPTB are summarized in Table 3.1 [13, 25].

Table 3.1 Pathological approaches for the diagnosis of PTB and EPTB

Types of samples used	Sensitivity and specificity	Advantages	Disadvantages
Sputum smear, smears made from fluid, fine-needle aspiration (FNA) smears, biopsy	Very low and high	Easy, inexpensive, fast, grading of disease	Species identification, viable bacteria & drug susceptibility not detected
Sputum smear, smears made from fluid, FNA smears, biopsy	Low and high	Slightly better sensitivity than ZN staining	Expensive and complex, species identification, feasible bacteria and drug susceptibility not detected.
Tissue and fluid samples	Low and very high	Gold approach, differentiating different species of mycobacteria, viable bacteria & drug susceptibility detection	Multifaceted, costly, lengthy, low sensitivity, need BSL3 facilities
Urine, cerebral fluid (CSF), pleural fluid, pus, or biopsy specimens	High and very high	Fast, robotic, identify species, vigorous disease,drug susceptibility	Costly, radioactive exposure, technically challenging
Sputum smear, tissue from FNA or biopsy and fluid samples	High and high	Easy, fast, vigorous, detect mycobacterialspecies	Cannot detect viable bacteria, drug sensitivity not detected
Tissue and biopsy sample	High and low	Helps in delineating from other granulomatous pathologies to a certain extent	Relentless, unusual lesion in HIV, necessary surgical facility, sinus tract and fistula formation

3.4 Differential Diagnosis of TB

PTB or EPTB can be challenging to diagnose due to its similarity to other diseases. When encountering a nodule on imaging, it is imperative to consider alternative diagnoses, like lung cancer and tuberculoma. The characteristics of the nodule and the structures around it, particularly the existence of pleural indentation or notching, are used to make the diagnosis. Currently, the drawback of imaging-based diagnosis is that some tuberculomas may exhibit the same characteristics as lung adenocarcinomas. Therefore, obtaining pathological or bacteriological evidence using appropriate techniques for tuberculosis differential diagnosis is critically needed [26–28]. The details of various infections that may mimic the infection of PTB and EPTB are summarized in Table 3.2. In brief, the following various differential diagnoses of tuberculosis are observed.

Table 3.2 The different infections or diseases that resemble PTB and EPTB

PTB	EPTB
• Sarcoidosis	• Sarcoidosis
• Blastomycosis	• Blastomycosis
• Tularemia	• Tularemia
• Actinomycosis	• Actinomycosis
• *M. avium*-intracellular	• Hidradenitis suppurativa
• *M. chelonae*	• Eosinophilic granuloma
• *M. fortuitum*	• *M. avium*-intracellular
• *M. gordonae*	• *M. chelonae*
• *M. kansasii*	• *M. fortuitum*
• *M. marinum*	• *M. gordonae*
• *M. xenopi*	• *M. kansasii*
Non Small Cell Lung Cancer	• *M. marinum*
	M. xenopi
	• Endemic syphilis
	• Erythema induratum (nodular vasculitis)
	• Erythema nodosum
	• Leishmaniasis leprosy
	• Cat scratch disease
	• Syphilis
	• Syringoma
	• Rheumatoid arthritis

3.4.1 Actinomycosis

Actinomycosis is a chronic to subacute bacterial infection caused by filamentous, gram-positive, non-acid-fast, and anaerobic to microaerophilic bacteria. It is distinguished by the infection's progressive spread, suppurative and granulomatous inflammation, the development of many abscesses, and the appearance of sinus tracts that could expel sulfur granules. These features are all present simultaneously.

3.4.2 Aspergillosis

Aspergillosis is caused by a type of mold(fungus) Chronic Pulmonary Aspergillosis is a long term infection of the lung usually but not exclusively caused by Aspergillus fumigatus. It mainly affects people who have weakened immune systems, as well as those who have underlying diseases, such as asthma or cystic fibrosis; have previous lung disease; or take long-term corticosteroids. Unlike tuberculosis, aspergillosis is not transmissible from person to person. However, the signs and symptoms of aspergillosis can resemble those of TB.

3.4.3 Bronchiectasis

Permanent dilatation of the bronchi and bronchioles is caused by the degeneration of elastic and mucosal tissues, which is frequently caused or exacerbated by chronic necrotizing bronchiole infection. Another cause of bronchiectasis could be TB. Nontuberculous bronchiectasis may resemble tuberculosis with necrotizing granulomas. Clinical and prognostic manifestations of posttubercular and nontubercular bronchiectasis differ markedly.

3.4.4 Fungal Pneumonia

Histoplasma capsulatum, *Coccidio desimmitis*, *Blastomyces dermatitidis*, and other endemic fungal diseases can infect both healthy hosts and people with weakened immune systems. Distinction must be made between the two, fungal pneumonia and tubercular pneumonia. Patients with inherited or acquired abnormalities in the host immune system are also more likely to contract pneumonia from opportunistic fungal pathogens like *Candida, Aspergillus*, and *Muco*r species. Therefore early detection of the causative organism is essential for timely correct treatment modalities.

3.4.5 Histoplasmosis

Histoplasma capsulatum is a fungus that has two different forms: yeast and mycelium. It grows as yeast inside mammals' bodies and survives at their normal temperature. Histoplasmosis and tuberculosis may coexist prior to HIV infection in HIV patients. Because of this, histoplasmosis symptoms and signs might be flawed for TB symptoms and vice versa, which can delay diagnosis and treatment leading complications and mortality.

3.4.6 Lung Abscess

Lung abscess due to tuberculous causes may mimic those of nontuberculous origin. The delay in differentiating can be fatal, especially in the case of aspiration lung abscess.

3.4.7 Nocardiosis

Nocardia is a typical saprophyte that spreads disease through direct cutaneous implantation or inhalation. It causes opportunistic lung infection in patients undergoing transplantation, chemotherapy, or immunosuppression or using steroids. Nocardiosis and PTB may both coexist and exhibit acid-fast bacilli in HIV-positive people. For the detection and diagnosis of such coinfections, rapid and sensitive techniques like PCR are recommended.

3.4.8 Non-small Cell Lung Cancer (NSCLC)

Recent studies have associated multiple diseases, including tuberculosis, with lung infections, indicating that they may be responsible for developing lung cancer. Therefore, the diagnosis of malignancy can be delayed in patients with a prior history of tuberculosis who present with radiological evidence of a new lesion.

3.4.9 Sarcoidosis

Sarcoidosis often manifests as a noncaseating epithelioid cell granuloma, although both sarcoidosis and TB are granulomatous diseases. Due to the high frequency of TB and the obvious clinico-radiological similarities between the two conditions, these patients get recurrent antitubercular treatment (ATT) as lung damage continues to progress [29].

3.5 Future Prospective for Pathology-Based Diagnosis of TB

Effective global TB control depends on the quick and accurate detection of all kinds of tuberculosis (TB), either PTB or EPTB. Clinicians who manage patients suspected of having active TB should conduct investigations using sputum smear microscopy and culture, adhering to TB diagnostic guidelines. It is essential that their diagnostic practices bring into line with the recommended methods for TB diagnosis. Modern techniques now make it feasible to identify active TB in people who have negative AFB sputum smear results and to quickly and accurately identify multiple drug-resistant *Mtb* strains from respiratory specimens [30]. Such novel methods are anticipated to be able to overcome some of the structural limitations of existing TB diagnostics, although future study is still required. The point-of-need testing technique encounters logistical, financial, and technical barriers in the majority of intermediately burdened TB settings, including the Indian population and globally. When attempting to integrate new technology and create fresh diagnostic procedures, it is imperative to take these limits into account [11]. Currently, a major concern for the capacity of TB laboratories is limited worldwide to dealing with the diagnostic challenges associated with PTB, EPTB, HIV-related TB, and MDR-TB [25]. The World Health Organization (WHO) and the National TB Control Program in endemic countries prioritize the expansion of laboratory networks and the strengthening of laboratory services. Recognizing the importance of integrating innovative, evidence-based diagnostic techniques into standard clinical practice, the WHO has emphasized the need for accelerated integration. In settings with limited resources, the Foundation for Innovative New Diagnostics (FIND) has actively promoted the research and evaluation of innovative diagnostic methods. Presently, pathological diagnostic research for PTB or EPTB showing a high degree of accuracy and specificity, it may offer a wide spectrum of diagnostic tests for PTB or EPTB. Also the pipeline of novel diagnostic instruments alreay expanded rapidly over the last 10 years [31]. Additionally, efficacy assessments and effectiveness studies are needed to determine the viability and sustainability of tuberculosis (TB) diagnostics in resource-constrained, TB-endemic countries. Pathological scores have proven to be accurate in the diagnostic setting in a number of investigations. It is critical to evaluate the baseline laboratory preparedness in order to determine the perceived needs, barriers, and facilitators for the implementation of novel TB diagnostic technologies in the future.

3.6 Conclusion

A rapid pathological diagnostic test for PTB and/or EPTB is required and it should be precise, quick, simple to use, long-lasting, and economical. In this histopathologic diagnosis, a prompt transfer from research laboratories to programmatic-level implementation is necessary to hasten the adoption of new technologies. In conclusion, more key research evidence is needed for accurate pathological diagnosis and evaluation of PTB and EPTB with high specificity and sensitivity for PTB and EPTB. This will yield important insights to enhance healthcare services.

References

1. WHO consolidated guidelines on tuberculosis: module 2: screening: systematic screening for tuberculosis disease [Internet]. [cited 2023 May 19]. https://www.who.int/publications-detail-redirect/9789240022676.
2. Attallah AM, Abdel Malak CA, Ismail H, El-Saggan AH, Omran MM, Tabll AA. Rapid and simple detection of a Mycobacterium tuberculosis circulating antigen in serum using dot-ELISA for field diagnosis of pulmonary tuberculosis. J Immunoassay Immunochem. 2003;24(1):73–87.
3. Pedersen JS, Clarke I, Mills J. Improved detection of mycobacteria species in formalin-fixed tissue sections. Histopathology. 2011;59(5):993–1005.
4. Mustafa T, Wiker HG, Mfinanga SGM, Mørkve O, Sviland L. Immunohistochemistry using a Mycobacterium tuberculosis complex specific antibody for improved diagnosis of tuberculous lymphadenitis. Mod Pathol. 2006;19(12):1606–14.
5. Logani S, Lucas DR, Cheng JD, Ioachim HL, Adsay NV. Spindle cell tumors associated with mycobacteria in lymph nodes of HIV-positive patients: "Kaposi sarcoma with mycobacteria" and "mycobacterial pseudotumor". Am J Surg Pathol. 1999;23(6):656–61.
6. Ulrichs T, Lefmann M, Reich M, Morawietz L, Roth A, Brinkmann V, et al. Modified immunohistological staining allows detection of Ziehl-Neelsen-negative Msycobacterium tuberculosis organisms and their precise localization in human tissue. J Pathol. 2005;205(5):633–40.
7. SomosKövi A, Hotaling JE, Fitzgerald M, O'Donnell D, Parsons LM, Salfinger M. Lessons from a proficiency testing event for acid-fast microscopy. Chest. 2001;120(1):250–7.
8. Goel MM, Budhwar P. Immunohistochemical localization of mycobacterium tuberculosis complex antigen with antibody to 38 kDa antigen versus Ziehl Neelsen staining in tissue granulomas of extrapulmonary tuberculosis. Indian J Tuberc. 2007;54(1):24–9.
9. Jain A. Extra pulmonary tuberculosis: a diagnostic dilemma. Indian J Clin Biochem [Internet]. 2011 [cited 2023 May 19];26(3):269–73. https://www.ncbi.nlm.nih.gov/pmc/articles/PMC3162957/.
10. Ahmed HGE, Nassar AS, Ginawi I. Screening for tuberculosis and its histological pattern in patients with enlarged lymph node. Patholog Res Int [Internet]. 2011 [cited 2023 May 19];2011:417635. https://www.ncbi.nlm.nih.gov/pmc/articles/PMC3108380/.
11. Corbett EL, Watt CJ, Walker N, Maher D, Williams BG, Raviglione MC, et al. The growing burden of tuberculosis: global trends and interactions with the HIV epidemic. AMA Arch Intern Med [Internet]. 2003 [cited 2023 May 19];163(9):1009–21. https://pure.johnshopkins.edu/en/publications/the-growing-burden-of-tuberculosis-global-trends-and-interactions-3.

12. Goel MM, Ranjan V, Dhole TN, Srivastava AN, Mehrotra A, Kushwaha MR, et al. Polymerase chain reaction vs. conventional diagnosis in fine needle aspirates of tuberculous lymph nodes. Acta Cytol. 2001;45(3):333–40.
13. Karimi S, Shamaei M, Pourabdollah M, Sadr M, Karbasi M, Kiani A, et al. Histopathological findings in immunohistological staining of the granulomatous tissue reaction associated with tuberculosis. Tuberc Res Treat. 2014;2014:858396.
14. Afsar I, Gunes M, Er H, Gamze SA. Comparison of culture, microscopic smear and molecular methods in diagnosis of tuberculosis. Rev Esp Quimioter. 2018;31(5):435–8.
15. dos Santos JB, Figueiredo AR, Ferraz CE, de Oliveira MH, da Silva PG, de Medeiros VLS. Cutaneous tuberculosis: diagnosis, histopathology and treatment - part II. An Bras Dermatol [Internet]. 2014 [cited 2023 May 19];89(4):545–55. https://www.ncbi.nlm.nih.gov/pmc/articles/PMC4148266/.
16. National Tuberculosis Advisory Committee. Guidelines for Australian mycobacteriology laboratories. Commun Dis Intell Q Rep [Internet]. 2006;30(1):116–28. PMID: 16637240.
17. S Nalin Kumar, T Srinivasa Prasad, P Anantha Narayan, J Muruganandhan. Granuloma with langhans giant cells: an overview. J Oral Maxillofac Pathol [Internet]. 2013 [cited 2023 May 19];17(3). https://pubmed.ncbi.nlm.nih.gov/24574664/.
18. Méndez-Samperio P. Diagnosis of tuberculosis in HIV co-infected individuals: current status, challenges and opportunities for the future. Scand J Immunol. 2017;86(2):76–82.
19. Zijenah L. The World Health Organization recommended TB diagnostic tools. In: Kayembe J-M, editor. Tuberculosis [Internet]. London: IntechOpen; 2018. https://doi.org/10.5772/intechopen.73070.
20. Centers for Disease Control and Prevention (CDC). Recommendations for use of an isoniazid-rifapentine regimen with direct observation to treat latent Mycobacterium tuberculosis infection. MMWR Morb Mortal Wkly Rep. 2011;60(48):1650–3.
21. Migliori GB, Tiberi S, Zumla A, Petersen E, Chakaya JM, Wejse C, et al. MDR/XDR-TB management of patients and contacts: challenges facing the new decade. The 2020 clinical update by the Global Tuberculosis Network. Int J Infect Dis. 2020;92S:S15–25.
22. Long R. Making a timely diagnosis of pulmonary tuberculosis. Can Respir J [Internet]. 2015 [cited 2023 May 19];22(6):317–21. https://www.ncbi.nlm.nih.gov/pmc/articles/PMC4676393/.
23. Lee JY. Diagnosis and treatment of extrapulmonary tuberculosis. Tuberc Respir Dis (Seoul) [Internet]. 2015 [cited 2023 May 19];78(2):47–55. https://www.ncbi.nlm.nih.gov/pmc/articles/PMC4388900/.
24. Noussair L, Bert F, Leflon-Guibout V, Gayet N, Nicolas-Chanoine M-H. Early diagnosis of extrapulmonary tuberculosis by a new procedure combining broth culture and PCR. J Clin Microbiol [Internet]. 2009 [cited 2023 May 19];47(5). https://pubmed.ncbi.nlm.nih.gov/19321729/.
25. Purohit M, Mustafa T. Laboratory diagnosis of extra-pulmonary tuberculosis (EPTB) in resource-constrained setting: state of the art, challenges and the need. J Clin Diagn Res. 2015;9(4):EE01–6.
26. Lewinsohn DM, Leonard MK, LoBue PA, Cohn DL, Daley CL, Desmond E, et al. Official American Thoracic Society/Infectious Diseases Society of America/Centers for Disease Control and Prevention clinical practice guidelines: diagnosis of tuberculosis in adults and children. Clin Infect Dis [Internet]. 2017 [cited 2023 May 19];64(2):e1–33. https://doi.org/10.1093/cid/ciw694.
27. American Thoracic Society. Targeted tuberculin testing and treatment of latent tuberculosis infection. Am J RespirCrit Care Med. 2000;161:S221–47.
28. Richeldi L. An update on the diagnosis of tuberculosis infection. Am J Respir Crit Care Med. 2006;174(7):736–42.
29. Badar F, Azfar SF, Ahmad I, Yasmeen S, Kirmani S. Diagnostic difficulties in differentiating sarcoidosis from tuberculosis. Oman Med J [Internet]. 2011 [cited 2023 May 19];26(3):210–1. https://www.ncbi.nlm.nih.gov/pmc/articles/PMC3191697/.

30. Rasool G, Khan AM, Mohy-Ud-Din R, Riaz M. Detection of Mycobacterium tuberculosis in AFB smear-negative sputum specimens through MTB culture and GeneXpert® MTB/RIF assay. Int J Immunopathol Pharmacol. 2019;33:2058738419827174.
31. Nahid P, Pai M, Hopewell PC. Advances in the diagnosis and treatment of tuberculosis. Proc Am Thorac Soc [Internet]. 2006 [cited 2023 May 19];3(1):103–10. https://www.ncbi.nlm.nih.gov/pmc/articles/PMC2658675/.

Chapter 4
Assessment of Haematological and Biochemical Parameters in the Diagnosis and Prognosis of Tuberculosis

Tushar Sehgal, Amit Singh, and Sudip Kumar Datta

Abstract Tuberculosis manifests as pulmonary, extra-pulmonary and sometimes miliary disease. Being a systemic disease, it is important to evaluate various biochemical and haematological parameters in patients with tuberculosis for diagnosis, prognostication, selection of appropriate treatment modalities and monitoring of the course of the disease. Tuberculosis, being a multi-system illness with a wide range of clinical presentations, frequently presents as diagnostic dilemmas for treating physicians. Often, these supporting diagnostic modalities help clinicians make a holistic decision regarding the diagnosis and treatment of the disease. Besides, once therapy is initiated, the regular monitoring of haematological and biochemical parameters helps in diagnosing early toxicity and treatment outcomes.

In this chapter, we describe the haematological and biochemical parameters crucial to the diagnosis and prognosis of tuberculosis.

Keywords Tuberculosis · Biochemical tests · Haematological tests · Adenosine deaminase (ADA) · Anaemia

---

T. Sehgal · S. K. Datta (✉)
Department of Laboratory Medicine, All India Institute of Medical Sciences, New Delhi, India
e-mail: doctusharsehgal@aiims.edu

A. Singh
Department of Gastroenterology & HNU, All India Institute of Medical Sciences, New Delhi, India

Department of Microbiology, Central University of Punjab, Bathinda, India
e-mail: amit.singh@cup.edu.in

A. Singh, D. Sharma (eds.), *Diagnosis of Mycobacterium*,
https://doi.org/10.1007/978-981-99-5624-1_4

4.1 Introduction

Tuberculosis (TB) is a multi-organ disease and can influence biochemical and haematological changes. These parameters are the routine investigations carried out to rule out the type of infections in patients. The haematological parameters provide important information about the types and numbers of blood cells in the blood, mainly white blood cells (WBCs), red blood cells (RBCs), platelets erythrocyte sedimentation rate (ESR) and haemoglobin [1]. Biochemical changes such as adenosine deaminase (ADA) and serum albumin often helps in diagnosis [2]. Tuberculosis remains a global health issue with high morbidity and mortality rates, especially in Asian and African countries, with estimated cases of 10.6 million annually and 1.6 million deaths globally [3]. TB is an infectious disease caused by the bacilli *Mycobacterium tuberculosis* (*Mtb*). It primarily affects the lungs but can also affect the meninges, bones, joints, intestines, lymph nodes, skin, endometrium, and other tissues of the body. Tuberculosis transmission is still a problem due to a variety of factors, such as social, economic and environmental. When TB patients cough or sneeze, aerosols are formed and spread to other persons, whose immunity is weakened. The World Health Organization (WHO) and National Tuberculosis Elimination Programme (NTEP) recommended a directly observed treatment short-course (DOTS) strategy standard therapy for tuberculosis treatment [4–6].

Combinational drug therapy was used for the treatment of tuberculosis and has been associated with increased incidences of various side effects. These side effects may be minor to lethal. The major side effects are hepatotoxicity, nephrotoxicity, hypercalcaemia, neuropsychiatric manifestations, hyperuricaemia, ototoxicity and hypokalaemia. Several other biochemical and haematological changes may occur in the level of electrolytes, magnesium, phosphorus, lipid profile, platelet count, haemoglobin and hematocrit of TB patients, respectively [1, 2, 7, 8].

Although tuberculosis is a common illness, atypical and various spectra of clinical presentations pose a diagnostic and therapeutic challenge to physicians. To monitor the diagnostics and progress of tuberculosis therapeutics, haematological and biochemical parameters are very important. On the basis of the current literature and knowledge available, this chapter aims to study the changes in various haematological and biochemical parameters in all forms of tuberculosis patients.

4.2 Haematological Parameters

The diagnosis of tuberculosis mainly relies on the detection of tuberculosis by microscopy, culture and the most advanced Nucleic Acid Amplification Test (NAAT)-based diagnostics. Active tuberculosis produces clinical symptoms and signs, including abnormalities in haematological and biochemical parameters. Changes in haematological parameters are seen during the infection. Complete blood count (CBC) is recommended for all patients as a routine investigation, irrespective of infection type, providing essential information in deciding treatment. In tuberculosis, changes in various haematological parameters are seen, including anemia, lymphopenia, monocytopenia, thrombocytopenia and also thrombocytosis, neutrophilia and monocytosis. Reversible peripheral blood abnormalities usually are associated with PTB [9]. The haematological changes sometimes act as helpful factors, providing supporting diagnostic modalities and helping clinicians make a holistic decision regarding the diagnosis, prognosis and treatment of the disease. Due to significant variability in the clinical presentation of tuberculosis, it poses challenges for diagnostics and therapeutics to the physicians. In the literature, numerous haematological abnormalities in association with tuberculosis have been described, and these changes in haematological parameters may act as biomarker for diagnosis, prognosis and response to antituberculosis therapy.

4.2.1 Anaemia

Anaemia is a common condition associated with the diagnosis of tuberculosis. The prevalence of anaemia in tuberculosis patients is much higher than in healthy individuals. The hemoglobin (Hb) level tends to decrease as acid-fast bacilli (AFB) smear positivity increases [10], and the presence of anaemia during the course of TB delays smear negativity. Anaemia is also associated with unfavourable TB outcomes [10], and has a higher risk of mortality among TB patients. Again, the prevalence of anaemia is significantly higher in HIV-TB (TB associated with Human Immunodeficiency virus infection) than in HIV-negative TB patients. Normochromic normocytic anaemia is the most common type, followed by microcytic anaemia. Anaemia in tuberculosis patients occurs as a consequence of chronic inflammation and without loss of blood or bone marrow suppression [11]. The blunted response of erythropoietin due to the release of tumour necrosis factor (TNF) or other cytokines has been reported. In anaemic patients, cell-mediated immune response and bactericidal capacity of leucocytes are significantly suppressed. Recent meta-analysis shows that anaemia is a risk factor of TB, and the risk is increased depending on the severity of the anaemia [12]. Anaemia has been associated with severe forms of TB, including meningeal and disseminated forms, suggesting that it may serve to understand disease severity in TB.

TB induces a systemic inflammatory response that stimulates the synthesis of hepcidin from hepatocytes and macrophages, the primary iron metabolism regulator [13]. Hepcidin modulates cellular iron export to plasma via ferroportin. Ferroportin is both the receptor for hepcidin and the only known iron exporter in vertebrates [13]. Ferroportin is expressed on cells that manage iron professionally in the body, including duodenal enterocytes that absorb dietary iron, macrophages in the liver and spleen that recycle old erythrocytes, hepatocytes that store iron and placental trophoblasts that transfer iron to the foetus during pregnancy [14]. The internalisation of ferroportin is triggered by the binding of hepcidin to ferroportin; once internalised, the hepcidin-ferroportin complex is degraded in lysosomes and cellular iron export ceases [14]. Anaemia caused by a deficiency in folic acid or vitamin B12, although uncommon, has also been described in patients with TB. It is primarily associated with malnutrition, an increase in folate consumption due to the disease and intestinal malabsorption syndrome due to the localization of disease in the ileocecal region [15]. Anaemia can also occur in the early stages of tuberculosis treatment. As treatment reduces inflammation, iron becomes available for haematopoiesis, causing an increase in reticulocytes and normal hematopoiesis is gradually restored.

4.2.2 Leucocyte Abnormalities

TB affects the white blood cells (WBC). Both leukopenia and leucocytosis are abnormalities associated with tuberculosis [16]. In some instances, a leukamoid reaction may be seen as resembling a neoplastic process [16]. Neutropenia or neutrophilia, lymphocytopenia or lymphocytosis and monocytopenia or monocytosis may all be revealed in TB [16]. Lin F.-S et al. demonstrated that the prevalence of leukopenia in TB patients was 10.4%, whereas it was 9.1% in previously treated TB patients [17]. Neutrophils may also be affected, with neutrophilia being more common than neutropenia. Neutrophils are seen to revert to normal after successful treatment [16]. Neutropenia is multifactorial and may be caused by the suppression of bone marrow granulopoiesis by activated T cells, a deficiency in folate and vitamin B12, bone marrow fibrosis or splenic sequestration [16]. Gelaw Y et al. demonstrated that female gender, elderly age group, more than 6 months of anti-TB treatment and antibiotics were risk factors for low white blood cell count or leukopenia [15].

4.2.3 Platelet

Thrombocytosis and thrombocytopenia are both associated with TB [18]. Thrombocytopenia is seen in severe forms, such as miliary and disseminated TB, whereas thrombocytosis is more prevalent in pulmonary TB (PTB) [18]. In the majority of patients with disseminated/miliary TB, normal haematopoietic elements in the bone marrow are severely suppressed. TB granulomas cause anaemia, monocytopenia and an increase in histiocytosis of the bone marrow [19]. Thrombocytosis is associated with the severity of inflammation and elevated interleukin 6 levels [19]. Thrombocytosis could be the cause of hypercoagulability and may precipitate thromboembolism and deep vein thrombosis (DVT) [19]. Rarely has TB-related immune thrombocytopenia (ITP) been described in the medical literature [15].

4.2.4 *Erythrocyte sedimentation rate (ESR)*

ESR is a blood test that measures the rate of fall of red blood cells in a column of anticoagulated blood in 1 h in millimetres per hour (mm/h) [20]. ESR is a low-cost and easily accessible examination, particularly in resource-limited regions where tuberculosis is prevalent. ESR is a general indicator of inflammation [20]. Anaemia, kidney failure, obesity, ageing and female gender are non-inflammatory conditions that can induce an elevated ESR. ESR is also elevated during menstruation and pregnancy in women [20]. In addition, ESR is elevated in autoimmune diseases such as rheumatoid arthritis and lupus, infections, certain renal diseases and neoplasms, including lymphoma and plasma cell disorders [20]. The determination of the rise in ESR in pulmonary tuberculosis is related to an increase in fibrinogen and plasma globulin levels associated with acute phase responses. If the patient was on anti-tuberculosis therapy, the non-specific ESR value can also be used to assess the disease's response [20].

4.2.5 *Rare Manifestations of TB*

See Table 4.1.

Table 4.1 Rare manifestations of tuberculosis, description and clinical features

Manifestation	Description	Clinical features	Other characteristics
Henoch–Schönlein purpura (HSP)	It is a vasculitic syndrome characterised by diffuse urticarial lesions and palpable purpura primarily affecting the lower extremities	It is associated with joint pain, abdominal pain and renal impairment	Seen in patients with pulmonary TB
Pancytopenia	Haematological abnormality due to bone marrow localisation of TB	It is usually characterised by fever, fatigue, loss of weight and massive splenomegaly. Diagnosis is made on bone marrow biopsy	Factors leading to pancytopenia include hypersplenism, histiocytic hyperplasia, maturation arrest of haematopoietic cells and infiltration of the marrow by granulomas and subsequent fibrosis of the bone marrow

(continued)

Table 4.1 (continued)

Manifestation	Description	Clinical features	Other characteristics
Hemolytic anaemia	Autoimmune hemolytic anaemia (AIHA)	It is characterised by high reticulocyte count, raised unconjugated bilirubin, reduced haptoglobin and blood smear showing normocytic to macrocytic anaemia with polychromasia and other findings of spherocytes or agglutination	Seen in patients with severe forms, such as disseminated TB
Myelofibrosis	It is a rare disorder characterised by extensive fibrosis of the bone marrow and a marked decrease in normal haematopoiesis	It may present as pancytopenia secondary to the myelosuppressive effects of TB	TB may stimulate a fibrotic reaction or may develop in patients with pre-existing myeloproliferative disorders
Hemophagocytic lymphohistiocytosis (HLH)	Occurs rarely in TB. It is an uncommon and potentially fatal disorder associated with cytokine storm and inflammation	It is characterised by fever, cytopenias, organomegaly and deranged laboratory parameters, like hypertriglyceridemia, transaminitis, elevated ferritin and coagulopathy	HLH is associated with high mortality
Leukemoid reaction	It is defined as an increased WBC count (>50,000 cells/μL), which is a physiological response to stress or infection. It is often associated with the presence of immature precursors of white cells and red blood cells	High leukocyte count in blood films with immature precursors of white cells and red blood cells, may resemble a leukaemic process	It may rarely occur in patients with extra-pulmonary or disseminated TB and is common in older males
Disseminated intravascular coagulation (DIC)	It is a rare and potentially fatal manifestation of TB	Occurs probably due to mycobacterial endo- or exo-toxins, capable of initiating the clotting cascade, or protein release into circulation during bacteraemia	Has a high mortality and is seen in patients with miliary TB
Thromboembolism	It is more common in patients with TB than in the general population	TB can be responsible for hypercoagulability as a result of chronic inflammation	There is an increase in the plasma level of factor VIII, fibrinogen and the plasminogen activator inhibitor as well as a reduction in natural anticoagulants such as protein C and antithrombin

4.2.6 *Drug-Induced Changes*

Many abnormalities seen in TB can also be induced by anti-tubercular therapy (ATT), and it makes the diagnosis problematic in a patient who has started therapy. Autoimmune haemolytic anaemia may develop due to the adverse effect of rifampicin, PAS and INH or may have been an underlying condition in these patients. Megalobasltic anaemia may happen due to PAS (malabsorption of vit B12), sideroblastic anaemia due to INH, cycloserine, PZA, pure red cell aplasia due to INH (immune mediated), agranulocytosis due to thioacatazone, PAS and streptomycin, autoimmune thrombocytopenia due to INH and PAS, aplastic anaemia due to streptomycin, PAS; disseminated intravascular coagulation (DIC) due to INH (factor XIII deficiency), rifampicin (induction of cyto P-450), PAS (hypothrombinemia) etc. [21–24] (Table 4.2).

4.3 Biochemical Parameters

Tuberculosis is a systemic disease affecting multiple organ-systems and thus, can impact biochemical parameters.

4.3.1 *Sodium*

According to a number of studies, patients with newly diagnosed tuberculosis have lower serum sodium levels than patients receiving treatment and healthy controls [2]. Hyponatraemia i.e. low serum sodium levels may be caused by local invasion of the hypothalamus, pituitary gland and adrenals or by the syndrome of inappropriate anti-diuretic hormone secretion in the presence of TB meningitis [2]. In one study involving 200 cases of tuberculosis, Jonaidi et al. found that new cases of TB had lower serum sodium levels than those on treatment and controls. In contrast, a study conducted by Olalekan et al. in southwestern Nigeria on electrolyte imbalance among tuberculosis (TB) patients on medication found that serum sodium was considerably lower among those on treatment compared to new cases [25]. The inclusion of patients with chronic kidney disease or other comorbidities such as HIV, etc. in their study group may account for the difference [25].

Table 4.2 Drug-induced changes in tuberculosis patients

Drug	Blood findings/other disorders
Rifampicin	RBC-hemolysis
	WBC-leukopenia, agranulocytosis
	Platelets-thrombocytopenia, thrombocytopenic purpura
	Clotting abnormalities-DIC
Isoniazid	RBC—Sideroblastic anaemia, red cell aplasia WBC-agranulocytosis
	Platelet—Thrombocytopenia
Ethambutol	RBC—Hemolytic anemia
	WBC—Neutropenia, eosinophilia
	Platelet—Thrombocytopenia
Pyrazinamide	RBC—Megaloblastic anemia, sideroblastic anaemia
	Platelet—Thrombocytopenia
Rifapentine	RBC—Aplastic anemia, AIHA
	WBC—Agranulocytosis, leukemoid reaction, leukopenia
	Platelet—Thrombocytopenia
	Others—TTP, acute porphyria
Rifabutin	WBC—Leucopenia
Cycloserine	RBC—Megaloblastic anemia, sideroblastic anaemia
Para-aminosalicylic acid (PAS)	RBC—Hemolytic anemia, methemoglobinemia
	WBC—Agranulocytosis
	Platelet—Thrombocytopenia
Streptomycin	WBC—Eosinophilia, leukopenia
	Platelet—Thrombocytopenic purpura
Amikacin	Causes drug rash with eosinophilia and systemic symptoms known as DRESS syndrome
Levofloxacin	RBC—Hemolytic anemia
	WBC—Leukopenia
	Platelet—Thrombocytopenia
Moxifloxacin	WBC—Neutropenia
Clofazimine	RBC—Hemolytic anemia, macrocytosis, elevated reticulocyte count
Ethionamide prothionamide	WBC—Agranulocytosis, neutropenia
Meropenem/imipenem	RBC—Hemolytic anemia
	WBC—Neutropenia
	Platelet—Thrombocytopenia
Linezolid	RBC—Anemia
	WBC—Neutropenia
	Platelet—Thrombocytopenia

4.3.2 Potassium

Potassium levels in newly diagnosed TB patients may be low or elevated. Ufoaroh, et al. found that patients with newly diagnosed TB had substantially higher serum potassium levels than the control group and those receiving treatment [2]. In contrast to the cases and the non-TB control group, patients on follow-up treatment had a significantly lower serum potassium concentration in their study. Olalekan et al. found that the serum potassium levels of TB patients on treatment were lower than those of newly diagnosed TB patients [25]. Similarly, Bhagyamma et al. found in their study that patients with a new TB infection had decreased potassium serum levels [26]. Salina et al. demonstrated that potassium-deficient solid media inhibited the growth of *Mycobacterium tuberculosis*, whereas potassium-rich solid media and liquid media promoted its growth, indicating that mycobacteria may require a potassium-rich environment to proliferate [27].

4.3.3 Calcium

Low serum calcium levels have been observed in newly diagnosed TB patients. Similar to the results of Ali-Gombe et al. and Rohini et al., Ufoaroh CU et al. found that patients with newly diagnosed TB had substantially lower serum calcium levels than those in the follow-up and non-TB control groups in their study on TB patients [2, 28, 29]. The granulomatous disease tuberculosis is associated with alterations in calcium metabolism, including hypercalcaemia and hypocalcaemia. Hypocalcaemia can be attributed to malnutrition, impaired absorption and vitamin D deficiency in patients with pulmonary tuberculosis [2].

4.3.4 Albumin

Serum albumin levels are also substantially low in newly diagnosed TB patients who had not yet begun treatment as compared to the non-TB control group [2, 30].

4.3.5 Liver Function Test (LFT)

TB is typically treated with a combination of isoniazid, rifampicin, pyrazinamide and ethambutol, according to standard recommendations [31]. Drug-induced liver injury (DILI) is a significant adverse event of anti-TB treatment, leading to non-adherence, treatment failure, or drug resistance development [31]. Anti-TB DILI is reported in 2–28% of patients worldwide [32]. Clinical manifestations of anti-TB DILI can range from asymptomatic to moderate elevations of aspartate aminotransferase (AST) and alanine aminotransferase (ALT) to acute hepatitis and even liver failure [31]. Hence regular monitoring of LFT during the course of the management is essential. *Mycobacterium tuberculosis* (*Mtb*) is the most prevalent opportunistic infection among those with HIV [33]. There is an antiretroviral therapy (ART) for

HIV infection that has resulted in a substantial decrease in AIDS-related deaths, but it has also elevated the DILI associated with ART [34]. The reported incidence of severe DILI associated with ART is approximately 10% [35]. Three of the four standard ATT medications, namely rifampicin, pyrazinamide and isonicotinic acid hydrazide (INH), as well as the majority of ART medications, are associated with DILI [33]. Rifampicin can inhibit bilirubin absorption, resulting in hyperbilirubinaemia without conjugation. Both hepatocellular damage and granulomatous hepatitis may be caused by pyrazinamide. Hepatocytes are toxic to mono-acetyl hydrazine, one of the primary metabolites of isoniazid. Co-trimoxazole prescribed for *Pneumocystis jirovecii* or toxoplasmosis prophylaxis or treatment is associated with both cholestatic jaundice and hepatic necrosis [33].

4.3.6 *Adenosine Deaminase (ADA)*

ADA is an enzyme that catalyses the hydrolytic deamination of adenosine to inosine and deoxyadenosine to deoxyinosine as part of purine metabolism [36]. ADA is crucial for the proliferation and differentiation of lymphocytes, specifically T lymphocytes [37]. Serum ADA is elevated in diseases associated with cellular stimulation, such as typhoid fever, infectious mononucleosis, liver disease, sarcoidosis, leukaemia, brucellosis, acute pneumonia, rheumatoid arthritis, cancer and tuberculosis [38]. Although serum ADA levels are higher in patients with pulmonary tuberculosis than in healthy individuals, ADA should not be used to distinguish between pulmonary TB and other pulmonary infections [39]. The serum ADA test has low sensitivity (range, 12–44%) and high specificity range (96–100%) [36]. Therefore, this test is not very useful for screening for tuberculosis. This test is beneficial for excluding tuberculosis in suspected cases with negative microbiological results based on the high specificity of the serum ADA level [36]. Tarhan and colleagues suggested that serum ADA levels can be used as a supplementary laboratory test in conjunction with clinical and laboratory findings to diagnose tuberculosis [40].

4.4 Conclusion

Various studies have shown that tuberculosis is associated with haematological and biochemical changes [2, 41, 42]. The diagnosis of tuberculosis (TB) is not always easy; thus, clinicians often need to rely on several parameters to diagnose TB. Biochemical and haematological paramets play important role in the diagnosis of tuberculosis as well as prognostication and monitoring treatment progression.

References

1. Morris CD, Bird AR, Nell H. The haematological and biochemical changes in severe pulmonary tuberculosis. Q J Med. 1989;73(272):1151–9.
2. Ufoaroh CU, Onwurah CA, Mbanuzuru VA, Mmaju CI, Chukwurah SN, Umenzekwe CC, et al. Biochemical changes in tuberculosis. Pan Afr Med J [Internet]. 2021 [cited 2023 May 6];38(66). https://www.panafrican-med-journal.com/content/article/38/66/full.
3. WHO. Global tuberculosis report 2021 [Internet]. 2021 [cited 2021 Nov 27]. https://www.who.int/publications-detail-redirect/9789240037021.
4. RNTCP. DOTS-Plus guidelines [Internet]. 2010. http://health.bih.nic.in/Docs/Guidelines/Guidelines-DOTS-Plus.pdf.
5. NTEP. Training module on extrapulmonary tuberculosis-standard treatment workflow [Internet]. 2023. p. 1–20. https://tbcindia.gov.in/showfile.php?lid=3681.
6. NTEP. Annual reports:: Central TB Division [Internet]. 2021 [cited 2021 Dec 7]. https://tbcindia.gov.in/index1.php?lang=1&level=1&sublinkid=4160&lid=2807.
7. Addington WW. The side effects and interactions of antituberculosis drugs. Chest. 1979;76(6):782–4.
8. Zaleskis R. Postgraduate course ERS Copenhagen 2005 - the side-effects of TB therapy. Breathe. 2005;2(1):69–73.
9. Sheetal MMR, Patwegar AR. Comparative study of hematological parameters in newly diagnosed tuberculosis patient's pre-att & after intensive phase of ATT. IP Arch Cytol Histopathol Res. 2020;3(4):185–91.
10. de Mendonça EB, Schmaltz CA, Sant'Anna FM, Vizzoni AG, Mendes-de-Almeida DP, Raquel de Vasconcellos Carvalhaes de Oliveira, et al. Anemia in tuberculosis cases: a biomarker of severity? PLoS One. 2021;16(2):e0245458.
11. Baynes RD, Flax H, Bothwell TH, Bezwoda WR, Atkinson P, Mendelow B. Red blood cell distribution width in the anemia secondary to tuberculosis. Am J Clin Pathol. 1986;85(2):226–9.
12. Gelaw Y, Getaneh Z, Melku M. Anemia as a risk factor for tuberculosis: a systematic review and meta-analysis. Environ Health Prev Med. 2021;26(1):13.
13. Hella J, Cercamondi CI, Mhimbira F, Sasamalo M, Stoffel N, Zwahlen M, et al. Anemia in tuberculosis cases and household controls from Tanzania: contribution of disease, coinfections, and the role of hepcidin. PLoS One. 2018;13(4):e0195985.
14. Nemeth E, Ganz T. The role of hepcidin in iron metabolism. Acta Haematol. 2009;122(2–3):78–86.
15. Minardi ML, Fato I, Di Gennaro F, Mosti S, Mastrobattista A, Cerva C, et al. Common and rare hematological manifestations and adverse drug events during treatment of active TB: a state of art. Microorganisms. 2021;9(7):1477.
16. Balepur SS, Schlossberg D. Hematologic complications of tuberculosis. Microbiol Spectr. 2016;4:6.
17. Lin FS, Wu MY, Tu WJ, Pan HQ, Zheng J, Shi JW, et al. A cross-sectional and follow-up study of leukopenia in tuberculosis patients: prevalence, risk factors and impact of anti-tuberculosis treatment. J Thorac Dis. 2015;7(12):2234–42.
18. Alghamdi AA, Awan FS, Maniyar IH, Alghamdi NA. Unusual manifestation of extrapulmonary tuberculosis. Case Rep Med. 2013;2013:353798.
19. Avasthi R, Mohanty D, Chaudhary SC, Mishra K. Disseminated tuberculosis: interesting hematological observations. J Assoc Physicians India. 2010;58:243–4.
20. Amilo GI, Meludu SC, Ele PU, Ezechukwu C, Onyenekwe C, Chukwu MI. Haematologic indices in pulmonary tuberculosis with or without HIV co-infection in south eastern Nigeria. Adv Life Sci Technol. 2013;11(1):1.
21. Akhtar AJ, Crompton GK, Schonell ME. Para-aminosalicylic acid as a cause of intestinal malabsorption. Tubercle. 1968;49(3):328–31.
22. Marseglia GL, Locatelli F. Isoniazid-induced pure red cell aplasia in two siblings. J Pediatr. 1998;132(5):898–900.

23. Piso RJ, Kriz K, Desax MC. Severe isoniazid related sideroblastic anemia. Hematol Rep. 2011;3(1):e2.
24. Yakar F, Yildiz N, Yakar A, Kılıçaslan Z. Isoniazid- and rifampicin-induced thrombocytopenia. Multidiscip Respir Med. 2013;8(1):13.
25. Olalekan AW, Oluwaseun FA, Oladele HA, Akeem AD. Evaluation of electrolyte imbalance among tuberculosis patients receiving treatments in southwestern Nigeria. Alexandria J Med. 2015;51(3):255–60.
26. Bhagyamma DSN, DrU S, Anuradha DR. Study of electrolyte changes in tuberculosis and human immune deficiency virus (HIV) co-infected with tuberculosis patients: a hospital based study. IOSR. 2016;15(9):28–31.
27. Salina EG, Waddell SJ, Hoffmann N, Rosenkrands I, Butcher PD, Kaprelyants AS. Potassium availability triggers Mycobacterium tuberculosis transition to, and resuscitation from, non-culturable (dormant) states. Open Biol. 2014;4(10):140106.
28. Ali-Gombe A, Onadeko BO. Serum calcium levels in patients with active pulmonary tuberculosis. Afr J Med Med Sci. 1997;26(1–2):67–8.
29. Rohini K, Bhat S, Srikumar PS, Mahesh Kumar A. Assessment of serum calcium and phosphorus in pulmonary tuberculosis patients before, during and after chemotherapy. Indian J Clin Biochem. 2014;29(3):377–81.
30. Ramakrishnan K, Shenbagarathai R, Kavitha K, Uma A, Balasubramaniam R, Thirumalaikolundusubramanian P. Serum zinc and albumin levels in pulmonary tuberculosis patients with and without HIV. Jpn J Infect Dis. 2008;61(3):202–4.
31. Zhao H, Wang Y, Zhang T, Wang Q, Xie W. Drug-induced liver injury from anti-tuberculosis treatment: a retrospective cohort study. Med Sci Monit. 2020;26:e920350-1–8.
32. Shang P, Xia Y, Liu F, Wang X, Yuan Y, Hu D, et al. Incidence, clinical features and impact on anti-tuberculosis treatment of anti-tuberculosis drug induced liver injury (ATLI) in China. PLoS One. 2011;6(7):e21836.
33. Puri P, Kaur N, Pathania S, Kumar S, Sharma PK, Sashindran VK. Antitubercular therapy induced liver function tests abnormalities in human immunodeficiency virus infected individuals. Med J Armed Forces India. 2017;73(1):12–7.
34. Reisler RB, Han C, Burman WJ, Tedaldi EM, Neaton JD. Grade 4 events are as important as AIDS events in the era of HAART. J Acquir Immune Defic Syndr. 2003;34(4):379–86.
35. Sulkowski MS, Thomas DL, Chaisson RE, Moore RD. Hepatotoxicity associated with antiretroviral therapy in adults infected with human immunodeficiency virus and the role of hepatitis C or B virus infection. JAMA. 2000;283(1):74–80.
36. Salmanzadeh S, Tavakkol H, Bavieh K, Alavi SM. Diagnostic value of serum adenosine deaminase (ADA) level for pulmonary tuberculosis. Jundishapur J Microbiol. 2015;8(3):e21760.
37. Russo M, Giancane R, Apice G, Galanti B. Adenosine deaminase and purine nucleoside phosphorylase activities in peripheral lymphocytes from patients with solid tumours. Br J Cancer. 1981;43(2):196–200.
38. al-Shammary FJ. Adenosine deaminase activity in serum and pleural effusions of tuberculous and non-tuberculous patients. Biochem Mol Biol Int. 1997;43(4):763–79.
39. Boloursaz M, Khalilzadeh S, Khodayari A, Hakimi S. Adenosine deaminase level as an indicator for differentiating between active pulmonary tuberculosis infection and other pulmonary infections. J Compr Pediatr. 2012;3(1):3–6.
40. Tarhan G, Gümüşlü F, Yilmaz N, Saka D, Ceyhan I, Cesur S. Serum adenosine deaminase enzyme and plasma platelet factor 4 activities in active pulmonary tuberculosis, HIV-seropositive subjects and cancer patients. J Infect. 2006;52(4):264–8.
41. Kurup R, Flemming K, Daniram S, Marks-James S, Roberts MR. Hematological and biochemistry profile and risk factors associated with pulmonary tuberculosis patients in Guyana. Tuberc Res Treat. 2016;2016:e6983747.
42. Shah AR, Desai KN, Maru AM. Evaluation of hematological parameters in pulmonary tuberculosis patients. J Family Med Prim Care. 2022;11(8):4424.

Chapter 5
Different Methods of Microscopic and Bacteriological Diagnosis of Tuberculosis

Rupal Rai, Praveen Gautam, Bijina J. Mathew, Chandan Kumar Dubey, Shadma Siddiqui, Sudheer Gupta, Shivendra K. Chaurasiya, and Anirudh K. Singh

Abstract Tuberculosis (TB) remains one of the largest preventable causes of death worldwide with an estimated 1.6 million deaths and ten million infections reported in 2021 (WHO, https://www.who.int/news-room/fact-sheets/detail/tuberculosis, 2022). While things have improved significantly over the last five decades and many developed nations have managed to control the disease, TB control in low- and middle-income countries (LMICs) remains a challenge. The rising incidence of drug resistance is further worsening the situation. One of the keys to effective control and eventual eradication of TB is timely and accurate diagnosis of the infection. While nucleic acid amplification tests (NAATs) are becoming more and more popular methods for the diagnosis of TB, the high running cost of the test and poor infrastructure, especially in resource-limited settings, make it difficult to adopt these methods for the routine diagnosis of TB. Smear microscopy and culture, despite their limitations thus, remain widely used diagnostic modalities for this disease. While the sensitivities of these methods are poor, especially in children due to the paucibacillary nature of the specimens, substantially lower cost weighs in favor of these methods. In this chapter, we give a detailed account of existing bacterio-

Rupal Rai and Praveen Gautam contributed equally with all other contributors.

R. Rai · B. J. Mathew · S. K. Chaurasiya
Department of Biological Science and Engineering, Maulana Azad National Institute of Technology, Bhopal, India

P. Gautam · S. Gupta
3B Blackbio Biotech India Ltd, Bhopal, India

C. K. Dubey
3B Blackbio Biotech India Ltd, Bhopal, India

School of Sciences, SAM Global University, Raisen, India

S. Siddiqui · A. K. Singh (✉)
School of Sciences, SAM Global University, Raisen, India

A. Singh, D. Sharma (eds.), *Diagnosis of Mycobacterium*,
https://doi.org/10.1007/978-981-99-5624-1_5

logical and microscopic methods used for the diagnosis of TB and provide a perspective on the use of these methods in the future of TB diagnosis.

Keywords Tuberculosis · MB-7H9 medium · MB-7H10 medium · Lowenstein-Jensen medium · Automated culture · Light microscopy · Ziehl-Neelsen staining · Fluorescent microscopy

5.1 Introduction

Mycobacterium tuberculosis (Mtb) is one of the most successful pathogens known to humans. It causes tuberculosis (TB), historically known as consumption. While pulmonary TB is the most common manifestation of Mtb infection and is a bigger cause of health concern globally, it can cause diseases in other sites of the body, such as meningitis, osteoarticular TB [1], and gastrointestinal TB [2]. Pulmonary TB is a highly contagious disease because of the airborne transmission of bacilli coughed out by symptomatic patients. Treatment of the disease is lengthy, and the currently used drugs have side effects. The situation is worsening by the emergence of multidrug-resistant, extremely drug-resistant, and totally drug-resistant Mtb [3, 4]. Timely diagnosis, isolation of symptomatic patients, treatment compliance, and monitoring of treatment response are some of the keys to reducing the burden of TB. TB diagnosis especially is central to the goal of a TB-free world. Confirmed TB diagnosis historically has been challenging due to the slow growth rate of Mtb and often the paucibacillary nature of the samples used for the culture. Direct staining of sputum samples from symptomatic patients using Zeil-Nelssen (ZN) staining and light microscopy has remained the mainstay of TB diagnosis, and due to its ease and quick turnaround time, it is still a preferred method in economically challenging settings despite its low specificity and sensitivity. Nucleic-acid-based tests, such as real-time polymerase chain reaction, loop-mediated isothermal amplification, and line probe assays (LPA), have not only shortened the turnaround time but are also more sensitive when compared to microscopy and culture. These tests are not only offering to identify the pathogen but also tell about the antibiotic susceptibility of the isolate [5, 6]. However, a high operational cost makes it difficult to bring the benefits of these tests to the people who need them the most. Furthermore, these tests can only detect resistance caused by known or selected mutations and fail to access the phenotypic resistance to drugs. This is where culture-based antimicrobial susceptibility tests (AST) play a crucial role in determining the treatment course for a patient [7]. Cost, utility, and/or ease of bacteriological and microscopic methods make these methods of TB diagnostics relevant.

5.2 *Mycobacterium tuberculosis* Culture

Mtb is a very slow-growing fastidious bacterium, which makes the culture of this important human pathogen difficult. Slow growth rate not only delays the diagnosis, but it also poses the problem of contamination by normal flora, making it even more difficult. Nonetheless, culture provides definitive diagnostics and is considered the "gold standard" of TB diagnosis [8]. With the availability of quick and rapid molecular diagnostics tests, culture may not be the preferred modality for the detection of Mtb, but it remains relevant for drug susceptibility testing (DST), genotyping, and the monitoring of patients' response to treatments [7, 9]. DST is used to determine which antibiotics are effective at killing Mtb. This is important because TB is often treated with a combination of several different antibiotics to reduce the risk of the bacteria developing resistance. DST involves growing the bacteria on a solid media and adding different concentrations of antibiotics to see which ones can inhibit the growth of the bacteria [10, 11]. Genotyping is a process used to identify the specific strain of Mtb present in a sample. This is important because different strains of the bacteria may have different drug susceptibilities or be more or less virulent. Genotyping is typically done using techniques such as polymerase chain reaction (PCR) or DNA sequencing [12]. The continuous monitoring of patients' response to treatment is very important for Mtb infections due to the emergence of various drug-resistant strains. After starting treatment, the patient's sputum (a mixture of saliva and mucus that is produced by the respiratory tract) is collected and cultured to see if the bacteria are still present. If the bacteria are still present, it may indicate that the treatment is not effective and the patient's treatment plan may need to be adjusted [13].

5.2.1 *Sample Collection and Processing*

The most commonly used sample for Mtb culture is sputum from patients with pulmonary TB. An appropriately collected sputum sample is key to the diagnosis of TB. It is imperative that the sputum is collected and not the oronasal secretion/saliva. Sputum, also called phlegm, is a secretion of the airway, specifically the lung, and is thick and sticky due to the presence of mucus. A sputum sample is collected in a sterile container and processed for growth as soon as possible but not beyond 48 h. In cases where the sample processing may take 48–72 h, the sample should be collected in 1% cetylpyridinium chloride and 2% sodium chloride solution. As the sample contains thick mucus and normal flora which grow faster than Mtb, it is critical to homogenize and decontaminate the sample before it is inoculated. The commonly used agents for decontamination and homogenization are 4% sodium hydroxide (NaOH) solution (Patroff's modified method) or *N*-acetyl-L-cysteine sodium hydroxide solution (NALC-NaOH) [14]. Once the sample is homogenized, the Mtb cells are concentrated by centrifugation and inoculated for growth on the medium of choice. As the methods for decontamination also affect the survival of Mtb, care should be taken to minimize the death of the bacilli while processing the samples.

5.2.2 *Culture on Solid Media*

Egg-based Lowenstein-Jensen (LJ) and agar-based Middlebrook 7H10/7H11 (MB7H10/MB7H11) are the most commonly used solid media for growing Mtb. LJ culture media is widely used in the laboratory for the isolation and identification of mycobacteria. LJ medium is made up of fresh eggs from country hens, glycerol, mineral salts, and malachite green. Glycerol in the medium favors the growth of Mtb, and malachite green in the medium inhibits the growth of other types of bacteria that may be present in a sample, making it selective for Mtb. The eggs provide nutrients in the form of proteins and lipids and work as solidifying agents. LJ medium is suitable for the growth of Mtb and allows for a visual inspection of the colony morphology of the isolate, making it easier to identify the culture [15]. It is also used for the antimicrobial susceptibility testing (AST) of first-line and second-line drugs for TB [11]. Some of the limitations of this medium are slower growth of Mtb compared to the liquid medium, which delays the diagnosis, and shorter shelf life when compared to agar base media, such as MB7H10 and MB7H11. The average time of growth of detectable macroscopic colonies of Mtb on the LJ medium is around 25.4 days [16]. As LJ is always prepared as a slope or slant in culture tubes/bottles, the effective surface area for the growth of mycobacteria is smaller than agar-based media, which may be prepared in Petri dishes as well.

MB7H10 is made up of a mixture of ingredients that provide the nutrients and conditions necessary for the growth of mycobacteria. This was developed by Middlebrook and Cohn as a replacement for egg-based agar media. This medium contains formulations of oleic acid and albumin as key components that help in the faster growth of mycobacteria and provide protection against toxic agents [17]. Reportedly, MB7H10 grows fewer contaminants in comparison to egg-based mediums like the LJ medium, and the average time of growth of detectable macroscopic colonies of Mtb on the Middlebrooke medium is around 23.6 days [16]. Various inorganic salts, like magnesium sulfate, ferric ammonium citrate, sodium citrate, etc., are present in this medium, which assist in the growth of the tubercle bacteria. Sodium citrate in the medium, when converted into citric acid, is responsible for retaining cations in the solution, and glycerol acts as a source of carbon and energy for the bacilli. The oleic acid present in the medium supports the metabolism of mycobacteria, while albumin protects the bacilli against free fatty acids in the medium, which can be toxic. MB7H10 also contains catalase, which protects the bacteria from toxic peroxidases that may be present in the medium [18]. MB7H11 has a similar composition as MB7H10, but it has one more addition of pancreatic digest of casein, which enhances the growth of fastidious, drug-resistant Mtb, which grows poorly on MB7H10 agar base. Both MB7H10 and MB7H11 are very clear and transparent media, which help in the observation and quantification of the bacterial colony. MB7H10 and MB7H11 media can be made selective against the growth of contaminating bacteria and fungi by the addition of a mixture of four antimicrobials: Polymyxin B, amphotericin B, carbenicillin, and trimethoprim lactate. The addition of these antimicrobial agents helps in recovering mycobacteria from samples containing mixed flora [19].

5.2.3 *Culture in Liquid Media*

The use of liquid media for mycobacteriology testing is not necessarily recommended as a standard practice for all laboratories. However, it may be used in some cases for the detection and cultivation of fast-growing mycobacterial species. It is possible that liquid media result in the increased recovery of mycobacteria and decreased time to detection compared to solid media. However, this can vary depending on the type of mycobacterial species and the specific methodology used. The choice of media and methodology used should be based on established guidelines and protocols, and it is important to consider factors such as sensitivity, specificity, and ease of use when making a decision. Liquid media are more prone to contamination compared to solid media, which can impact the accuracy of the results. As a result, the addition of antimicrobials to prevent contamination is often necessary when using liquid media for mycobacteriology testing. However, the specific type and amount of antimicrobial used should be based on established guidelines and protocols to ensure reliable results. The growth detection of an average growing bacteria can be around 15–16 days. There are several types of liquid media that can be used to culture and isolate Mtb, however, MB7H9 broth medium or its derivative are the most widely used media for Mtb.

MB7H9 broth is a type of liquid growth medium that is used to culture and isolate Mtb. The composition of this medium is similar to that of MB7H10 agar media. The presence of supplements like glycerol, oleic acid, albumin, and dextrose supports the growth of mycobacteria. Additional supplementation by polysorbate-80 (Tween-80) enhances the growth of mycobacteria in the medium. However, before the addition of Tween-80, the albumin in the media is heat-treated via autoclaving to avoid the release of free fatty acids from polysorbate 80 by lipase [20].

5.2.4 *Automated Liquid Culture*

Diagnosis by the conventional method of culturing bacteria either on solid media or liquid media is a time-taking process and involves lots of precision to avoid contamination, with a low recovery of bacilli from samples. The automated liquid culture not only improved the recovery of bacilli, i.e., it is a sensitive method for the detection of mycobacteria but was majorly adopted for its time-efficient protocol.

The implementation of automated liquid culture has increased the efficiency of the diagnosis of TB. The automated liquid culture requires expensive equipment, which can be a challenge in economically weaker settings [21, 22]. However, it is useful for DST against second-line drugs, including bedaquiline and linezolid, as well as for fluoroquinolones and amikacin when the genetic basis of the resistance is not known [23]. The system can use a variety of specimens/samples as inoculum to culture the bacilli and uses liquid media for growth. There are many systems available, with differences in the methods of detection. Currently, three FDA-cleared semi-automated liquid culture systems are being used for the diagnosis and AST of Mtb.

BACTEC 460 TB (B460) system by Becton Dickinson Microbiology Systems, Sparks, Md., USA, was the first automated culture system implemented for the diagnosis of TB. Despite having superior sensitivity, the use of radioisotopes in detection limited the use of this system [24]. The Mycobacteria Growth Indicator Tube (MGIT) system BACTEC MGIT 960 was then developed as an upgradation that utilizes a fluorescent indicator, ruthenium pentahydrate. The fluorescence is quenched by the oxygen in the medium, and as the bacteria grow, oxygen in the medium is depleted, resulting in the release of fluorescence. This fluorescence is detected by the instrument, and samples are labeled as positive for the presence of mycobacteria [25]. Prior to inoculation of the sample, each tube is supplemented with a growth supplement and an antibiotic mixture of polymyxin B, amphotericin B, nalidixic acid, trimethoprim, and azlocillin (PANTA) to avoid contamination. The tubes are incubated at 37 °C and automatically monitored each hour for up to 42 days or till the positive detection of fluorescence [26]. The average detection time is around 9.7 days in BACTEC MGIT 960 system [27]. Various studies have confirmed the increased sensitivity in the diagnosis of mycobacteria with this system as compared to the conventional culture detection method [28–32].

The VersaTREK system, formerly known as the ESP culture system II, can be used for Mtb growth and the detection of MDR TB. The system is based on evaluation pressure within a sealed bottle that is increased as the bacilli grow. It monitors the changes in gas production and consumption, mainly CO_2 due to microbial growth, every 24 min. A special algorithm also helps in the detection of very slow-growing mycobacteria. This system automatically incubates and continuously monitors culture bottles inoculated with specimens possibly containing mycobacteria. The bottles are incubated for up to 42 days or until detected positive [33]. The incubation in a sealed bottle reduced the chances of contamination when compared to other systems. Resistance toward antibiotics is detected by comparing the growth of bacilli in tubes with respective drugs, with growth observed in drug-free tubes. The system has been tested in comparison to BACTEC MGIT and showed similar efficiency as BACTEC MGIT 960 [34, 35]. However, when compared to the conventional solid culture, it significantly reduces the mean time to detection (TTD) rate and increases the sensitivity of detection [36].

BacT alert 3D by bioMérieux is used for the detection of various microorganisms, including bacteria, fungi, and yeast. BacT is a highly efficient system utilizing the detection method based on pressure and fluorescence. The samples are inoculated in sealed tubes containing liquid medium, mainly MB7H9 with revised reconstitution fluid-antibiotic supplement to avoid contamination. The colorimetric gas permeable sensor is embedded at the bottom of the tube, which changes color in response to changes in the pH of media due to CO_2 released by actively growing mycobacteria. The color change is monitored by a reflectometric detection unit contained inside each incubating drawer of the instrument. It is an automated, rapid, and less labor-intensive mycobacterial culturing system, which is an alternative to BACTEC-MGIT 960 system [37]. The tube used in BacT contains a modified MB7H9 broth supplemented with bovine serum albumin, catalase, and casein. Before inoculation, MB/BacT antibiotic supplement (amphotericin B, azlocillin, nalidixic acid, polymyxin B, trimethoprim, vancomycin) in reconstitution fluid (amaranth, glycerol, Tween-80, purified water) is added to the bottle. The tubes are incubated at 37 °C and read every 10 min. The results can be automatically read on the computer screen attached to the instrument (bioMérieux). The turnaround time is higher for BacT than the conventional culture methods [38, 39]. The BacT alert 3D equipment can be implemented for detection of mycobacteria in high burden countries.

5.3 Microscopic Examination

The current guidelines by the World Health Organization (WHO) have specifically determined microscopy as the initial investigation for patients suspected of TB. WHO and the International Union Against Tuberculosis and Lung Disease (UNION) pan the technical guidelines for microscopy-based diagnosis [40]. Particularly in pulmonary TB, the guidelines suggest collecting sputum samples at three different time points. The quality of sputum samples is crucial for microscopy. Sputum, a sticky and thick sample taken with spontaneous expectoration, is considered the ideal sample. Nasal, pharyngeal secretions or saliva with blood are not appropriate samples for microscopy. Bronchial lavage and gastric lavage are also taken as samples if medical professionals advise for it. On the other hand, in cases other than pulmonary TB, samples like cerebrospinal fluid (CSF); pleural, ascitic, pericardial, joint, and other fluids; biopsies; and resected material like pus, blood, and urine, are also tested initially with microscopy. The samples other than sputum have a low range of bacilli, therefore, are not relied on microscopy for the detection of TB [41, 42]. Microscopy is also considered for treatment monitoring as the prolonged treatment of TB requires continuous monitoring. The low level of bacilli or dead bacilli detection at the last stage of intensive treatment is further confirmed by the solid culture technique to determine the viability of bacilli. Treatment failure is determined with repeated highly positive smear microscopy results even at the intensive phase of treatment [8]. Sputum smear microscopy, thus, provides a simple, fast, and cheap diagnosis method for the proper monitoring of patients. However, microscopic examination can identify the bacilli but are unreliable to differentiate between live and dead bacilli, similarly unreliable to differentiate between Mtb and nontuberculous mycobacteria (NTM) [43]. Thus, other molecular diagnostic tests are further required for confirmation [44]. However, the results from microscopic examination determine the next step of molecular diagnosis. As per the guidelines, if the sputum sample is found to be positive for acid-fast bacilli (AFB), then DNA extracted from the processed sample is used for LPA to look for the presence of MTB complex and determine rifampicin and isoniazid resistance, as per the standard protocol [45]. Sputum-negative samples for AFB are further tested by the cartridge-based nucleic acid amplification test (CBNAAT) [46]. Irrespective of microscopic results, it is recommended that all samples are cultured on the LJ medium as per the standard protocol. The specificity and sensitivity of the microscopic examination further depend on the microscopy technique and staining methods used [47].

5.3.1 Light Microscopy: Ziehl-Neelsen Staining

The cell wall of mycobacteria contains mycolic acids and is rich in lipids, which are impermeable to many dyes. Thus, the primary stain used in Ziehl-Neelsen (ZN) staining is carbol fuchsin, which has an affinity toward mycolic acids and binds to lipids present in the cell wall of the bacilli. Fuchsin is soluble in phenol and is thus prepared using phenol. Heat and phenol help solubilize lipids present in the cell wall of mycobacteria, and phenol also acts as a mordant. Once stained, the bacilli resist decolorization by acid alcohol, thus giving mycobacteria their colloquial name, acid-fast bacilli. The commonly used counterstain is methylene blue. The artifacts and background thus appear blue, whereas the acid-fast bacilli appear pink and purple under bright field microscopy.

Depending on the use of heat during staining, ZN staining is further categorized into hot and cold staining [48]. Hot staining is widely used as the heating step helps in the solubilization of lipids present in the cell wall of mycobacteria and allows more penetration of the dye carbol fuchsin [49]. As per standard protocols, the smears are prepared directly from samples and dried. The slides are placed on a staining rack with smears facing up. The staining rack should have a gap of about 5 cm. The whole smear is gently covered with 0.3% carbol fuchsin reagent. Then the slides are heated over the flame of a Bunsen burner, in swinging movements until the first white steams are released. This is enough for carbol fuchsin to properly penetrate the bacilli and bind to lipids. Care should be taken to not overheat, and the t slides/smear should not dry. Each slide is then rinsed individually in a gentle stream of running water until all free stains are washed away. Then comes the decolorizing step, in which the slides are flooded with 25% H_2SO_4 or 3% hydrochloric acid in ethyl alcohol (decolorizing solution) for 2–3 min. Slides are again thoroughly rinsed with water, and excess water is drained from the slides. The slides are then counterstained by flooding the smear with 0.1% methylene blue for 30 s and thoroughly rinsed with water. Smears are allowed to air-dry and observed under the microscope (100×) for AFB [50]. In 1915, Kinyoun published a method, which is known as the "cold staining" method because the heating step was removed in favor of using a higher concentration of carbol-fuchsin primary stain and for a longer time of incubation of the smear with the primary dye [51].

5.3.2 Fluorescent Microscopy

Fluorescence microscopy was introduced in the 1930s in an attempt to improve the outcomes of smear microscopy [52]. Fluorescent microscopy offers similar specificity but higher sensitivity than bright field microscopy. One of the advantages of fluorescent microscopy is that smears are examined with an objective of acquiring lower augmentation (20× and 40×) than in the ZN technique (100×), which allows the scanning of a much bigger surface of the smear in less time. Fluorescence microscopy exploits fluorescent dyes, which absorb light rays of shorter wavelengths and emit light rays of longer wavelengths. The fluorescent microscope uses fluorescent bulbs containing inert gas within a glass casing. The fluorescent light is thus emitted quickly, releases trapped energy, and returns to stability. Fluorescent lights produce UV radiation and then convert it into visible light through a phosphor coating inside the bulb. However, the development of illumination systems based on LEDs, called LED fluorescence microscopy (LED-FM), has taken over conventional fluorescent microscopy as it utilizes electromagnetic radiation and does not waste energy by producing waste heat or nonvisible electromagnetic radiation (such as ultraviolet (UV)) [53]. A fluorescent dye, auramine O, is used in fluorescent microscopy for the detection of Mtb as auramine O has a specific affinity to mycolic acids. The dye forms a complex with mycolic acids present in the cell wall of mycobacteria, and the backgrounds, including the artifacts, are counterstained by potassium permanganate. The cells visualized under UV light appear as bright yellow or reddish-orange rods against a dark background [54, 55]. LED microscopy is also cost-effective; thus, it has been phased as an alternative to conventional ZN light microscopy and conventional fluorescent microscopy [56].

The smear prepared for fluorescent microscopy should be as thin as the smear for ZN staining. Slides are placed on a staining rack with smears facing up and flooded with 0.1% auramine O reagent for 10 min. Each slide is rinsed individually in a gentle stream of running water until all free stains were washed away. For decolorization, slides are further flooded with 3% acid alcohol (decolorizing solution) for 2–3 min and then thoroughly rinsed with water, and the slides are flooded with potassium permanganate, countered stain for 3–4 min, washed, air-dried, and observed under the microscope using 40x objective for AFB [57].

Table 5.1 Grading scales for light (Ziehl-Neelsen) and fluorescent microscopy

Union/WHO scale 100× field = HPF result	Bright field (1000× magnification)	Fluorescence (200–250× magnification)	Fluorescence (400× magnification)
NEGATIVE	0 AFB/100 OIF	0 AFB/100 HPF	0 AFB/100 HPF
SCANTY	1–9 AFB/100 OIF	1–29 AFB/100 HPF	1–19 AFB/100 HPF
1+	10–99 AFB/100 OIF	30–299 AFB/100 HPF	20–199 AFB/100 HPF
2+	1–10 AFB/1 OIF (on average)	10–100 AFB/1 HPF	05–50 AFB/1 HPF
3+	>10 AFB/1 OIF (on average)	>100 AFB/1 HPF	>50 AFB/1 HPF

Source: NTEP Training Modules [58]

5.3.3 *Reporting of Microscopy Results*

The results of microscopy are analyzed using the grading scales as per WHO guidelines (Table 5.1).

5.4 Advanced Microscopy for Mycobacterial Research

While not currently recommended for TB diagnosis, there are several advanced microscopy techniques, such as confocal laser scanning microscopy, fluorescent resonance energy transfer microscopy, single-molecule localization microscopy, saturated structure illumination microscopy, time-lapse microscopy, and bio-orthogonal corrective light electron microscopy, that are being used in TB research. These techniques offer higher resolution, and some of them can be used to detect Mtb in infected tissues. However, these methods require expensive equipment and highly skilled professionals to operate them. In wide-field microscopy, light/fluorescence above and below the focal plane of the objective lens will also be detected by the eyepiece or the detector. These signals from the out-of-focus regions add blurs to the image, thus reducing the resolution. Laser scanning confocal microscopy (LSCM) provides a solution to this shortcoming. LSCM rejects any out-of-focus light by focusing the illumination and detection optics on the same diffraction-limited spot in the sample, such that only one spot of the sample is illuminated at a time. The confocal image of the sample is built up by sweeping a laser beam across the sample via scanning galvanometer mirrors, and images are acquired point by point. These images are then assembled together, and a complete image of the sample/specimen is formed without any out-of-focus blurring. This technique can generate both 2D and 3D images. LSCM excludes all nonspecific fluorescence by analyzing the emission spectra of each fluorescence and omitting any fluorescence other than those that do not emit a spectrum similar to the specific fluorescent markers used in the study. This confirms the specificity of the immunolabeling.

Erokhina et al. have demonstrated that LCMS-assisted immunohistochemical detection of Mtb using anti-TB polyclonal antibodies provides better resolution and specificity for Mtb detection in comparison to wide-field immunofluorescent microscopy [59].

5.5 Conclusion

Bacteriologic and/or microscopic methods historically remained the mainstay of TB diagnosis until very recently. Only after the advent of highly sensitive and specific molecular diagnostics tests that international governing bodies have started discouraging the use of these methods because of the long turnaround time (culture-based diagnostics) and low sensitivity and specificity (culture/microscopy). It is critical that results from these tests are further confirmed by using other biochemical/molecular methods, and repeat tests must be performed in patients who demonstrate clinical symptoms but tested negative by microscopy or culture. Despite all these limitations, these classical methods remain relevant in resource-limited settings for the diagnosis of TB and are still the methods of choice for AST, surveillance, epidemiologic studies, and the assessment of treatment response. Efforts to improve currently used media or develop new media that support faster growth of Mtb and use of artificial intelligence guided evaluation of microscopic slides are innovations, which promise increase the utility of these methods in TB diagnosis and help in the management of this global health threat.

References

1. Pigrau-Serrallach C, Rodríguez-Pardo D. Bone and joint tuberculosis. Eur Spine J. 2013;22(4):556–66.
2. Choi EH, Coyle WJ. Gastrointestinal tuberculosis. Microbiol Spectr. 2016;4(6).
3. Catalogue of mutations in Mycobacterium tuberculosis complex and their association with drug resistance [Internet]. [cited 2023 Mar 19]. https://www.who.int/publications-detail-redirect/9789240028173.
4. Singh R, Dwivedi SP, Gaharwar US, Meena R, Rajamani P, Prasad T. Recent updates on drug resistance in *Mycobacterium tuberculosis*. J Appl Microbiol. 2020;128(6):1547–67.
5. Eddabra R, Ait BH. Rapid molecular assays for detection of tuberculosis. Pneumonia. 2018;10:4.
6. MacLean E, Kohli M, Weber SF, Suresh A, Schumacher SG, Denkinger CM, et al. Advances in molecular diagnosis of tuberculosis. J Clin Microbiol. 2020;58(10):e01582–19.
7. Schön T, Miotto P, Köser CU, Viveiros M, Böttger E, Cambau E. Mycobacterium tuberculosis drug-resistance testing: challenges, recent developments and perspectives. Clin Microbiol Infect. 2017;23(3):154–60.
8. Diagnosis of tuberculosis disease [Internet]. [cited 2023 Mar 17]. https://www.cdc.gov/tb/education/corecurr/pdf/chapter4.pdf.
9. Talbot EA, Adams LV, Fordham von Reyn C. The importance of culture for diagnosing tuberculosis. Clin Infect Dis. 2005;41(8):1213–4.

10. Canetti G, Froman S, Grosset J, Hauduroy P, Langerová M, Mahler HT, et al. Mycobacteria: laboratory methods for testing drug sensitivity and resistance. Bull World Health Organ. 1963;29(5):565–78.
11. Training manual M tuberculosis C DST [Internet]. [cited 2023 Mar 19]. https://tbcindia.gov.in/WriteReadData/l892s/6995271860Training%20manual%20M%20tuberculosis%20C%20DST.pdf.
12. Kato-Maeda M, Metcalfe JZ, Flores L. Genotyping of Mycobacterium tuberculosis: application in epidemiologic studies. Future Microbiol. 2011;6(2):203–16.
13. World Health Organization. Monitoring treatment response [Internet]. Companion handbook to the WHO guidelines for the programmatic management of drug-resistant tuberculosis. 2014 [cited 2023 Mar 19]. https://www.ncbi.nlm.nih.gov/books/NBK247422/.
14. Kubica GP, Dye WE, Cohn ML, Middlebrook G. Sputum digestion and decontamination with N-acetyl-L-cysteine-sodium hydroxide for culture of mycobacteria. Am Rev Respir Dis. 1963;87:775–9.
15. Lowenstein Jensen medium [Internet]. [cited 2023 Mar 19]. https://legacy.bd.com/europe/regulatory/Assets/IFU/Difco_BBL/244420.pdf.
16. Pfyffer GE, Wittwer F. Incubation time of mycobacterial cultures: how long is long enough to issue a final negative report to the clinician? J Clin Microbiol. 2012;50(12):4188–9.
17. Middlebrook G, Cohn ML. Bacteriology of tuberculosis: laboratory methods. Am J Public Health Nations Health. 1958;48(7):844–53.
18. Middlebrook 7H10 Agar [Internet]. [cited 2023 Mar 18]. https://www.bd.com/resource.aspx?IDX=9004.
19. Middlebrook 7H11 [Internet]. [cited 2023 Mar 18]. https://legacy.bd.com/europe/regulatory/Assets/IFU/US/8801671(0703)en.pdf.
20. Middlebrooke 7H9 [Internet]. [cited 2023 Mar 17]. https://legacy.bd.com/europe/regulatory/Assets/IFU/Difco_BBL/212352.pdf.
21. Lawson L, Emenyonu N, Abdurrahman ST, Lawson JO, Uzoewulu GN, Sogaolu OM, et al. Comparison of Mycobacterium tuberculosis drug susceptibility using solid and liquid culture in Nigeria. BMC Res Notes. 2013;6(1):215.
22. Park SK, Kim SC, Kim DM, Lee CW, Kim Y, Cho SN. Fully automated liquid culture system compared with Lowenstein-Jensen solid medium for rapid recovery of Mycobacteria in sputums. Tuberc Respir Dis. 2002;53(6):635.
23. World Health Organization. Technical manual for drug susceptibility testing of medicines used in the treatment of tuberculosis [Internet]. Geneva: World Health Organization; 2018 [cited 2023 Mar 19]. p. 39. Report No.: 9789241514842. https://apps.who.int/iris/handle/10665/275469.
24. Heifets L, Linder T, Sanchez T, Spencer D, Brennan J. Two liquid medium systems, mycobacteria growth indicator tube and MB redox tube, for *Mycobacterium tuberculosis* isolation from sputum specimens. J Clin Microbiol. 2000;38(3):1227–30.
25. Caulfield AJ, Wengenack NL. Diagnosis of active tuberculosis disease: from microscopy to molecular techniques. J Clin Tuberc Mycobact Dis. 2016;4:33–43.
26. Tortoli E, Cichero P, Piersimoni C, Simonetti MT, Gesu G, Nista D. Use of BACTEC MGIT 960 for recovery of mycobacteria from clinical specimens: multicenter study. J Clin Microbiol. 1999;37(11):3578–82.
27. Pfyffer GE, Welscher HM, Kissling P, Cieslak C, Casal MJ, Gutierrez J, et al. Comparison of the Mycobacteria Growth Indicator Tube (MGIT) with radiometric and solid culture for recovery of acid-fast bacilli. J Clin Microbiol. 1997;35(2):364–8.
28. Anargyros P, Astill DS, Lim IS. Comparison of improved BACTEC and Lowenstein-Jensen media for culture of mycobacteria from clinical specimens. J Clin Microbiol. 1990;28(6):1288–91.
29. Chien HP, Yu MC, Wu MH, Lin TP, Luh KT. Comparison of the BACTEC MGIT 960 with Löwenstein-Jensen medium for recovery of mycobacteria from clinical specimens. Int J Tuberc Lung Dis. 2000;4(9):866–70.

30. Diriba G, Kebede A, Yaregal Z, Getahun M, Tadesse M, Meaza A, et al. Performance of Mycobacterium growth indicator tube BACTEC 960 with Lowenstein–Jensen method for diagnosis of Mycobacterium tuberculosis at Ethiopian National Tuberculosis Reference Laboratory, Addis Ababa, Ethiopia. BMC Res Notes. 2017;10(1):181.
31. Kumari P, Thakur JK, Kumar P, Kumar R, Parekh D. Comparison of LJ medium and BACTEC MGIT 960 culture system for the diagnosis of tuberculosis. J Clin Diagn Res [Internet]. 2020 [cited 2023 Mar 17]. https://jcdr.net/article_fulltext.asp?issn=0973-709x&year=2020&volume=14&issue=12&page=DC09&issn=0973-709x&id=14304.
32. Palomino JC. Nonconventional and new methods in the diagnosis of tuberculosis: feasibility and applicability in the field. Eur Respir J. 2005;26(2):339–50.
33. Gravet A, Souillard N, Habermacher J, Moser A, Lohmann C, Schmitt F, et al. Culture and susceptibility testing of mycobacteria with VersaTREK. Pathol Biol (Paris). 2011;59(1):32–8.
34. Espasa M, Salvadó M, Vicente E, Tudó G, Alcaide F, Coll P, et al. Evaluation of the VersaTREK system compared to the Bactec MGIT 960 system for first-line drug susceptibility testing of Mycobacterium tuberculosis. J Clin Microbiol. 2012;50(2):488–91.
35. Yuksel P. Comparison of the VersaTrek and BACTEC MGIT 960 systems for the contamination rate, time of detection and recovery of mycobacteria from clinical specimens. Afr J Microbiol Res [Internet]. 2011 [cited 2023 Mar 17];5(9). http://www.academicjournals.org/ajmr/abstracts/abstracts/abstract%202011/4May/Yuksel%20et%20al.htm.
36. Falconi FQ, Infante Suárez L, López MJ, García SC. Comparison of the VersaTREK system and Löwenstein-Jensen medium for the recovery of mycobacteria from clinical specimens. Scand J Infect Dis. 2008;40(1):49–53.
37. Piersimoni C, Scarparo C, Callegaro A, Tosi CP, Nista D, Bornigia S, et al. Comparison of MB/BacT ALERT 3D system with radiometric BACTEC system and Löwenstein-Jensen medium for recovery and identification of mycobacteria from clinical specimens: a multicenter study. J Clin Microbiol. 2001;39(2):651–7.
38. Ängeby KAK, Werngren J, Toro JC, Hedström G, Petrini B, Hoffner SE. Evaluation of the BacT/ALERT 3D system for recovery and drug susceptibility testing of Mycobacterium tuberculosis. Clin Microbiol Infect. 2003;9(11):1148–52.
39. Martinez MR, Sardiñas M, Garcia G, Mederos LM, Díaz R. Evaluation of BacT/ALERT 3D system for mycobacteria isolates. J Tuberc Res. 2014;02(02):59–64.
40. WHO consolidated guidelines on tuberculosis: module 3: diagnosis: rapid diagnostics for tuberculosis detection, 2021 update [Internet]. [cited 2023 Mar 17]. https://www.who.int/publications-detail-redirect/9789240029415.
41. Desikan P. Sputum smear microscopy in tuberculosis: is it still relevant? Indian J Med Res. 2013;137(3):442–4.
42. Maynard-Smith L, Larke N, Peters JA, Lawn SD. Diagnostic accuracy of the Xpert MTB/RIF assay for extrapulmonary and pulmonary tuberculosis when testing non-respiratory samples: a systematic review. BMC Infect Dis. 2014;14(1):709.
43. Hepple P, Ford N, McNerney R. Microscopy compared to culture for the diagnosis of tuberculosis in induced sputum samples: a systematic review [review article]. Int J Tuberc Lung Dis. 2012;16(5):579–88.
44. Chawla K, Gupta S, Mukhopadhyay C, Rao PS, Bhat SS. PCR for M. tuberculosis in tissue samples. J Infect Dev Ctries. 2009;3(2):83–7.
45. Rufai SB, Kumar P, Singh A, Prajapati S, Balooni V, Singh S. Comparison of Xpert MTB/RIF with line probe assay for detection of rifampin-monoresistant Mycobacterium tuberculosis. J Clin Microbiol. 2014;52(6):1846–52.
46. Raizada N, Sachdeva KS, Nair SA, Kulsange S, Gupta RS, Thakur R, et al. Enhancing TB case detection: experience in offering upfront Xpert MTB/RIF testing to pediatric presumptive TB and DR TB cases for early rapid diagnosis of drug sensitive and drug resistant TB. PLoS One. 2014;9(8):e105346.
47. Tuberculosis: standard treatment guidelines [Internet]. [cited 2023 Mar 18]. https://www.nhm.gov.in/images/pdf/guidelines/nrhm-guidelines/stg/stg-tb.pdf.

48. Singhal R, Myneedu VP. Microscopy as a diagnostic tool in pulmonary tuberculosis. Int J Mycobacteriol. 2015;4(1):1–6.
49. Allen JL. A modified Ziehl-Neelsen stain for mycobacteria. Med Lab Sci. 1992;49(2):99–102.
50. RNTCP Lab Network Guidelines.pdf [Internet]. [cited 2023 Mar 18]. https://tbcindia.gov.in/WriteReadData/l892s/4234099618RNTCP%20Lab%20Network%20Guidelines.pdf.
51. Lahiri K, Chatterjee S. A simple cold staining method for acid fast bacilli. Med J Armed Forces India. 1994;50(4):256–8.
52. Riley PA. Principles of microscopy, culture and serology-based diagnostics. Medicine (Baltimore). 2017;45(10):639–44.
53. World Health Organization. Fluorescent light-emitting diode (LED) microscopy for diagnosis of tuberculosis: policy statement [Internet]. World Health Organization; 2011 [cited 2023 Mar 18]. Report No.: WHO/HTM/TB/2011.8. https://apps.who.int/iris/handle/10665/44602.
54. Lichtman JW, Conchello JA. Fluorescence microscopy. Nat Methods. 2005;2(12):910–9.
55. Young MR. Principles and technique of fluorescence microscopy. J Cell Sci. 1961;s3-102(60):419–49.
56. Reza LW, Satyanarayna S, Enarson DA, Kumar AMV, Sagili K, Kumar S, et al. LED-fluorescence microscopy for diagnosis of pulmonary tuberculosis under programmatic conditions in India. PLoS One. 2013;8(10):e75566.
57. Flourescence_Microscopy Manual.pdf [Internet]. [cited 2023 Mar 18]. https://tbcindia.gov.in/WriteReadData/l892s/7890638455Flourescence_Microscopy%20Manual.pdf.
58. NTEP training modules [Internet]. [cited 2023 Mar 19]. https://tbcindia.gov.in/WriteReadData/NTEPTrainingModules1to4.pdf.
59. Erokhina MV, Nezlin LP, Avdienko VG, Voronezhska EE, Lepekha LN. Immunohistochemical detection of Mycobacterium tuberculosis in tissues of consumptives using laser scanning microscopy. Biol Bull. 2016;43(1):21–5.
60. WHO. Tuberculosis (TB) [Internet]. 2022 [cited 2023 Mar 17]. https://www.who.int/news-room/fact-sheets/detail/tuberculosis.

Chapter 6
Molecular Diagnosis of Tuberculosis

Sanjay Singh Negi, Pushpendra Singh, and Kuldeep Sharma

Abstract Tuberculosis (TB) is a global health problem as it causes a significantly large number of new cases and mortality every year across the world. Its magnitude has further, been aggravated due to various factors including the emergence of drug resistance cases (multidrug, extensive and total drug resistance TB), overcrowding and poverty forcing poor nutrition and various precipitating infectious agents like HIV, HCV, leishmania enabling Mycobacterium tuberculosis (MTB) complex to sets in infection in patients coinfected with these agents. Thus, early, accurate, and specific diagnosis is required for the initiation of specific antitubercular treatment (ATT) for effective management. Although conventional clinical, microscopic, and radiological diagnostic approaches are indispensable, they have their limitations. Further, the low sensitivity of conventional diagnostic approach based on microscopy and culture and non-recommended serological tests has warranted the utmost need to develop more rapid and accurate tests for the diagnosis of MTB complex. The slow generation time of 16–18 hours of MTB restrict the utility of culture for diagnosis as it takes an average of 3-4 weeks to maximum of 6-8 weeks for confirmation. Conventional polymerase chain reaction (PCR), real-time PCR (RT-PCR), loop-mediated isothermal amplification (LAMP) assay, line probe assay (LPA), automated systems like GeneXpert and Truenat, and whole genome or target sequencing have made a tremendous advancement in the diagnosis of both pulmonary and extrapulmonary TB. The unique advantage has already been witnessed and reported in extrapulmonary TB, wherein the paucibacillary nature of the bacilli has made conventional approach almost impossible to diagnose it. This book chapter accordingly highlights the recent development witnessed in the molecular diagnosis of both TB and drug resistance.

Keywords Tuberculosis · PCR · Xpert · Truenat · LAMP · Drug resistance

The original version of this chapter has been revised. The correction to this chapter can be found at https://doi.org/10.1007/978-981-99-5624-1_15

S. S. Negi (✉) · P. Singh · K. Sharma
Department of Microbiology, All India Institute of Medical Sciences (AIIMS), Raipur, Chhattisgarh, India
e-mail: negidr@aiimsraipur.edu.in

A. Singh, D. Sharma (eds.), *Diagnosis of Mycobacterium*,
https://doi.org/10.1007/978-981-99-5624-1_6

6.1 Introduction

Tuberculosis (TB) is still being recognized as the leading infectious disease globally. The recent global TB report, 2022, has reported 1.6 million deaths and 10.6 million new cases worldwide due to TB in 2021 [1]. The situation is further aggravated by an increase in the number of cases of drug-resistant TB. The World Health Organization (WHO) has reported 450, 000 new cases of rifampicin (RIF)-resistant TB in 2021 [1]. Furthermore, isoniazid (INH), another important drug for treatment, was reported to have a resistance of 7.1% in new cases, along with 7.9% in previously treated TB cases [1].

TB is caused by *Mycobacterium tuberculosis* (*M. tuberculosis*) complex (MTB), which comprises *M. tuberculosis* (human tubercle bacillus, MTB), *M. bovis* (bovine tubercle bacillus), *M. canetti* (reported from a few cases in East Africa), *M. africanum* (reported in a few cases from West Africa), *M. caprae* (closely related to *M. bovis*), *M. microti* (vole bacillus, rare and less virulent), and *M. pinnipedii* (recently isolated from humans). Among all of them, MTB is the most common causative agent of TB. The most common manifestation of MTB is pulmonary TB (PTB), wherein the lung is the affected site. However, through a hematogenous route, it can disseminate to other organs to cause extrapulmonary TB (EPTB). Virtually, except hair, it can affect any organ. The commonest site is the lymph nodes (posterior cervical and supraclavicular lymph nodes), to cause tubercular lymphadenitis, pleura(pleural TB), upper airways (laryngx, pharynx, and epiglottis), renal TB and genital TB in which the fallopian tube and endometrium are commonly involved in females to cause infertility, while in males, the epididymis is the most affected site, skeletal TB, tubercular meningitis, gastrointestinal TB, tubercular pericarditis, skin tubercular lesin of scrofuloderma and lupus vulgaris, and military or disseminated TB. The management of TB is an uphill task since it requires multi-antimicrobial therapy for a longer period of time, ranging from a minimum of 6 months up to even 2 years. The management is severely hampered by the emergence of various drug-resistant forms of bacilli, like multidrug-resistant TB (MDR-TB), extensive drug-resistant TB (XDR-TB), and total drug-resistant TB (TDR-TB). Accordingly, better and appropriate management of TB utmost requires an early, sensitive, and specific diagnosis of MTB to initiate optimal treatment in positive cases to prevent its further spread and to reduce associated mortality and morbidity [2].

6.2 Laboratory Diagnosis

The diagnosis of TB includes conventional, radiographical, and molecular diagnostic modalities. Since this chapter focuses on the molecular diagnosis of TB, the conventional approach is very briefly discussed.

6.3 Conventional Methods for the Diagnosis of TB

Conventional diagnosis is the most widely used microscopy post Ziehl-Neelson (ZN) or fluorescent staining and culture for the isolation of MTB on either solid culture medium like Lowenstein Jensen (LJ Media) or liquid culture system like Mycobacteria Growth Indicator Tube (MGIT-Automated). Microscopy is a rapid test and may be completed within half an hour to detect acid-fast bacilli (AFB) in the smear examination. Since its limit of detection requires a minimum bacterial load of 10,000/mL of clinical sample, it lacks sensitivity and specificity [2]. It further does not differentiate between various *Mycobacterium* species [2]. Culture, on the other hand, is considered a gold standard as it provides specific growth identification features to identify MTB. However, although it is more sensitive than any other conventional methods, it is time-consuming. Culture using LJ media requires 4–8 weeks while the liquid culture system around 2–8 weeks. Further, it requires laboratory infrastructure and trained staff. The identification of antitubercular (ATT) drug sensitivity patterns requires another 4 to 6 weeks. Therefore, the utmost need for a new rapid, sensitive, and specific test was felt. The World Health Organization's (WHO's) End TB Strategy has set a goal of 95% reduction in TB death and 90% reduction in incidence rate by 2035. This ambitious plan for TB elimination highlights the critical role of laboratories in providing an early, sensitive, and specific diagnosis of TB and universal drug susceptibility testing to determine the MTB sensitivity pattern against antitubercular drugs. Various drug-resistant forms of TB, like monodrug resistant-, multidrug-resistant (MDR)-, XDR-, and TDR-TB, have further warranted the consistent demand for an upgrade in conventional drug-susceptibility testing (DST) to provide fast and reliable sensitivity patterns of the MTB [3].

6.4 Molecular Methods for the Diagnosis of TB

Molecular diagnosis, in the last one/two decades, has provided a revolutionary improvement in the field of both the detection and sensitivity testing of MTB. Presently, a variety of nucleic acid amplification test (NAAT)-based assays are being used to determine the drug susceptibility of MTB for the effective treatment of TB patients. DST-MTB, in relation to the "critical concentration," compares the growth of susceptible versus resistant strains [4]. The discrepancies among genotype and phenotype DST approaches, which are limited mainly to certain antibiotics, need to be resolved by more research into resistance mechanisms [5]. The phenotypic and genotypic methods should be used to define the antibiotic resistance profile to treat patients with MDR-TB and XDR-TB in the era of drug resistance MTB [6].

Rifampicin resistance is a suboptimal surrogate for MDR-TB. Resistance to rifampicin (RIF) of more than 95% in MTB is principally connected to changes within 81 bp (codons 507 to 533) RIF resistance-determining region (RRDR) in the *rpoB* gene [7, 8]. Earlier studies have shown that mutations in the *rpoB* RRDR are more probable to deliberate high levels of RIF resistance [8–10]. Single nucleotide changes are the domination mutations, resulting in mutations of particular amino acid residues at positions 516, 526, and 531. In-frame insertion and deletion changes also happen, although at lesser incidences at residue positions 511, 516, 518, and 522. In the *rpoB* gene, especially the RRDR region, one deletion, two insertions, five multiple mutations (with two or three concerned codons), and 16 single base mutations, were reported [11, 12] (Table 6.1).

Resistance to isoniazid (INH) is a complex process that generates mutations/substitutions/deletions in various genes, like *katG*, *inhA*, *ahpC*, *kasA*, and *ndh*, which have all been related to INH resistance [13]. The INH is a drug needing activation by the catalase/peroxidase protein that encode by *katG*. This activation of INH is severly hampered due to the frequently reported mutation S315T in the *katG* gene resulting in drug resistance against INH. Another mechanism contains mutations in the *inhA* promoter at C15T, which results in an excess expression of inhA, to exhibit low-level INH resistance [14]. Torres et al. exposed new changes in sequence that are capable of elucidating 98% of phenotypic INH resistance through *katG* and *inhA* promoter substitution [15] (Table 6.1).

Resistance to pyrazinamide (PZA) is primarily caused by mutations in the *pncA* gene of MTB. The mutations are remarkably situated and spread throughout the *pncA* gene, which encodes the pyrazinamidase protein or enzyme, which activates the drug [16]. Different studies on PZA resistance had reported the clustered mutations in the three conserved regions at amino acids positions 3–71, 61–85, and 132–142 most responsible for drug resistance against PZA [17–19] (Table 6.1). Groups of PZA-resistant MTB with identical *pncA* mutations have been reported but so far have been geographically restricted [19]. Studies have reported PZA resistance of more than 90% in XDR-TB isolates and about 40–50% in primary MDR-TB [20].

Ethambutol (EMB) is a bacteriostatic agent that acts against MTB by preventing membrane-related arabinosyl transferases, which are encoded by the *embCAB* operon (*embA*, *embB*, and *embC*) [21]. Earlier studies exhibited that common EMB-resistant clinical isolates that have mutations within *embB*, mainly at position 306, could be used as resistance markers for the fast detection of EMB resistance [22]. Some of the changes within the upstream region of *embA* are similarly connected with EMB resistance [21] (Table 6.1).

Molecular techniques have seen a revolutionary evolution in the last two decades for the diagnosis of MTB and the determination of drug resistance profile with the emergence of various molecular tests like polymerase chain reaction (PCR), real-time PCR, loop-mediated isothermal amplification (LAMP), TruNat, GeneXpert, specific target nucleotide sequencing, and next-generation sequencing (NGS).

This chapter has thus attempted to summarize the existing molecular diagnostic tests and the future aspects in the field of detection of MTB and Drug resistant (DR)-MTB. Instead of exhaustive detail of all the commercial NAAT tests, the chapter provides the basic priniciples of the tests along with their sensitivity, specificity, TAT (Turn Around Time), advantages and limitations.

Table 6.1 Anti-mycobacterial drugs and mechanisms of drug resistance [23]

S. no.	Agent	Mode of action	Target	Gene	Most prevalent mutation	Frequency in resistant strain
1	Isoniazid	Inhibition of mycolic acid synthesis	Mycolic acids	*katG*	Ser-315-Thr	60–70
		Drug target alteration	Enoyl-ACP reductase	*inhA*	Asn-146, Asn-268, and Asn-302	
		Drug target overexpression	Enoyl-ACP reductase promoter	*inhA* promotor	Cys-15-Thr	
2	Rifampicin	Inhibition of RNA polymerase	RNA polymerase	*rpoB*	Asp-516-Val, His-526-Tyr, Ser-531-Leu	Up to 70
3	Pyrazinamide	Inhibition of energy production and trans-translation	Fatty acid synthase-I, ribosomal protein S1	*pncA*	Asp-12-Ala/Asn, Leu-85-Pro	70–100
4	Ethambutol	Inhibition of arabinogalactan synthesis	Arabinosyl transferases	*embCAB*	Met-306-Val/Ile/Leu	69
5	Streptomycin	Binds to the small 16S rRNA of the 30S ribosomal subunit irreversibly, along with protein S12	Ribosomal protein S12	*rpsL*	Lys-43-Arg	60
6	Fluoroquinolone	Inhibition in cleaved complexes	DNA gyrase	*gyrA*	Ala-90-Val, Ser-91-Pro, and Asp-94-(Gly/Ala/His/Asn)	>90
7	Ethionamide	Inhibition in fatty acid biosynthesis pathway	Enoyl-ACP reductase	*inhA*	Asn-146, Asn-268, and Asn-302	<10
		Inhibits the cell wall synthesis	NADPH-specific FAD-containing monooxygenase	*EthA*	Val202-Leu	

The following molecular tests are presently being used in the routine diagnosis of TB.

6.4.1 Loop-Mediated Isothermal Amplification (LAMP)

LAMP is a molecular test based on the isothermal nucleic acid amplification technique developed by Notomi et al., [24]. In LAMP, amplification is characteristically carried out at one particular temperature to help avoid any requirement of a thermal cycler, which is mandatory for PCR. It employs a set of four specific primers targeting the six sites of selective DNA fragment and a strand displacing DNA polymerase to provide a rapid, sensitive, and specific amplification of the target DNA under isothermal conditions without requiring expensive instruments and reagents. The isothermal assay is an ultrasensitive nucleic acid amplification method and can produce large amounts of nucleic acid with minute quantities of DNA templates within a short time. The negative and positive results can easily be detemined by the color differentiation of the assay.

Several LAMP-based tests have been developed and is being used in the diagnosis of TB. It has been recommended by the World Health Organization (WHO) in August, 2016, as an alternate approach to smear microscopy for the diagnosis of pulmonary TB in adults [25]. It is recommended by WHO as a follow-up test in adult patients with signs and symptoms consistent with pulmonary TB, especially in smear-negative sputum specimens (Table 6.2). Various targets like *gyrB*, *rrs*, *rimM*, *IS6110*, *hspX*, *mpb64*, and *sdaA* genes, have been used in TB-LAMP assay [26–29]. Various earlier studies have found it around 92% to 100% more sensitive for smear-positive samples than smear-negative samples, where sensitivity is reported between 52.1% and 90.3% [30–33]. LAMP test has also shown promising result in extra pulmonary samples when compared to conventional microscopy, solid and liquid culture. In extrapulmonary samples its sensitivity of 95.6% was found statistically significant to smear microscopy, solid and liquid culture having sensitivity of 17.4%, 65.2%, and 69.6% [34]. It is however found with a shortcoming of exposure of reaction tubes to aerosol contamination, showing a false-positive result. WHO accordingly has excluded all data obtained from extrapulmonary samples to keep them still under investigation [25] (Fig. 6.1).

6.4.2 COBAS TaqMan TB

This test is developed by Roche Diagnostics, Tokyo, Japan, to replace the COBAS amplicon assay [35]. This is a real-time PCR assay that amplifies specific Taqman probes and primers to amplify the specific segment of 16S rRNA to detect the presence or absence of MTB complex in the clinical specimen in around 2.5 h for a maximum of 48 samples. This test is approved by the US Food and Drug

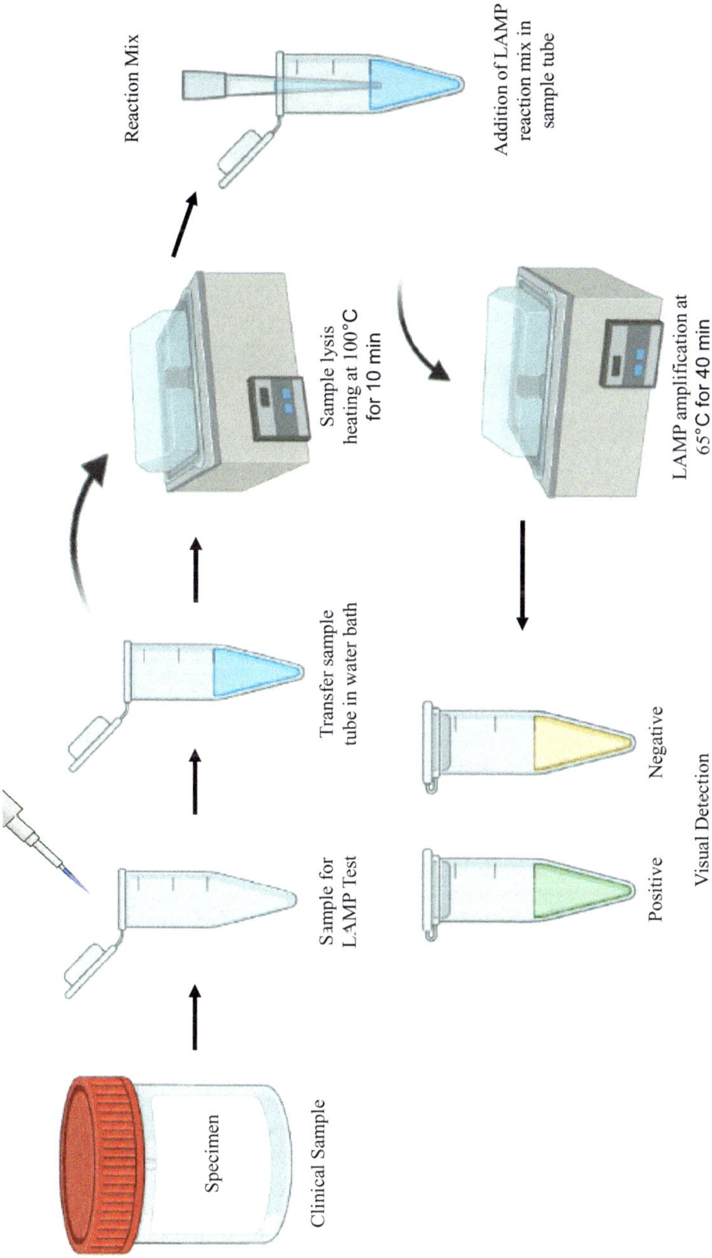

Fig. 6.1 Loop-mediated isothermal amplification (LAMP)

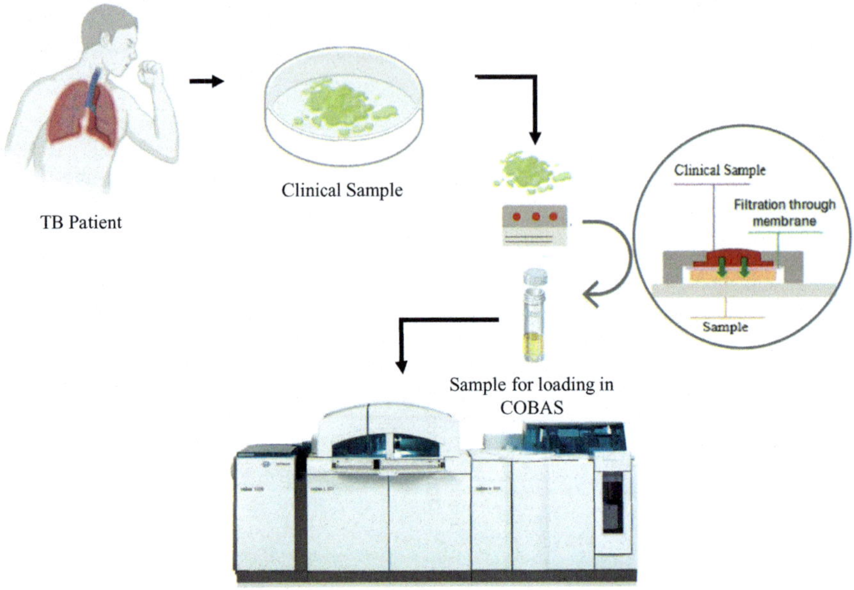

Fig. 6.2 COBAS TaqMan TB

Administration (FDA) for smear-positive and/or smear-negative pulmonary disease. Although this assay is limited by the manufacturer to respiratory specimens only, it has also been evaluated in extrapulmonary specimens and found its diagnostic sensitivity (63.6%) and specificity (94.6%) low for the extrapulmonary specimens in comparison to pulmonary specimens having the sensitivity of 88.4% and specificity of 98.8% [36, 37]. It was further found that this assay has higher sensitivity in smear-positive specimens than in smear-negative specimens. A sensitivity parameter was reported between 96.9% and 98% in smear-positive specimens and between 34.9% and 79.5% in smear-negative samples [36, 38, 39] (Fig. 6.2).

6.4.3 *GeneXpert MTB/RIF Assay*

This is an automated cartridge-based nucleic acid amplification test (CBNAAT) developed by Cepheid Inc., Sunnyvale, California, United States of America. This test simultaneously provides the result for the detection of MTB and rifampicin resistance/sensitive pattern within 2 h of the test, with the advantage of minimum human intervention [40]. It has also been approved by WHO and US FDA [41, 42] (Table 6.2). For this reason, this assay is also known as Xpert MTB/RIF test. This assay has been reported as a sensitive test for the rapid diagnosis of TB in comparison to conventional techniques [43]. Its sensitivity has been reported at 100% in smear-positive samples [43, 44]. The added advantage of the assay is

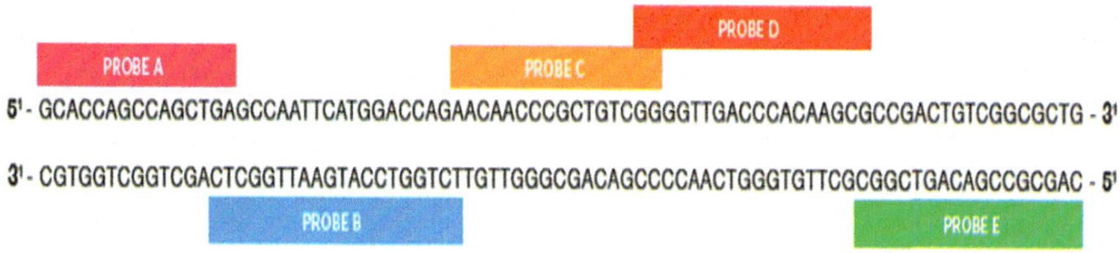

Fig. 6.3 The *rpoB* gene 81 bp RIF resistance determining region and Xpert MTB/RIF probes

providing results for the RIF sensitivity/resistance pattern to the tune of 94% sensitivity and 98% specificity. The biggest advantage of the GeneXpert has been observed in smear-negative specimens, wherein it has also been found to be a good tool for the diagnosis of TB. Earlier studies have shown a sensitivity of 47.8% to 73% in smear-negative pulmonary specimens while a sensitivity of 28.2% to 73.2% in smear-negative extrapulmonary specimens [43, 44]. It is, however, reported by some studies as less sensitive than liquid culture in children and adults for the diagnosis of TB [45, 46]. The mutations yielding the expected probe reactions are in different colors [47]. Further, in detecting rifampicin drug resistance, it has been reported that since its probe is specifically designed to detect mutation in 81 bp stretch of RRDR of the *rpo B* gene, it may fail to detect resistance against rifampicin if the resistance-conferring mutation occurs outside this region, like I491F mutation. It was confirmed in one of the earlier studies that it only detected rifampicin resistance in 38 (30%) multidrug resistance cases out of 125 cases in comparison to DNA sequencing [48]. Thus, the inability of the GeneXpert to detect resistance-conferring silent mutations within and outside of the RRDR region may render the treatment ineffective, causing a serious jolt to optimistic WHO TB elimination (Fig. 6.3).

6.4.4 Gene Xpert MTB/RIF Ultra Assay

To improve on these shortcomings, a new version of the test, known as Xpert Ultra, has been developed. Xpert Ultra has been provided with two amplification targets, IS6110 and IS1081, and by doubling the size of the DNA delivered to PCR and covering the RRDR region from codon positions 507 to 533 by probes for the detection of 27 different mutations, the limit of detection has been reduced from 112.6 CFU/mL in sputum for GeneXpert to 15.6 CFU/mL in sputum for Ultra [49]. The sensitivity of Xpert Ultra has been reported significantly higher than that of Xpert MTB/RIF in both smear-negative, culture-positive TB patients and HIV-infected patients in earlier published studies. Xpert Ultra has been reported for improved detection of rifampicin drug resistance by detecting mutants at codon 533, silent differentiated mutations at codon positions 513 and 514, and hetero-resistant mutants that were earlier missed by susceptibility testing as well as GeneXpert [49]. However, mutation outside the RRDR regions, like Ile491Phe, is

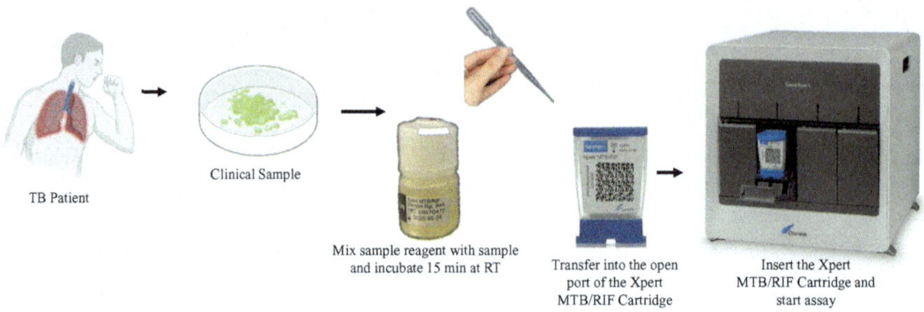

Fig. 6.4 GeneXpert MTB/RIF Ultra assay

still a concern as its not detectable by Xpert Ultra [50]. Nonetheless, due to its higher sensitivity in comparison with GeneXpert, Xpert Ultra is recommended by WHO to replace GeneXpert to improve the diagnosis of TB, especially the paucibacillary forms of TB, like extrapulmonary TB, pediatric TB, or HIV-associated TB [51] (Table 6.2) (Fig. 6.4).

6.4.5 Line Probe Assay (LPA)

Line probe assay (LPA) is a qualitative in vitro diagnostic test based on DNA-STRIP technology for the identification of the MTB complex and its resistance to RIF (mutations within the *rpoB* gene) and INH (mutations within the *katG* gene and the *inhA* promoter) [52]. The LPA test includes DNA extraction, master mix preparation and addition, multiplex amplification with biotinylated primers, and detection with reverse hybridization. The test is an aid in the rapid diagnosis of MDR-TB, which is a prerequisite for the appropriate treatment initiation [53, 54]. Resistance to rifampicin and isoniazid diagnosis is based on detected probe reactions for each individual mutation. Hence, mutated strains with low-level resistance and double or triple mutations, including noted delayed reactions, problems of hybridization, partial inhibition of color development, and mutations outside the region from RRDR, are not diagnosed [11, 55].

To characterize first-line drug-resistant MTB bacilli, WHO has endorsed the existing reverse hybridization-based line probe assay, namely the GenoType MTBDRplus assay for drug resistance [11] (Table 6.2). The drug resistance diagnosis by this assay depends on the amplified gene regions containing resistance-associated mutations determined by reverse hybridization to the immobilized mutated and wild-type sequences on the strip. In the presence of uncommon or rare mutations, the MTBDRplus assay shows its limitation during diagnosis. The GenoType MTBDRplus works on a very short piece of a targeted gene, which leads to failure in the proper diagnosis of resistant or susceptible MTB, mainly when novel or rare nonsynonymous mutations are outside the targeted drug resistance regions. Nucleic-acid-based assays, such as the Genotype MTBDRPlus assay, have

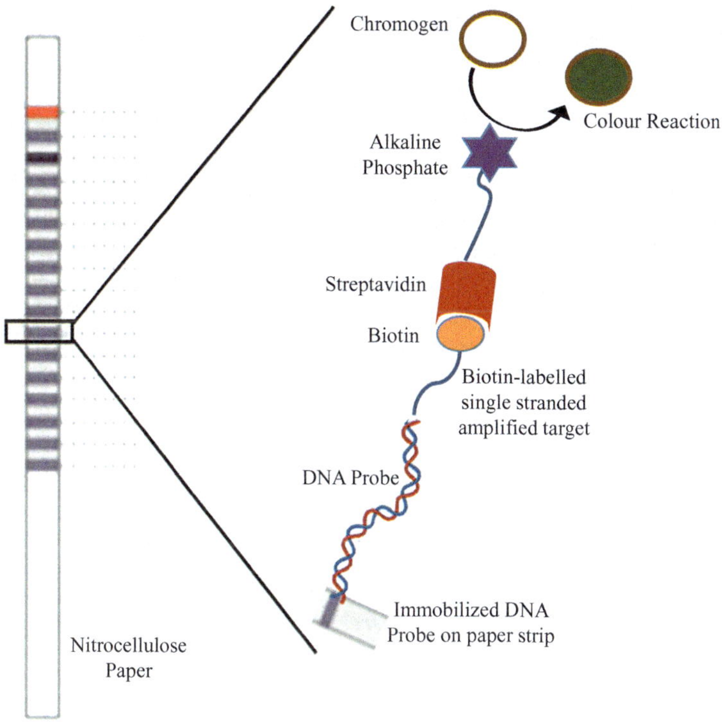

Fig. 6.5 Line probe assay (LPA)

failed to differentiate synonymous and nonsynonymous mutations that may or may not lead to phenotypic drug resistance [56].

GenoType MTBDRsl provides additional information on MDR strains to detect and characterize the resistant strains against second-line drugs such as fluoroquinolones (FLQ) and second-line injectable drugs (SLID) (capreomycin, kanamycin, or amikacin). This assay detects and differentiates the preextensively drug-resistant (pre-XDR) and extensively drug-resistant TB (XDR-TB). XDR-TB occurs when MDR-TB inherits additional resistance to any FLQ (ofloxacin, levofloxacin, gatifloxacin, and moxifloxacin) among any of the SLID aminoglycosides (capreomycin, amikacin, and kanamycin) [57–59]. GenoType MTBDRsl VER 2.0 is a redesigned and improved version of the previous one and bears the ability to detect the most common mutations at *gyr A* and *gyr B* for FLQ resistance and *rrs* and *eis* genes for SLID resistance. The *gyr A* probes target codons 85 to 97, rrs probes target the positions 1401 to 1484, codons 536 to 541 in the gyrB gene and − 10 to −14 position in the eis promoter (Fig. 6.5).

6.4.6 Truenat MTB-RIF Dx

Novel fast molecular diagnostics will possibly dramatically increase MTB detection and point of care, which are key mechanisms of the WHO End TB Strategy. Truenat MTB, Truenat MTB Plus, and Truenat MTB-RIF Dx assays have been manufactured by Molbio Diagnostics/Bigtec Labs, Goa/Bengaluru, India. Truenat assays are an innovative micro real-time PCR assay based on a chip that detects MTB in clinical samples within just about an hour [60] (Table 6.2). After getting positive assay results, one more additional chip should be required to identify RIF drug resistance, with an additional time period. The assay is ready to run in a Truenat lab system, which is a battery-powered preparation system (device for DNA extraction) and an automatic PCR machine, accessible in one, two, or four module configurations. The Truenat system sensitivity for MTB diagnosis is 68–100%, and linkage-to-care is 84–100%. The Truenat system is a very popular and widely used point-of-care (POC) machine that is cost-effective with >88% sensitivity and is decremental cost-effective when sensitivity is 74–88% and specificity is 80–100% (Fig. 6.6).

6.4.7 Sequencing Platforms for the Identification of Novel Mutations of Drug Resistance

Timely and accurate identification of TB diagnosis and drug-resistance mutational changes in sequences plays an important role in the effective management and elimination of TB. Many techniques for targeted or whole-genome sequencing, including the Sanger (capillary-based) and next-generation sequencer (NGS) and third-generation sequencing, are employed for the detection of drug resistance in MTB [61]. Notably, PCR-based targeted and whole-genome sequencing are two different approaches for the analysis of the targeted gene or genome sequence of MTB to reveal a clear picture of all mutations in a single assay [62]. Sequencing is a highly reliable and accurate method used as the gold standard assay for the detection of any mutational changes for the detection of resistance in both reported and novel mutations. Precise whole-genome sequence or targeted gene sequence data can be achieved within 48 h from an automated sequencer manufactured by Thermo Fisher (SeqStudio Flex Genetic Analyzers and Applied Biosystems 3500 Series Genetic Analyzer), Illumina Singapore (MiniSeq, MiSeq, NovaSeq), that use fluorescent chemistry methods. Except sequencing for detection of rifampicin drug resistance which cost reasonably low as it requires mainly identification of nucleotide changes in the RRDR region of 81 nucleotide sequences in the rpoB gene, the high cost involved in determining drug resistance for other antimycobaterial drugs wherein multiple genes/targets are required to unravel the mutational changes restrict their testing in daily routine [63]. The MDR-TB results through sequencing involve mainly an analysis of mutational sequential changes in specific drug target genes (Fig. 6.7 and Table 6.2).

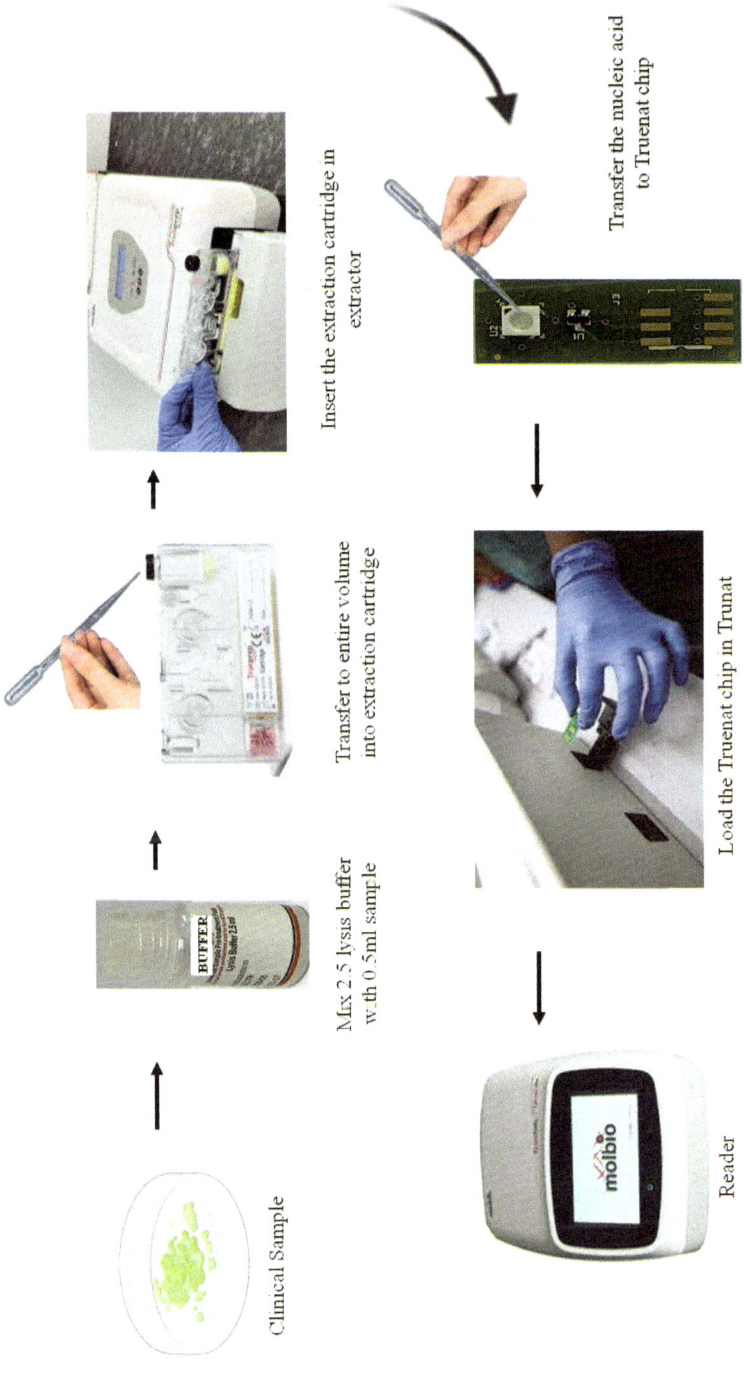

Fig. 6.6 Truenat MTB-RIF Dx

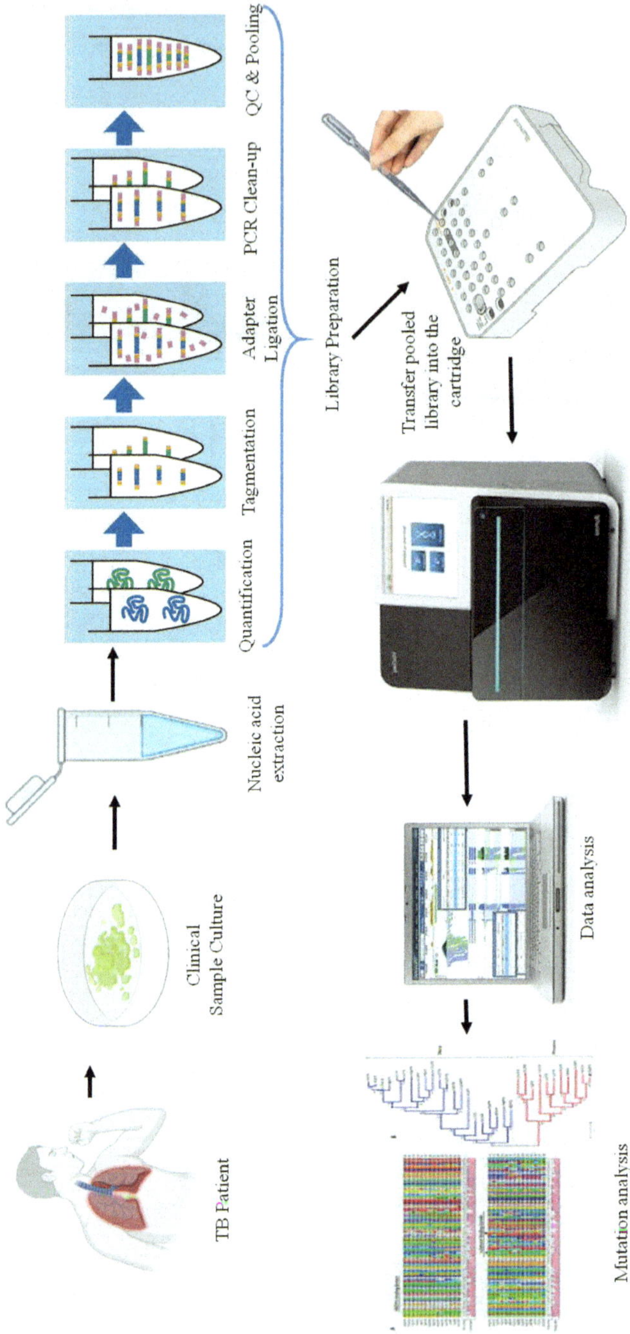

Fig. 6.7 Sequencing platforms for the identification of novel mutations of drug resistance

Table 6.2 WHO recommended various molecular diagnostic assays for the rapid detection of MTB and drug-resistance

S. no.	Molecular assay	Make	Principle	Purpose	Gene target	TAT (turn around time)	Advantages	Limitation	Efficiency of technique	
									Sensitivity	Specificity
1	LAMP	Eiken Chemical Co	Loop-mediated isothermal amplification (LAMP)	MTB diagnosis	rpoB, IS6110	1 h	• Minimal expertise required • Rapid and cost-efficient	• Aerosol contamination is high • Requires several manual steps	90–95%	95–98%
2	COBAS TaqMan TB	Roche Diagnostics	Realtime PCR assay	MTB diagnosis	16S rRNA	2.5 h	• Minimal expertise required • Expensive equipment is not required	• This assay is limited to respiratory specimen only	63.6–88.4%	94.6–98.8%
3	Gene Xpert MTB/RIF assay	Cepheid Inc	Automated cartridge-based nucleic acid amplification test (CBNAAT)	MTB diagnosis and RIF resistance	rpo B	2 h	• Rapid diagnosis of TB with rifampicin sensitivity/resistance pattern	• High costs and sophisticated equipment is required • Does not detect all mutations conferring resistance to anti-TB agents	85–96%	96–98%

(continued)

Table 6.2 (continued)

S. no.	Molecular assay	Make	Principle	Purpose	Gene target	TAT (turn around time)	Advantages	Limitation	Efficiency of technique	
									Sensitivity	Specificity
4	Gene Xpert TB Ultra	Cepheid Inc	Real-time PCR assay/ melting temperature analysis	MTB diagnosis and RIF resistance	IS6110 and IS1081	2 h	• Improves the diagnosis of TB, especially extrapulmonary TB, pediatric TB, or HIV- associated TB	• Low specificity from high-incidence countries	90–94%	96–99%
5	Line probe assay	Elabscience	Multiplex PCR + reverse hybridization	MTB diagnosis and RIF and INF resistance	rpoB, katG, inh genes	6 h	• High sensitivity for the detection of RIF resistance	• Low sensitivity for detecting INH resistance • False diagnosis of MTB complex in negative clinical samples	92–99% RIF 84–95% INF	94–99% RIF 94–98% INF
6	Trunat	Molbio Diagnostics	Micro RT-PCR	Chip-based RT- PCR test for RIF in resistant MTB	rpoB	1 h	• Microchip-based real-time PCR assay	• Low sensitivity to detect RIF resistance	80–84%	90–92%
7	Next-Generation Sequencing	Illumina	Sequencing based	All antibiotic resistance	Whole genome	24 h	• Next-generation-based sequencing	• High costs and sophisticated equipment required		

6.5 Conclusion

Tuberculosis (TB) is the world's most common and major public health concern, which worsens as it evolves into various drug-resistant strains, like multidrug/extensive / total drug resistant TB. The rapid detection of MTB and site-specific mutation associated with drug resistance are major challenges for the better management of drug resistant TB. A culture-based conventional drug sensitivity test, due to its longer turnaround time, may waste valuable treatment time while patients begin inappropriate antibiotic treatments that are little effective or increase the risk of mutations in relevant genes. In the last one to two decades, the development of various new molecular tools for the detection of TB and MDR-TB has significantly reduced the diagnostic time duration and, in particular cases, will possibly prove more easy to adapt and process than culture-based DST due to its fast turnaround time. The early and effective detection of drug resistance-conferring mutations plays a considerable role in the effective control of the disease. Most molecular assays are based on RRDR to differentiate RIF- susceptible and -resistant MTB, which, however, could fail when the mutation(s) occur outside this region. The molecular tests have obvious advantages over culture methods in detection of low-level rifampicin resistance which is often missed and undetected in culture based approaches. Resistance to other antitubercular drugs like INH, Streptomycin, EMB, pyrazinamide, fluoroquinolones, and second-line injectables can also be detected by molecular methods. Further improvements are, however, still needed to make the molecular test more accurate, sensitive, specific, and affordable to provide vital resistance information against both primary and secondary antitubercular antimicrobial for the better management of TB.

References

1. WHO. Global Tuberculosis Report. 2022. https://www.who.int/teams/global-tuberculosis-programme/tb-reports/global-tuberculosis-report-2022. Accessed 29 Dec 2022.
2. Eddabra R, Ait BH. Rapid molecular assays for detection of tuberculosis. Pneumonia. 2018;10:4.
3. WHO. Anti-tuberculosis drug resistance in the world: fourth global report fourth global report. In: WHO/IUATLD Global Project on Anti-Tuberculosis Drug Resistance Surveillance, 2002–2007; 2008. p. 1–153. https://www.who.int/publications/i/item/9789241563611.
4. Canetti G. Present aspects of bacterial resistance in tuberculosis. Am Rev Respir Dis. 1965;92:687–703.
5. Ahmad S, Mokaddas E, Al-Mutairi N, Eldeen HS, Mohammadi S. Discordance across phenotypic and molecular methods for drug susceptibility testing of drug-resistant mycobacterium tuberculosis isolates in a low TB incidence country. PloS One. 2016;11:e0153563.
6. Balasingham SV, Davidsen T, Szpinda I, Frye SA, Tønjum T. Molecular diagnostics in tuberculosis: basis and implications for therapy. Mol Diagn Ther. 2009;13:137–51.
7. Mboowa G, Namaganda C, Ssengooba W. Rifampicin resistance mutations in the 81 bp RRDR of rpoB gene in mycobacterium tuberculosis clinical isolates using Xpert® MTB/RIF in Kampala, Uganda: a retrospective study. BMC Infect Dis. 2014;14:481. https://pubmed.ncbi.nlm.nih.gov/25190040.

8. Schön T, Juréen P, Chryssanthou E, Giske CG, Kahlmeter G, Hoffner S, et al. Rifampicin-resistant and rifabutin-susceptible mycobacterium tuberculosis strains: a breakpoint artefact? J Antimicrob Chemother. 2013;68:2074–7. https://doi.org/10.1093/jac/dkt150.
9. Nusrath Unissa A, Hassan S, Indira Kumari V, Revathy R, Hanna LE. Insights into RpoB clinical mutants in mediating rifampicin resistance in mycobacterium tuberculosis. J Mol Graph Model. 2016;67:20–32. https://www.sciencedirect.com/science/article/pii/S1093326316300560.
10. Jamieson FB, Guthrie JL, Neemuchwala A, Lastovetska O, Melano RG, Mehaffy C. Profiling of rpoB mutations and MICs for rifampin and rifabutin in Mycobacterium tuberculosis. J Clin Microbiol. 2014;52:2157–62. https://pubmed.ncbi.nlm.nih.gov/24740074.
11. Sinha P, Srivastava GN, Tripathi R, Mishra MN, Anupurba S. Detection of mutations in the rpoB gene of rifampicin-resistant mycobacterium tuberculosis strains inhibiting wild type probe hybridization in the MTBDR plus assay by DNA sequencing directly from clinical specimens. BMC Microbiol. 2020;20:284. https://doi.org/10.1186/s12866-020-01967-5.
12. Singh A, Gupta AK, Singh S. In: Saxena SK, Khurana SMP, editors. Molecular mechanisms of drug resistance in mycobacterium tuberculosis: role of nanoparticles against multi-drug-resistant tuberculosis (MDR-TB) BT - NanoBioMedicine. Singapore: Springer; 2020. p. 285–314. https://doi.org/10.1007/978-981-32-9898-9_12.
13. Almeida Da Silva PE, Palomino JC. Molecular basis and mechanisms of drug resistance in mycobacterium tuberculosis: classical and new drugs. J Antimicrob Chemother. 2011;66:1417–30. https://doi.org/10.1093/jac/dkr173.
14. Cohen KA, Bishai WR, Pym AS Molecular basis of drug resistance in mycobacterium tuberculosis. Microbiol Spectr. 2014;2(3). https://doi.org/10.1128/microbiolspec.MGM2-0036-2013.
15. Torres JN, Paul LV, Rodwell TC, Victor TC, Amallraja AM, Elghraoui A, et al. Novel katG mutations causing isoniazid resistance in clinical M. tuberculosis isolates. Emerg Microb Infect. 2015;4:e42. https://pubmed.ncbi.nlm.nih.gov/26251830.
16. Scorpio A, Zhang Y. Mutations in pncA, a gene encoding pyrazinamidase/nicotinamidase, cause resistance to the antituberculous drug pyrazinamide in tubercle bacillus. Nat Med. 1996;2:662–7. https://doi.org/10.1038/nm0696-662.
17. Stoffels K, Mathys V, Fauville-Dufaux M, Wintjens R, Bifani P. Systematic analysis of pyrazinamide-resistant spontaneous mutants and clinical isolates of mycobacterium tuberculosis. Antimicrob Agents Chemother. 2012;56:5186–93. https://pubmed.ncbi.nlm.nih.gov/22825123.
18. Cuevas-Córdoba B, Xochihua-González SO, Cuellar A, Fuentes-Domínguez J, Zenteno-Cuevas R. Characterization of pncA gene mutations in pyrazinamide-resistant mycobacterium tuberculosis isolates from Mexico. Infect Genet Evol. 2013;19:330–4. https://www.sciencedirect.com/science/article/pii/S1567134812003954.
19. Sengstake S, Bergval IL, Schuitema AR, de Beer JL, Phelan J, de Zwaan R, et al. Pyrazinamide resistance-conferring mutations in pncA and the transmission of multidrug resistant TB in Georgia. BMC Infect Dis. 2017;17:491. https://pubmed.ncbi.nlm.nih.gov/28697808.
20. Chang KC, Yew WW, Zhang Y. Pyrazinamide susceptibility testing in Mycobacterium tuberculosis: a systematic review with meta-analyses. Antimicrob Agents Chemother. 2011;55:4499–505. https://pubmed.ncbi.nlm.nih.gov/21768515.
21. Li M, Chen R, Lin S, Lu Y, Liu H, Li G, et al. Detecting ethambutol resistance in mycobacterium tuberculosis isolates in China: a comparison between phenotypic drug susceptibility testing methods and DNA sequencing of embAB [internet]. Front Microbiol. 2020;11:781. https://doi.org/10.3389/fmicb.2020.00781.
22. Alcaide F, Pfyffer GE, Telenti A. Role of embB in natural and acquired resistance to ethambutol in mycobacteria. Antimicrob Agents Chemother. 1997;41:2270–3. https://pubmed.ncbi.nlm.nih.gov/9333060.
23. Nasiri MJ, Haeili M, Ghazi M, Goudarzi H, Pormohammad A, Imani Fooladi AA, et al. New insights in to the intrinsic and acquired drug resistance mechanisms in mycobacteria. Front Microbiol. 2017;8:681. https://pubmed.ncbi.nlm.nih.gov/28487675.

24. Notomi T, Okayama H, Masubuchi H, Yonekawa T, Watanabe K, Amino N, et al. Loop-mediated isothermal amplification of DNA. Nucleic Acids Res. 2000;28:E63.
25. WHO. The use of loop-mediated isothermal amplification (TB-LAMP) for the diagnosis of pulmonary tuberculosis: policy guidance. Geneva: World Health Organization; 2016. https://apps.who.int/iris/handle/10665/249154.
26. Aryan E, Makvandi M, Farajzadeh A, Huygen K, Bifani P, Mousavi S-L, et al. A novel and more sensitive loop-mediated isothermal amplification assay targeting IS6110 for detection of mycobacterium tuberculosis complex. Microbiol Res. 2010;165:211–20.
27. Bi A, Nakajima C, Fukushima Y, Tamaru A, Sugawara I, Kimura A, et al. A rapid loop-mediated isothermal amplification assay targeting hspX for the detection of mycobacterium tuberculosis complex. Jpn J Infect Dis. 2012;65:247–51.
28. Balne PK, Barik MR, Sharma S, Basu S. Development of a loop-mediated isothermal amplification assay targeting the mpb64 gene for diagnosis of intraocular tuberculosis. J Clin Microbiol. 2013;51:3839–40.
29. Nimesh M, Joon D, Varma-Basil M, Saluja D. Development and clinical evaluation of sdaA loop-mediated isothermal amplification assay for detection of mycobacterium tuberculosis with an approach to prevent carryover contamination. J Clin Microbiol. 2014;52:2662–4.
30. Ou X, Li Q, Xia H, Pang Y, Wang S, Zhao B, et al. Diagnostic accuracy of the PURE-LAMP test for pulmonary tuberculosis at the county-level laboratory in China. PloS One. 2014;9:e94544.
31. Bojang AL, Mendy FS, Tientcheu LD, Otu J, Antonio M, Kampmann B, et al. Comparison of TB-LAMP, GeneXpert MTB/RIF and culture for diagnosis of pulmonary tuberculosis in The Gambia. J Infect. 2016;72:332–7.
32. Gray CM, Achilles K, Pratibha N, Jorge G, Carlos Z, Moses J, et al. Feasibility and operational performance of tuberculosis detection by loop-mediated isothermal amplification platform in decentralized settings: results from a multicenter study. J Clin Microbiol. 2016;54:1984–91. https://doi.org/10.1128/JCM.03036-15.
33. Kaku T, Minamoto F, D'Meza R, Morose W, Boncy J, Bijou J, et al. Accuracy of LAMP-TB method for diagnosing tuberculosis in Haiti. Jpn J Infect Dis. 2016;69:488–92.
34. Ghosh PK, Chakraborty B, Maiti PK, Ray R. Comparative evaluation of loop-mediated isothermal amplification and conventional methods to diagnose extrapulmonary tuberculosis. Ann Trop Med Publ Health. 2017;10:160–4.
35. Leylabadlo HE, Kafil HS, Yousefi M, Aghazadeh M, Asgharzadeh M. Pulmonary tuberculosis diagnosis: where we are? Tuberc Respir Dis. 2016;79:134–42.
36. Bloemberg GV, Voit A, Ritter C, Deggim V, Böttger EC. Evaluation of Cobas TaqMan MTB for direct detection of the mycobacterium tuberculosis complex in comparison with Cobas Amplicor MTB. J Clin Microbiol. 2013;51:2112–7.
37. Jönsson B, Lönnermark E, Ridell M. Evaluation of the Cobas TaqMan MTB test for detection of mycobacterium tuberculosis complex. Infect Dis. 2015;47:231–6. https://doi.org/10.3109/00365548.2014.987162.
38. Horita N, Yamamoto M, Sato T, Tsukahara T, Nagakura H, Tashiro K, et al. Sensitivity and specificity of Cobas TaqMan MTB real-time polymerase chain reaction for culture-proven mycobacterium tuberculosis: meta-analysis of 26999 specimens from 17 studies. Sci Rep. 2015;5:18113.
39. Jönsson B, Lönnermark E, Ridell M. Evaluation of the Cobas TaqMan MTB test for detection of mycobacterium tuberculosis complex. Infect Dis. 2015;47:231–6.
40. Steingart KR, Schiller I, Horne DJ, Pai M, Boehme CC, Dendukuri N. Xpert® MTB/RIF assay for pulmonary tuberculosis and rifampicin resistance in adults. Cochrane Database Syst Rev. 2014;2014:CD009593.
41. WHO. Automated real-time nucleic acid amplification Technology for Rapid and Simultaneous Detection of tuberculosis and rifampicin resistance: Xpert MTB/RIF assay for the diagnosis of pulmonary and Extrapulmonary TB in adults and children. Geneva: WHO; 2013.

42. US Food and Drug Administration. FDA permits marketing of first US test labeled for simultaneous detection of tuberculosis bacteria and resistance to the antibiotic rifampin. Clin Infect Dis. 2013;57:i.
43. Iram S, Zeenat A, Hussain S, Wasim Yusuf N, Aslam M. Rapid diagnosis of tuberculosis using Xpert MTB/RIF assay—report from a developing country. Pak J Med Sci. 2015;31:105–10.
44. Ozkutuk N, Surucüoglu S. Evaluation of the Xpert MTB/RIF assay for the diagnosis of pulmonary and extrapulmonary tuberculosis in an intermediate-prevalence setting. Mikrobiyol Bul. 2014;48:223–32.
45. Nicol MP, Workman L, Isaacs W, Munro J, Black F, Eley B, et al. Accuracy of the Xpert MTB/RIF test for the diagnosis of pulmonary tuberculosis in children admitted to hospital in Cape Town, South Africa: a descriptive study. Lancet Infect Dis. 2011;11:819–24.
46. Rachow A, Clowes P, Saathoff E, Mtafya B, Michael E, Ntinginya EN, et al. Increased and expedited case detection by Xpert MTB/RIF assay in childhood tuberculosis: a prospective cohort study. Clin Infect Dis. 2012;54:1388–96.
47. Lawn SD, Nicol MP. Xpert® MTB/RIF assay: development, evaluation and implementation of a new rapid molecular diagnostic for tuberculosis and rifampicin resistance. Future Microbiol. 2011;6:1067–82. https://pubmed.ncbi.nlm.nih.gov/21958145.
48. Sanchez-Padilla E, Merker M, Beckert P, Jochims F, Dlamini T, Kahn P, et al. Detection of drug-resistant tuberculosis by Xpert MTB/RIF in Swaziland. N Engl J Med. 2015;372:1181–2. https://doi.org/10.1056/NEJMc1413930.
49. Chakravorty S, Simmons AM, Rowneki M, Parmar H, Cao Y, Ryan J, et al. The new Xpert MTB/RIF ultra: improving detection of mycobacterium tuberculosis and resistance to rifampin in an assay suitable for point-of-care testing. MBio. 2017;8:8.
50. Dorman SE, Schumacher SG, Alland D, Nabeta P, Armstrong DT, King B, et al. Xpert MTB/RIF ultra for detection of mycobacterium tuberculosis and rifampicin resistance: a prospective multicentre diagnostic accuracy study. Lancet Infect Dis. 2018;18:76–84.
51. WHO. Use of Xpert MTB/RIF and Xpert MTB/RIF ultra on GeneXpert 10-colour instruments: WHO policy statement. Geneva: WHO; 2021. https://www.who.int/publications/i/item/9789240040090.
52. Madhuri K, Deshpande S, Dharmashale S, Bharadwaj R. Utility of line probe assay for the early detection of multidrug-resistant pulmonary tuberculosis. J Glob Infect Dis. 2015;7:60–5. https://pubmed.ncbi.nlm.nih.gov/26069424.
53. Meaza A, Kebede A, Yaregal Z, Dagne Z, Moga S, Yenew B, et al. Evaluation of genotype MTBDRplus VER 2.0 line probe assay for the detection of MDR-TB in smear positive and negative sputum samples. BMC Infect Dis. 2017;17:280. https://pubmed.ncbi.nlm.nih.gov/28415989.
54. Yadav RN, Singh BK, Sharma SK, Sharma R, Soneja M, Sreenivas V, et al. Comparative evaluation of GenoType MTBDRplus line probe assay with solid culture method in early diagnosis of multidrug resistant tuberculosis (MDR-TB) at a tertiary Care Centre in India. PloS One. 2013;8:e72036. https://doi.org/10.1371/journal.pone.0072036.
55. Rando-Segura A, Aznar ML, Moreno MM, Espasa Soley M, Sulleiro Igual E, Bocanegra Garcia C, et al. Molecular characterization of rpoB gene mutations in isolates from tuberculosis patients in Cubal, Republic of Angola. BMC Infect Dis. 2021;21:1056. https://doi.org/10.1186/s12879-021-06763-8.
56. Singhal R, Myneedu VP, Arora J, Singh N, Sah GC, Sarin R. Detection of multi-drug resistance & characterization of mutations in Mycobacterium tuberculosis isolates from North- Eastern States of India using GenoType MTBDRplus assay. Indian J Med Res. 2014;140:501–6. https://pubmed.ncbi.nlm.nih.gov/25488443.
57. Deguchi T, Yasuda M, Asano M, Tada K, Iwata H, Komeda H, et al. DNA gyrase mutations in quinolone-resistant clinical isolates of Neisseria gonorrhoeae. Antimicrob Agents Chemother. 1995;39:561–3.
58. Matrat S, Veziris N, Mayer C, Jarlier V, Truffot-Pernot C, Camuset J, et al. Functional analysis of DNA gyrase mutant enzymes carrying mutations at position 88 in the a subunit found

in clinical strains of mycobacterium tuberculosis resistant to fluoroquinolones. Antimicrob Agents Chemother. 2006;50:4170–3.
59. Wang J-Y, Lee L-N, Lai H-C, Wang S-K, Jan I-S, Yu C-J, et al. Fluoroquinolone resistance in mycobacterium tuberculosis isolates: associated genetic mutations and relationship to antimicrobial exposure. J Antimicrob Chemother. 2007;59:860–5.
60. Lee DJ, Kumarasamy N, Resch SC, Sivaramakrishnan GN, Mayer KH, Tripathy S, et al. Rapid, point-of-care diagnosis of tuberculosis with novel Truenat assay: cost-effectiveness analysis for India's public sector. PloS One. 2019;14:e0218890.
61. Ezewudo M, Borens A, Chiner-Oms Á, Miotto P, Chindelevitch L, Starks AM, et al. Integrating standardized whole genome sequence analysis with a global mycobacterium tuberculosis antibiotic resistance knowledgebase. Sci Rep. 2018;8:15382. https://doi.org/10.1038/s41598-018-33731-1.
62. Fluit AC, Visser MR, Schmitz FJ. Molecular detection of antimicrobial resistance. Clin Microbiol Rev. 2001;14:836–71. https://pubmed.ncbi.nlm.nih.gov/11585788.
63. Parsons LM, Somoskövi A, Gutierrez C, Lee E, Paramasivan CN, Abimiku A, et al. Laboratory diagnosis of tuberculosis in resource-poor countries: challenges and opportunities. Clin Microbiol Rev. 2011;24:314–50. https://pubmed.ncbi.nlm.nih.gov/21482728.

Chapter 7
The Diagnosis and Challenges of Pediatric Tuberculosis

Alkesh Khurana and Bhavna Dhingra

Abstract The diagnosis of tuberculosis (TB) is fraught with numerous challenges, and more so in children, mainly due to the paucibacillary nature of the disease and the difficulty in accessing a suitable sample for diagnosis. Upfront universal drug sensitivity testing for Mycobacterium Tuberculosis is now recommended as per the National Tuberculosis Elimination Programme (NTEP) approved Nucleic Acid Ampliication Test (NAAT). The lack of exact estimates of the burden of latent TB in the community; grossly inadequate reporting of TB cases; and the empirical and inappropriate use of antitubercular drugs, contributing to the widespread resistance and lack of easily available child-friendly drug formulations, are some of the main challenges of pediatric TB. Improved diagnosis, notification and the development of newer effective drugs, as well as the availability of child-friendly preparations, will go a long way in helping to achieve the goal of TB elimination.

Keywords Tuberculosis · Pediatric TB · Lymph node tuberculosis · Pleural effusion · Abdominal tuberculosis · CNS tuberculosis · Bone and joint tuberculosis

7.1 Introduction

In high-endemic countries, high clinical suspicion of tuberculosis is kept in the presence of the following constitutional symptoms:

- Fever of unknown origin documented for 2 weeks or more and/or,
- Persistent cough for 2 weeks or more and/or,

A. Khurana
Department of Pulmonary Medicine, AIIMS, Bhopal, India

B. Dhingra (✉)
Department of Pediatrics, AIIMS, Bhopal, India
e-mail: bhavna.pediatrics@aiimsbhopal.edu.in

A. Singh, D. Sharma (eds.), *Diagnosis of Mycobacterium*,
https://doi.org/10.1007/978-981-99-5624-1_7

- Unexplained documented weight loss of more than 5% in the last 3 months in spite of optimal nutrition.

The proper documentation of the constitutional symptoms increases the positive predictive value of these symptoms in the diagnosis of tuberculosis [1, 2].

History of contact with an index case should prompt the physician to investigate for active disease if clinical features are suggestive of the disease. Any patient with lung/airway involvement has the potential to spread the mycobacteria through the droplet route and is hence considered infectious. Although the risk is higher in sputum-positive cases as compared to negative cases, the latter is not completely devoid of infectious transmission. Sometimes extrapulmonary TB patients may have concomitant pulmonary involvement, and the risk of transmission may therefore be masked. Hence, contact with a patient with any form of tuberculosis should not be ignored. Figure 7.1 depicts the algorithm for an approach to the diagnosis of tuberculosis in children as per the Revised National TB Control Programme.

The following demographic features increase the risk of tuberculosis:

- Residing in a TB-endemic area.
- Recent contact with a patient with active pulmonary tuberculosis.
- Residence where there is community exposure, such as a nursing home, where occult exposure to tuberculosis might have occurred, or
- Underlying immunosuppression due to HIV infection or the administration of immunosuppressive drugs, which reduce immunity [3–5].

Failure to consider tuberculosis in the differential diagnosis of patients at risk leads to delays in the diagnosis and administration of antitubercular drugs.

7.2 Laboratory Diagnosis of TB

7.2.1 Microbiological Tests

Sample collection for microbiological diagnosis in children has always been an area of concern when compared to adults for obvious reasons. For pulmonary tuberculosis, gastric lavage and induced sputum are usual substitutes in younger children vis-à-vis early morning sputum in adults [6]. Bronchoalveolar lavage can be done but only in some referral institutes in India. Extrapulmonary samples, e.g., lymph node aspirates and pleural fluid aspirates, are relatively easier to obtain, although not as easily done as in adults. Getting a sample to detect MTB from other extrapulmonary sites, e.g., central nervous system (CNS) tuberculosis, bone and joint tuberculosis remains a challenge in both children as well as adults. Because of these constraints in children, it is recommended that children be subjected to chest radiography first as a screening modality and then to targeted sample collection.

Whether pulmonary or extrapulmonary, all samples are subjected to the detection of mycobacteria and its drug resistance pattern by using the National

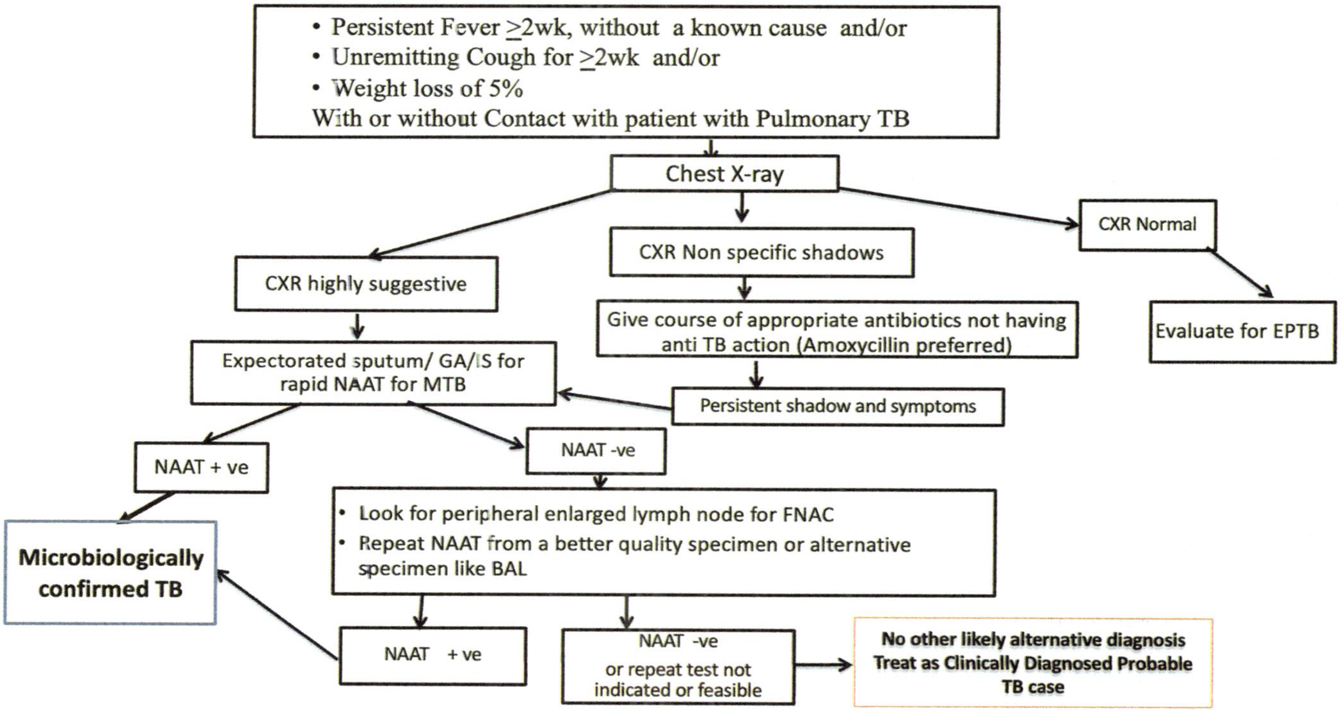

Fig. 7.1 Algorithm for an approach to pediatric pulmonary tuberculosis [1, 2]. (Algorithm adapted from IAP NTEP Pediatric TB guidelines 2019 and 2021, central TB division MOHFW New Delhi India)

Tuberculosis Elimination Programme (NTEP)-approved NAAT (nucleic acid amplification test) assay, cartridge-based NAAT (CBNAAT), or TruNAT. The NAAT assay works under the principle of automated amplification of *Mycobacterium tuberculosis* (MTB) DNA, and the test itself takes only 2 h to get completed and generate the report. NAAT detects *Mycobacterium tuberculosis* and the status of resistance to rifampicin. The NAAT assay has now replaced conventional microscopy and culture as the initial investigation of choice in the diagnosis of tuberculosis in children. In the next step, line probe assays (LPAs) are a part of the guidelines. Smear-positive samples are subjected to LPA directly. First-line LPA detects *Mycobacterium tuberculosis* and also resistance to both isoniazid and rifampicin. Second-line LPA detects resistance to quinolones and second line injectables [7]. Although the yield of a single specimen is quite good in these molecular tests, two samples are taken, even for NAAT. The second sample is usually used either in cases of some unexpected logistic issues or when there is discordance between the NAAT assay and LPA. It is worth mentioning that these molecular assays perform poorly in culture-negative TB cases, and up to half of pediatric tuberculosis cases are culture negative. Commercial serodiagnostics are not recommended for the diagnosis of pulmonary or extrapulmonary tuberculosis in either children or adults, irrespective of their HIV status since 201 [6].

7.2.2 TST and IGRAs

The Mantoux test, also called the tuberculin skin test (TST), is the oldest and the most traditional test for detecting latent TB infection. In children where obtaining a tissue sample is always difficult, TST has been used, along with clinical and radiological pictures, to decide on whether to make a diagnosis of TB disease as well. Two TU of PPD 23 antigen has been recommended for use intradermally to carry out this test, which requires repeat reading after 48–72 h to measure the induration on the forearm [6]. The test is usually taken as positive if the induration is >10 mm, except in immunosuppressed conditions, e.g., HIV, where the cut-off may be reduced to 5 mm. Interferon-gamma release assays (IGRAs) provide the same information but in an in vitro environment. Two tests are currently approved by the World Health Organization (WHO), namely, QuantiFERON TB Gold and the TB spot. The main advantage of IGRAs over TST is that they neither require a repeat visit to evaluate the result nor cross-react with Bacillus Calmette–Guérin (BCG) vaccination. The blood lymphocytes are exposed to TB-specific antigens, namely ESAT-6 and CFP-10, and thereafter the release of interferon γ is measured as a marker of TB infection [8].

7.3 Site or Type of Tuberculosis

7.3.1 Lymph Node Tuberculosis

One of the most common extrapulmonary sites for tuberculosis, enlarged neck nodes are a common site in pediatric population. Tubercular lymph nodes are usually not very large in size initially, are frequently nontender/mildly tender, and may or may not have fluctuations. Whenever there is clinical suspicion, fine-needle aspiration (FNA) should be done and subjected to smear and CBNAAT for tuberculosis diagnosis. FNA corresponds to histopathology in up to 90% cases, the latter being the gold standard for diagnosis of lymph node pathologies [6]. If FNA is unyielding of TB despite being clinically suspected, an excision biopsy for histopathology should be advised to find out the etiology. The presence of epithelioid granulomas despite having negative CBNAAT sometimes warrants starting antituberculous treatment (ATT) by the treating physician. However, it must be mentioned that children frequently have enlarged neck nodes because of recurrent upper respiratory infections/tonsillitis with a positive TST; thus, this scenario alone does not fulfill the mandate of starting ATT [9]. Although the final decision on starting ATT rests with the treating physician, every attempt must be made to obtain a microbiological diagnosis first.

7.3.2 Pleural Effusion

Another common site for extrapulmonary tuberculosis is the pleural cavity. After a clinical suspicion, the presence of pleural fluid should be confirmed, which is usually done through radiology (chest radiograph, ultrasonography (USG), or both). Pleural fluid analysis via pleurocentesis is almost always required to comment upon the etiology. A lymphocytic exudative picture is suggestive of tuberculosis in the background of relevant clinical history. Pleural fluid CBNAAT yield is not more than 5%. Also, the measurement of pleural fluid adenosine deaminase (ADA) in children is not as reliable as in adults. In cases where a microbiological diagnosis of TB has not been made, it is more daunting to carry out thoracoscopy in children than in adults because of the lack of expertise and lack of equivalent equipment for children. Therefore, pleural biopsy needles (Abraham's and Cope's biopsy needles), which are now becoming outdated in adults, are still used in children [6, 10]. For all these reasons, making a confirmatory/microbiological diagnosis of tuberculous pleural effusion is far more difficult in children than in adults, similar to the involvement of other sites in children.

7.3.3 Abdominal Tuberculosis

Making a diagnosis of abdominal tuberculosis is probably one of the most difficult tasks in this huge spectrum of tuberculosis illness, especially in children. This form of tuberculosis is much less common in children than in adults [11]. The initial investigation is most of the time ultrasonography as abdominal radiographs are rarely helpful and hence not advised routinely. Some suggestive findings include the thickening of the bowel wall, especially in the ileocecal region, with or without accompanying mesenteric lymph nodes. Hypoechoic areas in the lymph nodes indicating necrosis within are a pointer toward tuberculosis. Contrast-enhanced computed tomography (CECT) can clarify these suggestive findings to a greater extent. However, obtaining a tissue diagnosis from the abdomen requires invasive methods, e.g., gastrointestinal (GI) endoscopy or sometimes image-guided FNA/biopsy from mesenteric lymph nodes [6]. These facilities are available only at tertiary referral centers, and hence, making a diagnosis of abdominal tuberculosis at the peripheral centers is highly dependent on the physician's clinical acumen and judgment.

7.3.4 CNS Tuberculosis

Tubercular meningitis (TBM) is the most severe form of tuberculosis and can be life-threatening in children if not treated timely. CECT head can identify various signs suggestive of tubercular etiology. CT can show basal meningeal enhancement, tuberculomas, hydrocephalus, or even infarcts sometimes. Magnetic resonance imaging (MRI) can be done when CT is inconclusive. Unlike adults, when performing these imaging diagnostics, especially MRI, children need sedation and bedside supervision by a trained physician, making the relatively easier task much more difficult. The next most important investigation is the cerebrospinal fluid (CSF) examination. Again, as in imaging, getting a CSF sample in a child requires highly skilled hands. Low CSF glucose (<40 mg/dL), elevated protein levels (>100 mg/dL), and a predominantly lymphocytic picture are all indicative of a tuberculous etiology. CBNAAT in CSF fluid can be positive in 16–40% of cases [6, 12]. Therefore, like other extrapulmonary sites, more than half of CNS TB cases are treated without any microbiological evidence, and hence there is a lack of a drug susceptibility pattern.

7.3.5 Bone and Joint Tuberculosis

TB of bones and joints account for 5–15% of all extrapulmonary TB cases. If untreated in children, up to 6% of the children with primary infection may later develop bone and joint tuberculosis [13]. The spectrum includes Pott's spine, dactylitis, arthritis, reactive arthritis (Poncet's disease), and osteomyelitis. Dactylitis, also called spina ventosa, affects children much more commonly than adults. Short tubular bones of the hands and feet are more commonly affected. Typically, it involves the proximal phalanges of the middle and index fingers and metacarpals of the middle and ring fingers. Diagnosis is attempted by X-ray, which initially shows a diaphyseal expansile lesion and sclerosis in the later stages. In Pott's spine, the most commonly involved site is the thoracic vertebrae, initially suspected on the basis of clinical signs only. X-rays reveal an abnormality only in the later stages when a significant bone erosion has occurred, and reduced disk space is the commonest X-ray finding. MRI is the radiological investigation of choice, which not only reveals the true extent of the bone and paravertebral involvement but also tells the radiological findings consistent with Pott's spine [6]. However, as is true with all other sites of involvement, an attempt for microbiological diagnosis should always be made. In this case, In this case, image-guided biopsy from paravertebral lesions or accompanying soft tissue abscess, if any, should be tried by a skilled surgeon taking full care of the adjoining neural tissue.

7.4 Challenges in the Elimination of Tuberculosis

The diagnosis of TB is fraught with various challenges, which get further compounded in children due to the fact that children do not bring out sputum, and obtaining an appropriate sample for diagnostic purposes is difficult. The various challenges can be grouped into different categories as depicted in Fig. 7.2.

7.4.1 Real Burden of the Disease and Notification

Over the last few years, the epidemiology of tuberculosis has been described in terms of estimated prevalence and incidence. Real figures concerning the burden of the disease have been lacking and hence have prevented us from taking appropriate and targeted measures accordingly. In 2012, the Government of India made tuberculosis a notifiable disease [4]. In 2013, there was nil notification from the private sector in Delhi, and in 2016, around 40% of TB cases were calculated to be missing from the total 10.2 million cases worldwide, and India accounted for one-fourth of these missing cases [5, 6]. In a study carried out in a tertiary care hospital in South India, TB notification was a meager 23% in 2018. Low rates of TB notification are more common in children and in the case of sputum-negative and extrapulmonary tuberculosis [7]. Notification attempts to bring the missing cases into the main frame and has been instrumental in the approach to epidemiology as it takes into account the real burden of the disease and the follow-up of these patients. Digitalization (the use of Nikshay software) has helped overcome a lot of barriers in the notification process [8]. However, despite this improvisation, there is huge

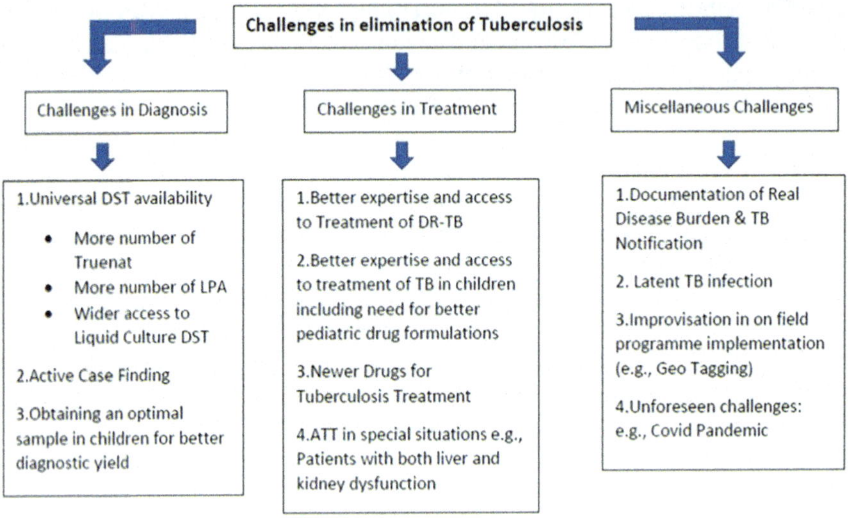

Fig. 7.2 Various challenges in the elimination of TB

variability in notification rates across the country and also within the districts of the same state.

7.4.2 *The Burden of Latent TB Infection*

Latent TB infection (LTBI) with persistent microbial viability has the possibility of reactivation anytime. The identification and eradication of this pool is a crucial step toward achieving the goal of TB elimination. A quarter of the global population is estimated to be TB infected, with a 10% lifetime risk of progressing to active disease, thereby creating a large pool of potential TB sufferers and spreaders. This risk is maximum in under five children (a relative risk of 22.9 vs. 8.2 in the 5–14 years vs. 13.4 in the >15 years age group). With the aim of eliminating tuberculosis, the treatment of LTBI has now found its place in the program guidelines, and all contacts of tuberculosis patients are now being prescribed treatment for latent TB infection. It has been reported that exposed infants who had not received preventive therapy had up to 18% risk of developing the disease in the next 2 years after exposure. Children between 2 and 5 years of age are estimated to have a 2-year cumulative tuberculosis risk of 19% [9].

7.4.3 Active Case Finding

As described above, special measures need to be taken to diagnose and treat the huge proportion of missing TB cases worldwide. Both WHO and the Revised National Tuberculosis Control Programme (RNTCP) have endorsed active case finding (ACF) and the screening of the household contacts of TB patients for wider coverage, as well as the implementation of TB control programs. ACF has led to higher and early case detection in areas where there was underreporting prior to this activity [10]. The results from various trials on ACF have yielded variable results. Whereas, on one hand, Mani et al. detected only one presumptive and one confirmed case of tuberculosis among 6606 persons (approximately 16 cases/lac population) who underwent ACF, on the other hand, Fox et al. detected additional 1084 cases/1 lac population in Vietnam with the help of ACF [11, 12]. It may be noted that the former was a community-based cross-sectional study, whereas the latter was a longitudinal study with a 2-year follow-up among household contacts of TB patients. Community-based studies are underway to evaluate the long-term and wider economic benefits of ACF to see if it reduces the high treatment costs imposed on the government and also mitigates the stigma of TB in households [13]. The target of TB elimination by 2025 cannot be achieved unless this strategy of ACF is implemented diligently all across India.

7.4.4 Prevention of the Emergence/Spread of Drug Resistance

Apart from encouraging early diagnosis and appropriate treatment, certain other practices need to be curbed to prevent the emergence of drug resistance. First and foremost is the mandatory implementation of directly observed therapy (DOTS). Individualized regimes should only be prescribed by certified specialists entitled to do so.

Another aspect is the application of enhanced infection control measures in hospital settings. The outbreak of extensively drug-resistant (XDR) TB in KwaZulu-Natal is a classical, infamous example from which to learn the importance of appropriate infection control practices [14]. Both indoor patients and healthcare workers were affected by the outbreak at the time. Last but not least is the administration of preventive therapy to contacts of multidrug-resistant TB (MDR TB) patients. Preventive therapy to contacts of drug-sensitive TB patients is mostly followed well, especially in low-endemic countries, but the practical implementation of the same to contacts of MDR TB patients needs to be followed strictly, especially in high-endemic countries [15]. The efficacy of chemoprophylaxis to contacts of MDR TB patients is well documented in the literature and also helps break the chain of transmission of DR TB [16, 17]. These measures should be undertaken seriously in addition to a routine upscaling of drug susceptibility testing (DST) and the treatment of drug-resistant (DR) TB on the path toward TB elimination.

7.4.5 Universal Drug Susceptibility Testing

The culture and drug susceptibility testing (DST) of MTB involves phenotypic and genotypic methods. Phenotypic culture methods are based on the ability of MTB to grow in culture media containing a particular critical concentration of anti-TB drugs. Solid cultures traditionally take 6–7 weeks to yield results, whereas liquid cultures do so in around 10 days. Phenotypic DST results for isoniazid, rifampicin, second-line injectables, and quinolones are generally reliable and reproducible. The availability and implementation of DST to other drugs, e.g., newer quinolones (moxifloxacin), ethionamide, pyrazinamide, and ethambutol, have become increasingly important in view of the management of drug-resistant tuberculosis. Because of the rapidity of results and the standardization of testing, genotypic methods have virtually replaced the phenotypic methods at the field level for the programmatic management of DR TB. The most common genotypic methods adopted in the program are CBNAAT/GeneXpert, which detects MTB DNA and resistance to rifampicin, and line probe assay (LPA), which detects not only MTB DNA in smear-positive samples but resistance to isoniazid as well. The detection of rifampicin resistance by mutations in the rpo gene by GeneXpert/LPA correlates well with the phenotypic methods using solid or liquid culture media [18].

It seems prudent and has been proven already that the implementation of a universal DST program is instrumental in the management of patients with DR TB. In Taiwan, the proportion of TB cases with drug susceptibility testing results increased from 24.2% in 2007 to 97.9% in 2016. During the same period, the prevalence of MDR TB decreased from 19.4 cases per million in 2007 to 8.4 cases per million in 2016 [19]. In 2018, India had only 624 GeneXpert machines and 74 RNTCP-certified laboratories for DST testing. Also, whereas India carried out over 10 lac GeneXpert/Rif resistance tests in 2017, it did less than 1 lac LPA and only around 26,000 second-line DSTs [20]. Although the number of GeneXpert and Truenat machines has gone up to 1268 and 1879 by 2020 [21], still there is a significant gap between the existing and the required infrastructure for DSTs, and this gap needs to be filled early and optimally as a major step toward TB elimination. Figure 7.3 depicts the methodology for the workup of tissue samples to establish drug resistance/sensitivity.

7.4.6 Management of Drug-Resistant Tuberculosis

In 2017, an estimated 5.58 lakh people had rifampicin-resistant TB (RR TB), and 82% of them had multidrug-resistant TB (MDR TB). Further analysis revealed that 3.5% of the new TB cases and 18% of the previously treated cases had MDR/RR TB [17]. For proven MDR tuberculosis, WHO and RNTCP now recommend a longer regimen or a shorter regimen. Once rifampicin resistance is confirmed, the patient is subjected to first-line LPA, second-line LPA, and liquid culture DST. If no

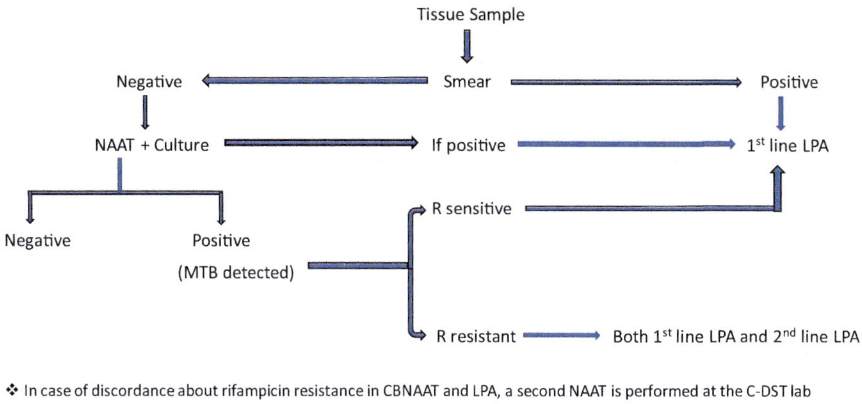

Fig. 7.3 Flow of sample workup for the diagnosis of TB and drug sensitivity testing

additional resistance is detected or even if isoniazid resistance is detected with only Inh A or Kat G mutation, a shorter all-oral MDR regimen is given to the patient. This consists of 4–6 months of bedaquiline, high-dose isoniazid, ethionamide, levofloxacin, clofazimine, pyrazinamide, and ethambutol, followed by 5 months of the latter four drugs [(4–6) $BDQ_{6m,}$ Lfx, Cfz, Z, E, Hh, Eto, followed by (5) Lfx, Cfz, Z]. An alternative shorter MDR regime of 9–12-month duration was also approved by WHO for patients who have not been previously exposed for more than 1 month to second-line ATT drugs and in whom resistance to second-line drugs had been excluded. This regimen consists of 4–6 months of intensive phase comprising kanamycin, ethionamide, high-dose isoniazid, moxifloxacin, clofazimine, pyrazinamide, and ethambutol, followed by 5 months of the latter four drugs [(4–6) Mfx, Km/Am, Eto, Cfz, Z, Hh, E/ (5) Mfx, Cfz, Z, E]. The new all-oral regimen is supposed to replace the injectable-containing shorter regimen gradually all over the country very soon. Disseminated MDR TB and CNS TB are not indications for the shorter regimen.

On the other hand, if isoniazid resistance is detected with both Inh A and Kat G mutations or if quinolone resistance is also detected, the all-oral longer regimen is given to the patient, which includes levofloxacin, bedaquiline, linezolid, clofazimine, and cycloserine [20] [(18–20) Lfx, Bdq (6 months or longer), Lzd, Cfz, Cs]. The total duration is recommended to be 18–20 months with no differentiation into an intensive or continuation phase. The standard treatment for isonicotinic acid hydrazide (INH) monoresistance is 6 months of rifampicin, ethambutol, pyrazinamide, and levofloxacin [22]. However, not all patients fit into monoresistance or MDR TB. Lots of patients will have varying combinations of polydrug resistance, and hence individualized and customized regimens based on drug susceptibility patterns to both first-line and second-line drugs are the need of the hour and are a challenging task to be implemented all across the country.

7.4.7 Management of Pediatric Tuberculosis

Pediatric TB has long been considered an offshoot of its adult counterpart and has been denied the due attention it deserves. India has the highest burden of pediatric TB among the high-burden countries [23]. Since, the childhood form of the disease has different and varied clinical presentations, has nonspecific symptoms that overlap with a host of other clinical conditions and is much more difficult to diagnose due to its paucibacillary nature, restricted availability of an appropriate sample and suboptimal performance of diagnostic tests for diagnosis, the adult strategies are not appropriate to control this menace in children. Estimating the actual burden of the disease in children is a challenge due to the complexity of diagnosis, the absence of mandatory reporting, and poor healthcare surveillance networks. It is estimated to represent at least 10–20% of the entire disease burden in high-prevalence settings [24].

Universal upfront testing for drug resistance in children through the inclusion of CBNAAT (cartridge-based nucleic acid amplification test) in the RNTCP diagnostic algorithm has led to a threefold increase in the diagnostic yield for pulmonary TB and has shown good to moderate sensitivity for extrapulmonary specimens, except pleural and ascitic fluid [25]. The availability and adoption of line probe assays (LPA) to detect resistance to INH and second-line drugs have helped prevent the amplification of drug resistance [26]. The availability of sensitivity testing has led to a paradigm shift toward optimized drug regimens in place of standard treatment regimens in children. Currently, less than 5% of children get a timely and appropriate diagnosis of MDR-TB [27]. A crucial step toward the elimination of TB is the timely identification and initiation of appropriate management for children with MDR TB as every year, about 30,000 children are diagnosed with MDR TB [28].

Community-wide screening approaches for children and the integrated linking of TB control programs with various other maternal and child health services are the need of the hour to help improve the strategies for TB control in children.

7.4.8 Better and Effective Implementation of NTEP

We also need to devise novel mechanisms of improvisation for better and more effective implementation of the program at the ground level simultaneously to achieve this goal of TB elimination. One such extremely cheap and innovative task was done by Singh et al. to ensure near real-time monitoring of the visits done by the senior treatment supervisor (STS). After the disease is notified, the patients are supposed to receive Public Health Action Support (PHAS) under the program. Since there has been no parameter to ensure this important participation of STS delivering PHAS components and having observed the lack of visits by program personnel, Singh et al. developed a very cost-effective web form to ensure near real-time monitoring of the visits done by the STS. By ensuring the "geotagging" of the visits by the STS, the investigators found that for 1772 notified TB patients, home visits took place for 669 new patients and 1381 visits for earlier notified patients [29]. This practically indicated that PHAS is being delivered to all notified patients. Even a typical pattern of a surge in visits just before monthly review meetings was observed. Such cost-effective and innovative strategies need to be considered as a novel opportunity and should be tried to be implemented all over the country for better results.

7.4.9 New Drugs for Tuberculosis

Since the disease refuses to die down despite so many advancements in diagnosis and therapeutics, there is a continuous need for better and more effective drugs to be made. However, tremendous advancements have been made in the last decade whereby, on one hand, newly synthesized drugs, like bedaquiline and delamanid, have already made their way into the programs and, on the other hand, already existing classes of drugs, namely, fluoroquinolones, carbapenems, and oxazolidinones, have also been extrapolated for use in tuberculosis. There is always a scope and need for more effective drugs to control the menace of drug-resistant tuberculosis. Some drugs in preclinical and phase 1 trials include PTBZ 169, BTZ 043, GSK 070, TBA 7371, TBI 166, Q-203, etc. [30]. Pyrroles (BM 212 and LL3858) and benzothiazinones are some other new classes of drugs that have shown synergistic activity with other first-line drugs in murine experimental models. Research related to newer drug development in tuberculosis needs continuous push and support to achieve the daunting task of TB elimination.

7.4.10 Covid-19 and Tuberculosis

Covid-19 has taken the world by storm in the last couple of years. There is a growing apprehension, especially in high TB-burden countries, about the possible impact of Covid-19 on old/active TB patients. Both diseases will have a tendency to spread more in overpopulated areas [31]. With the available data, it is still not clear whether Covid-19 has a higher predisposition in patients with tuberculosis or whether Covid-19 survivors will have a higher predisposition toward developing TB in the near future. A recent United Nations study has anticipated that the Covid-19 pandemic could lead to an increase in number of people living in poverty by up to half a billion, mainly in South Africa, Southeast Asia, and South America [32]. The possible socioeconomic impact of this hypothesis seems scary in the near future. We can just hope that these two public health problems do not lead to any synergism and that we do not have to face a new challenge in the background of so many existing challenges.

7.5 Conclusions

1. All efforts should be directed toward obtaining an appropriate sample for the microbiological confirmation of the diagnosis of TB in children.
2. Universal DST should be done in all samples subjected to the diagnostic confirmation of TB.
3. Improved notification and the timely treatment of latent TB will help achieve the goal of TB elimination.
4. Newer drugs with child-friendly formulations are the need of the hour.

References

1. Draft updated IAP RNTCP Ped TB guidelines. 2019. Central TB Division MOHFW New Delhi, India; p 21.
2. Sharma SK, Ryan H, Khaparde S, Sachdeva KS, Singh AD, Mohan A, Sarin R, Paramasivan CN, Kumar P, Nischal N, Khatiwada S, Garner P, Tharyan P. Index-TB guidelines: guidelines on extrapulmonary tuberculosis for India. Indian J Med Res. 2017;145(4):448–63. PMID: 28862176; PMCID: PMC5663158. https://doi.org/10.4103/ijmr.IJMR_1950_16.
3. Centers for Disease Control and Prevention (CDC). Tuberculosis associated with blocking agents against tumor necrosis factor-alpha--California, 2002–2003. MMWR Morb Mortal Wkly Rep. 2004;53(30):683–6.
4. Keane J, Gershon S, Wise RP, Mirabile-Levens E, Kasznica J, Schwieterman WD, Siegel JN, Braun MM. Tuberculosis associated with infliximab, a tumor necrosis factor alpha-neutralizing agent. N Engl J Med. 2001;345(15):1098–104. https://doi.org/10.1056/NEJMoa011110.
5. Okur E, Yilmaz A, Saygi A, Selvi A, Süngün F, Oztürk E, Dabak G. Patterns of delays in diagnosis amongst patients with smear-positive pulmonary tuberculosis at a teaching hospital in Turkey. Clin Microbiol Infect. 2006;12(1):90–2. https://doi.org/10.1111/j.1469-0691.2005.01302.x.
6. Pediatric TB. Management guidelines. https://tbcindia.gov.in/index1.php?sublinkid=4200&level=2&lid=2848&lang=1. Accessed 1 Feb 2023.
7. Guidelines for programmatic management of drug resistant tuberculosis in India. 2021. https://tbcindia.gov.in/. Accessed 2 Feb 2023.
8. Scarpellini P, Tasca S, Galli L, Beretta A, Lazzarin A, Fortis C. Selected pool of peptides from ESAT-6 and CFP-10 proteins for detection of mycobacterium tuberculosis infection. J Clin Microbiol. 2004;42(8):3469–74. PMID: 15297485; PMCID: PMC497651. https://doi.org/10.1128/JCM.42.8.3469-3474.2004.
9. Reddy MP, Moorchung N, Chaudhary A. Clinico-pathological profile of pediatric lymphadenopathy. Indian J Pediatr. 2002;69(12):1047–51. https://doi.org/10.1007/BF02724385.
10. Koegelenberg CF, Bolliger CT, Theron J, Walzl G, Wright CA, Louw M, Diacon AH. Direct comparison of the diagnostic yield of ultrasound-assisted Abrams and Tru-cut needle biopsies for pleural tuberculosis. Thorax. 2010;65(10):857–62.
11. Lal SB, Bolia R, Menon JV, Venkatesh V, Bhatia A, Vaiphei K, Yadav R, Sethi S. Abdominal tuberculosis in children: a real-world experience of 218 cases from an endemic region. JGH Open. 2019;4(2):215–20.
12. Rai A, Prasad R, Das BK, Anupurba S, Singh UK. Cerebrospinal fluid gene XPERT (CBNAAT) in children with tuberculous meningitis. J Clin Tuberc Other Mycobact Dis. 2021;24:100255.
13. Rahangdale A. Outcome of bone tuberculosis in children in rural India—a case series. Pediatr Oncall J. 2022;19:19.

14. Global TB Report. 2019. https://www.who.int/teams/global-tuberculosis-programme/global-report-2019. Accessed 16 Jan 2023.
15. WHO end TB Strategy. https://www.who.int/tb/strategy/end-tb/en/. Accessed 16 Jan 2023.
16. Fox GJ, Schaaf HS, Mandalakas A, Chiappini E, Zumla A, Marais BJ. Preventing the spread of multi-drug resistant tuberculosis and protecting contacts of infectious cases. CMI. 2017;23:147–53.
17. WHO consolidated guidelines on drug-resistant tuberculosis treatment. 2019. https://www.who.int/tb/publications/2019/consolidated-guidelines-drug-resistant-TB-treatment/en/. Accessed 19 Jan 2023.
18. Gilpin C, Korobitsyn A, Weyer K. Current tools available for the diagnosis of drug resistant tuberculosis. Ther Adv Infect Dis. 2016;3:145–51.
19. Lee PH, Chan PC, Peng YT, Chu PW, Wu MH, Jou R, et al. Impact of universal drug susceptibility testing and effective management of multidrug resistant tuberculosis in Taiwan. PloS One. 2019;14:e0214792.
20. Chatterjee S, Poonawala H, Jain Y. Drug-resistant tuberculosis: is India ready for the challenge? BMJ Glob Health. 2018;3(4):e000971. PMID: 30116597; PMCID: PMC6089296. https://doi.org/10.1136/bmjgh-2018-000971.
21. National Strategic plan to eliminate TB BY. 2025. https://tbcindia.gov.in/WriteReadData/NSP%20index%2020.02.2017%201.pdf. Accessed 16 Oct 2020.
22. Saravu K, Pai M. Drug-resistant tuberculosis: progress towards shorter and safer regimens. Lung India. 2019;36:373–5.
23. Global tuberculosis report 2017. World Health Organization, Geneva, Switzerland. www.who.int/tb/publications/globalreport/en/. Accessed 16 Jan 2023.
24. Guidance for TB notification in India. 2012. http://www.tbcindia.nic.in/WriteReadData/l892s/2362168570Guidance%20tool%20for%20TB%20notification%20in%20India.pdf. Accessed 17 Jan 2023.
25. Lawn SD, Zumla AI. Diagnosis of extrapulmonary tuberculosis using the Xpert® MTB/RIF assay. Exp Rev Anti-Infect Ther. 2012;10:631–5.
26. World Health Organization. The use of molecular line probe assays for the detection of resistance to second-line anti-tuberculosis drugs: policy guidance. 2016. http://www.who.int/iris/handle/10665/246131. Accessed 20 Jan 2023.
27. Bates M, O'Grady J, Maeurer M, et al. Assessment of the Xpert MTB/RIF assay for diagnosis of tuberculosis with gastric lavage aspirates in children in sub-Saharan Africa: a prospective descriptive study. Lancet Infect Dis. 2013;13:36–42.
28. Martinez L, Cords O, Horsburgh CR, Andrews JR, Pediatric TB Contact Studies Consortium. The risk of tuberculosis in children after close exposure: a systematic review and individual-participant meta-analysis. Lancet. 2020;395(10228):973–84.
29. Singh AR, Shewade HD, Pandey R. Near real-time supervision of home visits for patients with tuberculosis. Int J Tuberc Lung Dis. 2020;24:260–1.
30. Khurana AK, Dhingra B. What is new in management of pediatric tuberculosis? Indian Pediatr. 2019;56(3):213–20.
31. Saunders MJ, Evans CA. COVID-19, tuberculosis and poverty: preventing a perfect storm. Eur Respir J. 2020;56(1):2001348. PMID: 32444399; PMCID: PMC7243392. https://doi.org/10.1183/13993003.01348-2020.
32. Valensisi G. COVID-19 and global poverty: are LDCs being left behind? Eur J Dev Res. 2020;32(5):1535–57. Epub 2020 Oct 21. PMID: 33100600; PMCID: PMC7575865. https://doi.org/10.1057/s41287-020-00314-8.

Chapter 8
Diagnosis of Latent Tuberculosis

Parul Singh and Aishwarya Govindaswamy

Abstract Latent tuberculosis infection (LTBI) is defined as a state of persistent immune response to stimulation by *Mycobacterium tuberculosis* antigens without evidence of a clinically manifested active tuberculosis (TB) disease. One third of the world's population is affected by *M. tuberculosis*, among which a majority of them may not have active TB disease; around 5 to 10% of them can develop active TB disease after 5 years of initial infection. The risk factor for progression to active TB disease depends on the immunological status of an individual. Testing for LTBI can be done by tuberculin skin test (TST) and interferon-gamma release assays (IGRAs); however, these tests can neither differentiate LTBI from active TB nor predict if an individual with LTBI can progress to active TB. The treatment options include 3 months of weekly regimen of isoniazid plus rifapentine (3HP) or 3 months of daily regimen of isoniazid plus rifampicin (3HR) or 6/9 months of daily isoniazid (6H/9H). Short-course regimens 3HP and 4R are effective and safe and have higher completion rates over long 6H/9H regimens. The treatment of LTBI, therefore, by preventing active TB disease and developing newer diagnostic tests better than the current ones is crucial to achieving the World Health Organization's (WHO's) End TB target and reducing global TB incidence to 80% by 2030.

Keywords Tuberculosis · LTBI · Latent tuberculosis · TST · IGRA · Manteaux test · Quantiferon test

---

P. Singh
Department of Microbiology, All India Institute of Medical Sciences, Gorakhpur, India

A. Govindaswamy (✉)
Department of Microbiology, Apollo Proton Cancer Centre, Chennai, India

A. Singh, D. Sharma (eds.), *Diagnosis of Mycobacterium*,
https://doi.org/10.1007/978-981-99-5624-1_8

8.1 Introduction

Tuberculosis (TB) has been an infectious disease causing significant morbidity and mortality worldwide until the coronavirus (COVID-19) pandemic happened, which causes 1.6 million deaths every year. One third (around 10.6 million) of the world population is affected by TB, among which 2.6 million are in India, the country with the highest TB burden, as per the World Health Organization (WHO) [1]. An average of 5–10% of the affected population can develop active TB disease within the first 5 years of acquiring the initial infection, while the rest (95%) of them can develop latent infection and have risks of reactivation [2]. Latent tuberculosis infection (LTBI) is defined as a state of persistent immune response to stimulation by *Mycobacterium tuberculosis* antigens without evidence of clinically manifested active tuberculosis (TB) disease [3]. The high burden of latent TB infection can serve as a major barrier to the WHO End TB strategy [4]. It aims to reduce TB incidence by 80%, lessen TB deaths by 90%, and attain 0% catastrophic costs incurred by TB-affected families by 2030. These TB goals aim to end the global TB epidemic as a part of the Sustainable Development Goals (SDG) for 2016 to 2030 [5]. The above goals can be accomplished by improving diagnostic modalities, followed by completion of the treatment course, latent TB detection, and treatment.

8.2 Epidemiology

The estimation of the true burden of LTBI is quite challenging due to the absence of a gold-standard test for diagnosing latent TB. As per the literature, the global prevalence of LTBI is around 24.8% based on IGRAs (interferon-γ release assays) and 21.2% based on TST (tuberculin skin test), representing a huge reservoir of potential TB disease [6]. The prevalence of LTBI varies between different geographical locations. Regions with high prevalence based on IGRA are Southeast Asia (36%), followed by Africa (33.6%), Eastern Mediterranean (24%), Western Pacific (20.7%), America (13.7%), and Europe (12.2%) [6, 7]. Based on age, LTBI rates were found to be 10% in those aged 5 to 9 years and 17.9% in the age group 10–14 years in a study conducted in Bangladesh [8]; 30.3% in those aged 4–18 years in South Africa [9]; in Vietnam, 16.7% for age 6 to 14 years and 36.8% among adults [10]; and 19% in a study conducted in China for those aged >5 years [11]. In relation to the high incidence of TB in males as per WHO [12], studies have shown a high prevalence of LTBI in males than females in most countries [11, 13, 14].

8.3 Etiology of Tuberculosis

Tuberculosis is caused by members of the *Mycobacterium tuberculosis* complex, which comprises around nine species belonging to the genus *Mycobacterium*, family Mycobacteriaceae, and order Actinomycetales. The common species causing infection in humans worldwide is *Mycobacterium tuberculosis* sensu stricto *(M. tuberculosis)*. Others include *Mycobacterium bovis (M. bovis)*, *Mycobacterium africanum*, and *Mycobacterium canettii*. The species affecting animals are *Mycobacterium caprae (goats)*, *Mycobacterium microti (voles)*, *Mycobacterium pinnipedii (seals)*, *Mycobacterium mungi (mongooses)*, and *Mycobacterium orygis (antelope)* [15, 16]. Humans are the only reservoir for *M. tuberculosis*; however, animals can also be susceptible to the infection. *M. tuberculosis* is an aerobic, acid-fast, nonmotile, non-spore-forming bacilli with a cell wall comprising high molecular weight lipids. The bacteria grow slowly, with a generation time of around 15 to 20 h, and it takes 3 to 8 weeks for visible growth to occur in solid media. The organism tends to grow in parallel groups as serpentine cords [17].

8.3.1 Risk Factors for Latent Tuberculosis Infection

The risk factors for the reactivation of latent TB infection to active TB disease depend on various hosts and bacterial and environmental factors. Individuals with latent TB infection serve as a major reservoir for active TB disease [18]. The risk factors for LTBI reactivation are classified into high risk, moderate risk, and low risk (Table 8.1).

8.3.1.1 High-Risk Factors

The most potent risk factors are individuals with acquired immunodeficiency syndrome (AIDS) and human immunodeficiency virus infection (HIV). People with HIV and LTBI have more than a 100-fold risk of developing an active TB disease [19–21]. Patients undergoing organ transplants are at a high risk of developing infections due to drugs causing immunosuppression. As per a study conducted in Spain, recipients of liver, kidney, and heart transplants had a TB incidence of 0.8%, which is 20 times higher in comparison to the general population [22]. A study from India has reported an incidence of 11.8% among kidney transplant recipients, 70 times higher than in the general population [23]. In patients with silicosis, around 25–30% are at risk of developing active TB; the relative risk for TB is 2.8 in patients with silicosis in contrast to the general public [24, 25]. There is a high risk (>15 times) of TB reactivation in individuals recently (<2 years) infected with TB, and people who are in close contact with active TB disease patients have a high chance of acquiring infection within 2 years [21, 26]. Tumor necrosis factor-alpha (TNF-α)

Table 8.1 Risk factors associated with latent TB reactivation in individuals infected with *M. tuberculosis*

	TB risk
High risk	
Acquired immunodeficiency syndrome (AIDS)/human immunodeficiency virus infection (HIV)	10–110
Transplant, chemotherapy, and other immunocompromised state	20–70
Silicosis	30
Close contact with pulmonary TB patients	15
Renal failure requiring dialysis	6.9–52.5
Head and neck cancer, leukemia, lymphoma	16
Recent infection with TB (<2 years)	15
Tumor necrosis factor (TNF) alpha inhibitors	1.6–25.1
Silicosis	2.8
Moderate risk	
Apical fibronodular changes depicting healed TB in chest X-ray	6–19
Immigrants from high TB-burden countries	2.9–5.3
Health care workers	2.55
Acquiring infection at a young age (<4 years)	2.2–5
Prisoners, homeless, illicit drug users	–
Low risk	
Diabetes mellitus	1.6–7.83
Underweight individuals (with body mass index ≤20%)	2–3
Cigarette smoking	2–3.4
Treatment with systemic corticosteroids	2.8–7.7

has an important role in regulating inflammatory response in the body. TNF-α antagonists like infliximab, adalimumab, etanercept, golimumab, and certolizumab are being used in various clinical trials. First randomized clinical trials (RCTs) on infliximab have shown a fourfold risk of TB infection [27, 28], and several other studies from the literature have also shown a similarly high risk of TB infection in the TNF-α group, in contrast to the placebo group, with a relative risk of 1.6 to 25.1 [29]. In individuals with chronic renal failure and on hemodialysis, there is a 6.9- to 52.5-fold increase risk of TB [30]. Also, the diagnosis of TB in dialysis is difficult; the sensitivity of TST is decreased to 50% during chronic renal failure and hemodialysis, thus requiring IGRA or other invasive modalities for diagnosis [31].

8.3.1.2 Moderate Risk Factors

As per the literature, there is a six- to 19-fold increased risk of TB in individuals with old inactive TB lesions as shown in chest X-rays and had not received adequate treatment, and clinical trials have shown a 65% decrease in the incidence of TB in patients with fibrotic lesions after 6 months of isoniazid monotherapy, thus warranting prophylaxis in this group of the population [32, 33]. Developed nations have low TB prevalence, and immigrants from heavy TB-burden countries always pose a risk for TB; thus, the screening for and treatment of TB and LTBI are necessary for foreign nationals [34, 35]. However, not all countries uniformly perform screening for TB and LTBI, and the cutoff for diagnosing LTBI also varies between countries [19]. Health care workers are always at a high risk of developing hospital-acquired TB, in contrast to the general population not working in such settings, which can cause secondary outbreaks in hospitals if not treated properly [36, 37]. It could be a result of malfunctioning air conditioners causing the recirculation of contaminated air; inadequate usage of personal protective equipment during procedures like bronchoscopy, intubation, and assisted ventilation; the emergence of HIV epidemic; and immigrant patients from high TB-burden countries [38, 39]. Populations like prisoners, homeless individuals, and illicit drug users are usually underprivileged, and HIV coinfection is more common in them; thus, treating these population groups is quite a challenge [40]. In addition to this, prisoners are a main risk group in spreading drug-resistant TB as well [41]. Several studies have recommended both the screening and treatment of LTBI for the above risk groups [3].

8.3.1.3 Low-Risk Factors

For individuals with diabetes, the relative risk of TB varies between 1.16 and 7.83 [42, 43]. However, this does not warrant screening or testing for LTBI. This could be due to the low risk of TB acquisition, and there are no large randomized controlled trials (RCTs) on this topic. Moreover, TB risk is related to glycemic control in patients; a study has shown that poor glycemic control has more risk of TB reactivation [44]. Smoking tobacco can cause alterations in the immune responses of the lungs, making them more susceptible to acquiring infection. However, it does not warrant prophylaxis for LTBI due to health and financial constraints since in low- and middle-income countries, approximately 50% of men and 8% of women are indulged in tobacco smoking [45]. Other risk factors, like alcohol usage, underweight, and usage of corticosteroids, do not require LTBI testing until and unless they are accompanied by any of the high-risk factors [3].

8.4 Pathogenesis of Latent TB Infection

Most individuals who have a prime *Mycobacterium tuberculosis (Mtb)* infection develop specific acquired cell-mediated immunity that prevents mycobacteria from growing but does not completely eradicate them. Tuberculosis bacilli still exist in such a person but are inactive. For immune-competent subjects, the lifetime chance of latent TB reactivation is in the range of 10% [46, 47]. The primary defense mechanism of an organism against TB is cell-mediated immunity. The establishment of a Th1 immune response is essential for the management of *Mtb* infection in people who are resistant to treatment. This type of response involves the participation of resident alveolar macrophages, dendritic cells, T lymphocytes (TCD4+, TCD8+, Tγδ), and the release of proinflammatory cytokines, interferon-γ (IFN-γ), interleukin-2 (IL-2), IL-12, IL-18, tumor necrosis factor-α (TNF-α), chemokines (IL-8, monocyte chemoattractant protein-1 (MCP-1)), and macrophage inflammatory protein-1 alpha (MIP-1α) [48, 49]. They all contribute significantly to the recruitment of extra cells to the infection site for the development of granulomas, which confine and eradicate tuberculosis bacilli while also offering the long-term niche required for LTBI [50, 51].

In reaction to pulmonary inflammation brought on by the stimulation of host cells with mycobacterial antigens, the granuloma is a structural organization of various types of immune cells, including macrophages, T cells, B cells, dendritic cells, neutrophils, natural killer (NK) cells, and a fibroblast. Localized macrophages that phagocytose bacteria and produce proinflammatory cytokines, like TNF-, to draw in more cells, are what cause granuloma to form. In the granuloma, macrophages either fuse to create multinucleated giant cells or differentiate into epithelioid cells. A ring of lymphocytes, including CD4 T cells of the adaptive immune response, surrounds the aforementioned cells and may increase the bactericidal activity of macrophages by releasing IFN-γ. Later, the granuloma is covered in a compact layer of fibroblasts [52, 53]. The outcome of *Mtb* infection is determined by the adapted cell-mediated immune response and the correct formation of granulomas. The host response is adequate to stop the TB disease in 90% of *Mtb*-infected people. Changes in bacterial metabolism and host metabolism that are partially driven by *Mtb* effector proteins and glycolipids are associated with the persistence of TB bacilli in the granuloma. The persistent bacilli in the granuloma are susceptible to a variety of stressful situations, including hypoxia, nutrient deficiency, acidic pH, and nitric-oxide-inhibited respiration. All of these elements cause the genes to be expressed, which causes *Mtb* to enter a dormant state [54].

By reducing their metabolic and replicative activity as well as by obstructing their growth and development, the inactive bacilli can affect the activity of these organisms. They develop immune system resistance and escape being destroyed by immune cells [54]. It has been suggested that in *Mtb* infections, proper granuloma formation is essential for preventing the spread of mycobacteria and tissue damage, two aspects of active TB illness. The localization of lymphocytes that are particular for an antigen within the lungs may be hampered by an inadequate upregulation of adhesion molecules on circulating lymphocytes. As a consequence, the ability of the correct granuloma to stop *Mtb* growth is compromised. The processes that target antigen presentation affect the strength and consistency of the T-cell response within granulomas [55, 56]. Many *Mtb* components, including 19-kDa lipoprotein, mannose-capped lipoarabinomannan (Man-LAM), trehalose dimycolate (cord factor), and others, can influence how mycobacterial protein and glycolipid antigens are processed and presented by MHC class I, MHC class II, and CD1 molecules [52]. In this manner, TB bacilli may prevent macrophages from presenting antigens to T lymphocytes. The lack of activation of effector CD4+, CD8+, and /T lymphocytes; CD1-restricted T cells; and cytotoxic T cells impairs the ability of macrophages to kill bacteria, and altered immune cell activity, including that of cells involved in the inflammatory response, causes tissue injury and spread [48].

8.5 Clinical Presentation Latent TB Versus TB Disease

Tuberculosis disease caused by *M. tuberculosis* spreads from person to person via the air. The disease usually affects the lungs; however, it can also infect other organs, like the brain, spine, and kidneys. When an individual with infectious TB sneezes or coughs droplets, nuclei with *M. tuberculosis* are released into the air, and another person inhaling these droplet nuclei can get infected. However, not all who are infected get sick. Thus, two different TB-related conditions, latent TB infection and TB disease, exist [57]. The differentiation between these two conditions is important to decide on individuals' preventive therapy for latent TB [3]. The differences between latent TB infection and TB disease are illustrated in Table 8.2.

Table 8.2 Differences between latent TB infection and TB disease

Latent TB infection	TB disease
There are no symptoms	May have the following symptoms: Cough lasting for 3 weeks or longer, chest pain, weakness fatigue, coughing up sputum or blood, loss of appetite, loss of weight, fever, chills, night sweats
Individual does not feel sick	Individuals usually feel sick
Cannot spread TB bacilli to others around	May spread the TB bacilli to others
Chest X-ray is normal and sputum smear is negative	Chest X-rays are abnormal and have a positive sputum smear or culture
May require treatment for latent TB infection to prevent TB disease	Requires treatment for TB disease

8.5.1 Latent TB Infection

TB bacilli can reside in the human body without making an individual sick, causing latent TB infection. Individuals with good immune responses are able to fight the infection and thus may not exhibit any signs and symptoms. They may not feel sick and would not spread the infection to others. They can have a positive TB skin test or a positive test for TB in their blood. These individuals may not develop TB disease, remaining inactive for a lifetime. However, in people with a weakened immune system, the TB bacilli may become active and replicate, causing TB disease [58].

8.5.2 TB Disease

TB bacilli can replicate in the body if the body's immune system cannot fight the bacteria, thus causing TB disease. Patients with TB disease are sick and spread the bacilli to people around them. Many with latent TB infection may not develop TB disease, while some can get TB disease within weeks of acquiring the infection even before the immune system can fight the TB bacilli. Others can get TB disease after years of TB infection due to various reasons that may cause a weak immune system [58].

8.5.3 Individuals Requiring LTBI Testing and Treatment

The identification of at-risk populations that require testing and preventive treatment of LTBI is of the utmost importance due to the risks and costs entailed in the treatment. Thus, preventive treatment is recommended only for individuals with the highest risk for active TB progression, who would benefit the most from preventive treatment [3]. The WHO recommendation for systematic LTBI testing of the high-risk population is illustrated in Table 8.3.

8.6 Diagnosis of Latent Tuberculosis

The immunological markers of the immune reaction are used to identify this infection. To assess the adaptive cell-mediated immune response of *Mtb*-exposed asymptomatic hosts, two clinically applicable techniques have been developed. Standard immunologic diagnostic methods for LTBI include interferon-gamma release assays (IGRA) and the tuberculin skin test (TST) [59, 60].

Table 8.3 Population requiring systemic LTBI testing and treatment

Adults and adolescents living with HIV infection at any stage of illness
Infants and children living with HIV
Household contacts of HIV-negative children (<5 years) with bacteriologically confirmed pulmonary TB
Household contacts of adults, adolescents, and children with bacteriologically confirmed pulmonary TB in countries with low TB incidence
Household contacts of adults, adolescents, and children (<5 years) with bacteriologically confirmed pulmonary TB in countries with high TB incidence
Patients receiving anti-TNF treatment
Patients on dialysis
Patients undergoing organ or hematological transplant
Patients with silicosis
Prisoners, immigrants from high TB-burden countries, health care workers, users of illicit drugs, homeless in low TB-incidence countries

8.6.1 *Tuberculin Skin Test*

The tuberculin skin test (TST) was the only way to diagnose LTBI prior to 2001. The test was called after Charles Mantoux and Clemens von Pirquet, who put it into use in 1907, and was first introduced by Robert Koch in 1890. PPD (purified protein derivative) is a precipitate of nonspecies-specific antigens derived from filtrates of mycobacterial cultures, and tuberculin is a glycerol extract of mycobacteria.

The skin's induration is currently measured in TST 48 to 72 h after intradermal administration of PPD (the Mantoux procedure). A delayed-type hypersensitivity response is visible 2–8 weeks after infection if a person is infected. Using the Mantoux method, the purified protein derivative (PPD) solution containing 5 tuberculin units is injected intradermally into the skin. A PPD-driven classical T-cell-mediated delayed-type hypersensitivity response (DTH) is brought on by the test [60].

8.6.1.1 The Criteria for a Positive TST Result

A TST reaction of ≥5 mm induration is considered positive for:

HIV-positive individuals, recent encounters with infected TB patients, individuals whose chest X-ray results point to a history of TB illness, organ donation recipients, and other immunosuppressed patients, such as those receiving long-term corticosteroid therapy (e.g., those taking TNF-alpha antagonist or corticosteroid therapy equal to or higher than 15 mg of prednisone per day).

A TST reaction of ≥10 mm of induration is considered positive for:

Those born in nations with a high prevalence of TB illness, such as Mexico, the Philippines, Vietnam, India, China, Haiti, and Guatemala; drug abusers; mycobacteriology lab personnel; people with certain medical conditions that put them at risk for TB (e.g., silicosis, diabetes mellitus, severe kidney disease, certain types of cancer, or certain intestinal conditions); people who reside or labor in high-risk congregate environments (such as nursing homes, homeless shelters, or correctional facilities); individuals with low body weight; infants under the age of 5; and kids and teenagers who are exposed to adults who fall under the high-risk categories.

A TST reaction of ≥15 mm of induration is considered positive for:

People without established TB risk factors.

8.6.2 Interferon-Gamma Release Assays (IGRAs)

The discovery of a region of difference (RD1), which is missing from bacille Calmette-Guérin (BCG) strains and the majority of atypical mycobacteria but present in all *Mtb* and pathogenic *M. bovis* strains, was made possible by our understanding of the full genome sequences of mycobacteria. Because it appears to be involved in increasing the mycobacterial virulence by encoding two secretory proteins, ESAT-6 (early secretory antigenic target-6) and CFP-10 (culture filtrate protein-10), this region is one of the most intriguing genomic loci of the tubercle bacilli [63]. Both antigens have been discovered to be potent inducers of rodent delayed-type hypersensitivity (DTH) development and IFN production. It was effective to use rESAT-6 derived from *Escherichia coli* as an antigen for the human TB diagnostic skin test. In the guinea pig model, it was discovered that the combo of rESAT-6 and CFP-10 was more accurate than PPD for determining *Mtb* infection both in vitro and in vivo. Therefore, it appears that the antigenic complex is a very promising candidate for LTBI identification and may eventually take the position of PPD as a skin test reagent. Interferon-gamma release assays were developed on the basis of the effective antigenicity of ESAT-6 and CFP-10. They are diagnostic procedures that gauge the amount of IFN-gamma that T lymphocytes release in reaction to antigens unique to *M. tuberculosis* [61, 62].

There are two T-cell-dependent tests: QuantiFERON-TB Gold (QFT-G) and T-SPOT.TB assay. The tests have been approved by the Food and Drug Administration (FDA) for the diagnosis of infection with *M. tuberculosis* [64]. The QuantiFERON-TB Gold (QFT-G) contains two antigens of the *Mtb* RD1 region, ESAT-6 and CFP-10. In these tests, ELISA is used to detect the IFN response to mycobacterial antigens.

The T-SPOT.TB test, an enzyme-linked immunospot (ELISPOT) assay used in TB testing, measures the number of T lymphocytes that produce IFN-gamma after being stimulated with ESAT-6 and CFP-10. In this test, anti-IFN-gamma monoclonal antibodies that have been immobilized on a plate bind precisely to IFN-gamma molecules secreted by cells responding to the antigens. All *M. tuberculosis*-specific T cells that produce IFN-gamma are visible as dark spots, which can be identified using a microscope or an automatic reader [65, 66]. Laboratories provide both qualitative and quantitative results. Qualitative results are reported as positive, negative, indeterminate, invalid, or borderline.

8.6.2.1 Interpretations of TB Blood Test Results

POSITIVE: *M. tuberculosis* infection is likely.

NEGATIVE: Although rare, *M. tuberculosis* infection cannot be ruled out, particularly if (a) the patient exhibits symptoms typical of TB disease. (b) Following infection with *M. tuberculosis*, the patient is at significant risk of developing TB disease (e.g., the patient is immunosuppressed).

INDETERMINATE (QFTPlus only) or INVALID (T-Spot only): The probability of *M. tuberculosis* infection was not accurately predicted by the test. A second TB blood test or a TST could be beneficial.

BORDERLINE (T-Spot only): Repeating a TB blood test or performing a TST may be useful.

In the quantitative results, responses to the TB antigen and the two mitogen and nil controls are included, along with numerical values. When combined with risk factors, quantitative findings may be helpful for clinical decision-making in specific cases [67]. The advantages and limitations of TB blood tests are illustrated in Table 8.4.

8.6.3 *Other Diagnostic Considerations*

8.6.3.1 Chest Radiograph

Chest radiographs (X-rays) help differentiate between LTBI and pulmonary TB disease in individuals with positive tests for TB infection [68].

Table 8.4 TB blood tests—advantages and limitations

Advantages	Limitations
• Only one patient visit is necessary to perform the test • Not a source of the booster phenomenon • The results are available in 24 h • The TB blood tests are not affected by the bacille Calmette Guérin (BCG) vaccine and most of the nontuberculous mycobacteria (NTM) • Not impacted by the prejudices and mistakes related to TST location and reading	• Blood samples are supposed to be processed within 8 to 32 h of collection • The precision of TB blood tests can be reduced by mistakes made during the collection, transport, running, and interpretation of the test • Expensive

8.6.3.2 Sputum Examination for Acid-Fast Bacilli (AFB) Smear and Culture

People with positive TB test findings who have either an abnormal chest radiograph or respiratory symptoms should have a sputum evaluation (even when the chest radiograph is normal). Even with a typical chest radiograph, sputum evaluation for HIV-positive patients should be taken into consideration. Chest radiographs taken in the presence of advanced HIV can look normal [68].

8.6.3.3 Physical Examination and Medical History

Individuals with positive TB test results should have a physical examination, as well as a medical history that includes knowing about any prior TB infection tests that were positive, any prior treatment for LTBI or TB disease, and a risk assessment for liver disease. A patient's verbal history is not adequate proof of a prior positive TB blood test or TST result; written documentation is necessary [68].

8.6.4 Special Considerations in Testing for TB Infection

8.6.4.1 Bacille Calmette-Guérin (BCG) TB Vaccine

In nations with an active TB epidemic, the World Health Organization (WHO) advises giving infants the BCG vaccine. A false-positive TST response could result from the BCG vaccination. A positive TST response brought on by the BCG vaccine cannot be reliably distinguished from one brought on by an actual TB infection. Blood tests for TB are performed using *M. tuberculosis*-specific antigens, which do not cross-react with BCG and do not result in false-positive results in BCG patients. The recommended test for those who have received the BCG vaccine is a TB blood test [67, 68].

8.6.4.2 HIV Infection

One of the main causes of death among HIV-positive individuals worldwide is tuberculosis (TB). For people with both LTBI and untreated HIV infection, the annual chance of progression from LTBI to TB disease ranges from 7% to 10%. Those with LTBI who are not HIV positive run a 10% lifetime chance. As a result, those with HIV have a ten times higher chance of developing TB disease. With effective antiretroviral therapy (ART) for HIV, this risk is significantly decreased; however, it is still greater than it is for those with LTBI who are HIV negative. Regardless of their epidemiological risk of TB exposure, all HIV-positive individuals should undergo LTBI testing at the time of HIV diagnosis [67, 68].

8.6.4.3 Booster Phenomenon

The booster phenomenon is one element that may have an impact on the baseline TST's accuracy. Depending on how long it has been since the individual contracted *Mtb*, certain people with the infection may have a negative TST reaction. They might, however, exhibit a positive response upon retesting within a year of the initial test, suggesting a likely TB infection. This is so because their capacity to respond to the test is stimulated or amplified by the first test. Commonly known as the "booster phenomenon," this could be mistakenly understood as a skin test conversion (going from negative to positive) [67, 68].

8.6.4.4 Contacts

People who have been exposed to a confirmed or suspected TB patient are considered contacts (e.g., pulmonary or laryngeal TB with a positive sputum smear). When the initial TB blood test or TST result is negative for contacts of an individual with infectious TB disease, retesting 8–10 weeks after the end of exposure is advised [68].

8.6.4.5 Pregnancy

Pregnant women should be targeted for TB testing only if they have a specific risk factor for *Mtb* infection or progression from LTBI to TB disease. The TST is both safe and reliable throughout the course of pregnancy. TB blood tests are also safe to use during pregnancy, but they have not been fully evaluated for diagnosing TB infection in pregnant women [68].

8.7 Treatment of Latent Tuberculosis Infection

8.7.1 Treatment Regimens for Latent Tuberculosis Infection

The antimicrobial regimens available for LTBI include isoniazid monotherapy, rifampicin plus isoniazid, rifapentine and isoniazid combination, and rifampicin monotherapy, as described in Table 8.5. Isoniazid monotherapy for 6–12 months are in use for several years, and the efficacy for preventing the progression to TB disease is around 90% [3, 69]. However, its overall efficacy is reduced due to low adherence and increased hepatotoxicity. There are shorter rifamycin-based regimens with good completion rates and reduced hepatoxicity that are being used. The adverse effects of LTBI drugs are illustrated in Table 8.6. There are also studies showing that the above regimens do not cause any isoniazid- or rifamycin-resistant TB disease post treatment of LTBI [70, 71].

Table 8.5 Treatment options for latent tuberculosis infection

Treatment regimen	Duration	Frequency	Dose	Maximum dose	Total doses
Isoniazid monotherapy(6H/9H)	6–9 months	Daily	Adults: 5 mg/kg Children: 10 mg/kg	300 mg	180 (6H) 270 (9H)
Isoniazid plus rifampicin (3HR)	3–4 months	Daily	Rifampicin Adults:10 mg/kg Children: 15 mg/kg Isoniazid Adults: 5 mg/kg Children: 10 mg/kg	Rifampicin, 600 mg Isoniazid, 300 mg	90
Isoniazid and rifapentine (3HP)	3 months	Weekly	Isoniazid: 15 mg (individuals aged ≥12 years) Isoniazid: 25 mg (individuals aged 2–11 years) Rifapentine: 10.0–14.0 kg = 300 mg, 14.1–25.0 kg = 450 mg, 25.1–32.0 kg = 600 mg, 32.1–50.0 kg = 750 mg, >50 kg = 900 mg	Isoniazid, 900 mg Rifapentine, 900 mg	12
Rifampicin (4R)	3–4 months	Daily	Adults: 10 mg/kg Children: 15 mg/kg	600 mg	120

Table 8.6 Adverse effects of drugs for LTBI treatment

Drug	Adverse effects and considerations
Isoniazid	Hepatitis, elevation of aminotransferases, rash, peripheral neuropathy Pyridoxine can be administered to patients at risk of neuropathy Close follow-up in patients with baseline liver disease
Rifamycins (rifapentine, rifampin, rifabutin)	Rash, hematological abnormalities (thrombocytopenia), hepatitis, elevation of aminotransferases, flu-like illness, body fluids becoming orange Can cause potential drug interactions with anticonvulsants, antihypertensives, antiarrhythmics, warfarin, opioids, hormonal contraceptives, azole antifungals, HIV antiretroviral drugs Isoniazid-rifapentine not recommended for pregnant women or women who are expecting to be pregnant during the course of LTBI treatment Isoniazid-rifapentine not recommended for children <2 year old

8.7.1.1 Daily Isoniazid Monotherapy for 6 to 9 Months (6H/9H)

Isoniazid (INH) acts by inhibiting the mycolic acid synthesis of *M. tuberculosis*, which is an important component of the cell wall. Isoniazid monotherapy has been used for decades for LTBI preventive therapy, with several systemic reviews demonstrating its preventive efficacy. Systemic reviews of randomized controlled trials in people living with HIV have shown a 33% reduction in TB incidence and preventive efficacy of 64% for individuals with a positive TST. Also, the efficacy of the 6-month regimen was not significantly different in comparison to the 12-month regimen [72]. However, a reanalysis of US Public Health Service (USPHS) trials has demonstrated that isoniazid given for 9 to 10 months has maximum benefits [73]. Therefore, the recommended duration of isoniazid monotherapy has been 6 to 9 months. The Centers for Disease Control and Prevention (CDC) recommends 6H/9H regimens if other short-course regimens cannot be given due to drug-drug interactions with rifamycins. 6H/9H is recommended for HIV-positive and HIV-negative adults and children of all ages. However, efficacious treatment completion rates are low, and liver toxicity is high with 6H/9H regimens [74]. The adverse effects of isoniazid include hepatitis, which can cause a loss of appetite, fatigue, abdominal discomfort, jaundice, nausea, and peripheral neuropathy. With increasing age and alcohol consumption, the risk of hepatitis is high. Close monitoring is required in patients with preexisting chronic liver disease or hepatitis C infection. Drug interactions are observed with the following drugs: sodium valproate, phenytoin, carbamazepine, warfarin, monoamine oxidase inhibitors, and acetaminophen [75].

8.7.1.2 Rifampicin Daily for 3 to 4 Months (4R)

Rifampicin (RIF) acts by binding to the RNA polymerase, thereby inhibiting the RNA synthesis of mycobacteria. In a multicenter trial study program where the research participants were randomized into two groups, those receiving isoniazid daily for 9 months versus those taking rifampicin daily for 4 months have found that rifampicin monotherapy is noninferior to isoniazid for the prevention of LTBI. The treatment completion rates were high with rifampicin than with isoniazid (78.8% vs. 63.2%), and hepatoxicity was also less [76]. The CDC recommends the 4R group for HIV-negative adults, children of all age groups, and individuals who cannot tolerate INH or exposed to INH-resistant TB [74] . The adverse effects of rifampicin include gastrointestinal symptoms like nausea, abdominal pain, anorexia, skin rash, hypersensitivity, hematological side effects, and hepatoxicity. The body fluids and urine can have orange discoloration. Rifampicin is a potent inducer of CYP3A and thus can affect the metabolism of drugs like anticonvulsants, antiretrovirals, warfarin, methadone, azole antifungals, and immunosuppressants like cyclosporine [75].

8.7.1.3 Daily Rifampicin Plus Isoniazid for 3–4 Months (3HR)

A systematic review has shown that daily rifampicin plus isoniazid for 3–4 months versus daily isoniazid for 6 months has similar efficacy and safety profile [77]. There are randomized controlled trials conducted in HIV-positive and HIV-negative groups to compare the efficacy of 6 to 9 months of daily isoniazid and daily rifampicin plus isoniazid for 3–4 months. The studies have shown similar efficacy and prevention rates of TB [78, 79]. As per the CDC, the 3HR is one of the preferentially recommended short-course rifamycin-based regimens for children of all ages, adults, and HIV-positive and HIV-negative individuals [74].

8.7.1.4 Weekly Once Rifapentine and Isoniazid for 3 Months (3HP)

Rifapentine (RPT), which is a derivative of rifamycin, has a longer shelf life and is more potent than rifampin. PREVENT TB clinical trail 7731 individuals with positive TST consisting of people living with HIV/AIDS (PLWHA), contacts of pulmonary TB patients, recent TST convertors, chest X-ray showing fibrosis were included in the study. The above participants were randomized to receive 900 mg of isoniazid for 3 months of directly observed therapy (DOT) and 900 mg of rifapentine once weekly or 300 mg of isoniazid daily monotherapy for 9 months. The study participants were followed up for 33 months, and it was found that the isoniazid rifapentine combination was noninferior to isoniazid monotherapy for preventing TB disease. The adverse effects of hepatotoxicity were also less for the isoniazid rifapentine combination in contrast to the isoniazid group (0.4% vs. 2.7%). The isoniazid rifapentine group had good treatment completion rates in comparison to the isoniazid monotherapy group (82.1% vs. 69%) [80]. On the basis of PREVENT TB Trail, initially, the isoniazid rifapentine combination was only administered as DOT. However, a study based on DOT versus self-administered therapy (SAT) showed similar safety and treatment completion rates with SAT; thus, the regimen was recommended as SAT by the Centers for Disease Control and Prevention (CDC) [81]. The CDC recommends the 3HP regimen for children older than 2 years and for persons living with HIV/AIDS (PLWHA) who are on antiretroviral medications like efavirenz and raltegravir. The 3HP regimen is not recommended for children less than 2 years, pregnant and expecting mothers, and individuals infected with INH or RIF-resistant *M. tuberculosis*. Rifapentine is also a potent inducer of CYP3A, drug interactions and toxicity profiles are similar to as rifampin [82].

8.7.2 Monitoring of Adverse Effects During LTBI Treatment

Individuals undergoing treatment for LTBI are recommended to meet the health care professionals on a monthly basis for the monitoring of treatment adherence and for signs of drug toxicity. Patients experiencing symptoms like nausea, vomiting, anorexia, abdominal discomfort, weakness, fatigue, pale stools, orange color urine, and jaundice can consult the doctor anytime. Laboratory investigations are usually not recommended during follow-up visits until there are any high-risk factors, like the use of alcohol, history of liver disease, chronic liver disease, pregnancy, postpartum within 3 months of delivery, HIV infection, and age of more than 35 years. Vitamin B6 supplements should be given to at-risk individuals with malnutrition,

HIV infection, alcohol addiction, diabetes, renal failure, pregnancy, and breast feeding mothers taking isoniazid regimen who can develop peripheral neuropathy [3].

8.7.3 Special Considerations for LTBI Treatment

8.7.3.1 Persons Living with HIV/AIDS (PLWHA)

HIV/AIDS has a high risk of TB disease progression. In PLWHA, in the absence of antiretroviral therapy (ART) among patients with HIV and LTBI coinfection, 10% have the risk of progression to active TB disease every year [83]. Both LTBI treatment and ART are recommended in this group since it decreases progression to active TB disease. Isoniazid monotherapy is the recommended regimen for its proven efficacy [74]. Isoniazid monotherapy has reduced TB incidence in 64% of PLWHA having a positive TST [72]. In PLWHA who have not started ART or on ART regiments containing efavirenz or raltegravir once-weekly isoniazid plus rifapentine for 3 months regimen have been proven to be noninferior to isoniazid monotherapy [80]. A recent clinical trial has also shown that an ultra-short daily isoniazid 300 mg and rifapentine 300–600 mg daily for 1 month was noninferior to 9 months of isoniazid monotherapy in PLWHA for the prevention of TB. The ultra-short regimen also had fewer adverse effects and high completion rates [84].

8.7.3.2 Pregnant Women

The CDC recommends the treatment of pregnant women with high risk factors, such as HIV, or who had recent contact with infectious TB. In pregnancy when there is no high risk of progression to TB disease, LTBI treatment can be delayed until 2 to 3 months postpartum due to a high risk of hepatotoxicity. Isoniazid monotherapy is the preferred regimen, but rifampin can also be used as an alternative. Supplements, like 25–50 mg/day of pyridoxine (vitamin B6), are necessary to reduce the side effects of isoniazid. The isoniazid-rifapentine regimen is not recommended during pregnancy and in women expecting to be pregnant during the treatment course since its safety profile has not yet been studied [69, 74].

8.7.3.3 Breastfeeding Mothers

For women taking INH and RIF separately, breastfeeding is not a contraindication. Vitamin B6 supplements must be given to breastfeeding mothers on INH. An orange discoloration of breast milk is harmless for mothers on RIF [74].

8.7.3.4 Infants and Children

Infants and children are more vulnerable than older children and adults to developing serious forms of TB disease. The CDC recommends the treatment of children less than 5 years of age who were in contact with infectious TB, even if the TST or blood test is negative, for less than 8 to 10 weeks upon their exposure to infectious TB. The second TST or IGRAs after 8 to 10 weeks exposure to infectious TB if positive the LTBI treatment can be continued, if negative it can be discontinued. The recommended regimen is 3HP, 4R, 3HR, 6H, or 9H. The CDC recommends a 3HP regimen for adolescents and children older than 2 years, which can be given as DOT or SAT [74].

8.7.3.5 Contacts of Multidrug-Resistant (MDR) Tuberculosis

Contacts of MDR TB patients are at a high risk of developing MDR TB disease. Due to the lack of RCT, there are no standardized recommendations for TB preventive therapy. Preventive therapy is indicated only for household contacts at a high risk, which include children, PLWHA, and people on immunosuppressive medications. A confirmation of LTBI infection should be done before starting treatment. The drugs should be based on the susceptibility profile of the source case. The CDC recommends later-generation fluoroquinolone-based regimens for treatment [3, 74].

References

1. Global Tuberculosis Reports. https://www.who.int/teams/global-tuberculosis-programme/tb-reports. Accessed 10 Jan 2023.
2. Xiao X, Chen J, Jiang Y, Li P, Li J, Lu L, et al. Prevalence of latent tuberculosis infection and incidence of active tuberculosis in school close contacts in Shanghai, China: baseline and follow-up results of a prospective cohort study. Front Cell Infect Microbiol. 2022;12:1000663. https://doi.org/10.3389/fcimb.2022.1000663.
3. World Health Organization. Latent tuberculosis infection: updated and consolidated guidelines for programmatic management. Geneva: World Health Organization; 2018. https://apps.who.int/iris/handle/10665/260233. Accessed 3 Jan 2023.
4. Behr MA, Kaufmann E, Duffin J, Edelstein PH, Ramakrishnan L. Latent tuberculosis: two centuries of confusion. Am J Respir Crit Care Med. 2021;204(2):142–8.
5. The End TB Strategy [Internet]. https://www.who.int/publications-detail-redirect/WHO-HTM-TB-2015.19. Accessed 30 Nov 2022.
6. Cohen A, Mathiasen VD, Schön T, Wejse C. The global prevalence of latent tuberculosis: a systematic review and meta analysis. Eur Respir J. 2019;54(3):1900655. https://erj.ersjournals.com/content/54/3/1900655.
7. Houben RMGJ, Dodd PJ. The global burden of latent tuberculosis infection: a re-estimation using mathematical modelling. PLoS Med. 2016;13(10):e1002152.
8. Hossain S, Zaman K, Banu S, Quaiyum MA, Husain MA, Islam MA, et al. Tuberculin survey in Bangladesh, 2007–2009: prevalence of tuberculous infection and implications for TB control. Int J Tuberc Lung Dis. 2013;17(10):1267–72.

9. Shanaube K, Sismanidis C, Ayles H, Beyers N, Schaap A, Lawrence KA, et al. Annual risk of tuberculous infection using different methods in communities with a high prevalence of TB and HIV in Zambia and South Africa. PloS One. 2009;4(11):e7749.
10. Hoa NB, Cobelens FGJ, Sy DN, Nhung NV, Borgdorff MW, Tiemersma EW. First national tuberculin survey in Viet Nam: characteristics and association with tuberculosis prevalence. Int J Tuberc Lung Dis. 2013;17(6):738–44.
11. Gao L, Lu W, Bai L, Wang X, Xu J, Catanzaro A, et al. Latent tuberculosis infection in rural China: baseline results of a population-based, multicentre, prospective cohort study. Lancet Infect Dis. 2015;15(3):310–9.
12. Global Tuberculosis Report 2022 [Internet]. https://www.who.int/teams/global-tuberculosis-programme/tb-reports/global-tuberculosis-report-2022. Accessed 16 Jan 2023.
13. Kizza FN, List J, Nkwata AK, Okwera A, Ezeamama AE, Whalen CC, et al. Prevalence of latent tuberculosis infection and associated risk factors in an urban African setting. BMC Infect Dis. 2015;15(1):165.
14. Marks GB, Nhung NV, Nguyen TA, Hoa NB, Khoa TH, Son NV, et al. Prevalence of latent tuberculous infection among adults in the general population of Ca Mau, Viet Nam. Int J Tuberc Lung Dis. 2018;22(3):246–51.
15. Junghanss T, Kang G, Lalloo D. Manson's tropical diseases. Amsterdam: Elsevier Health Sciences; 2014. p. 1337.
16. Bennett JE, Dolin R, Blaser MJ. Mandell, Douglas, and Bennett's principles and practice of infectious diseases. Amsterdam: Elsevier Health Sciences; 2014. p. 5095.
17. Tille PM. Bailey & Scott's diagnostic microbiology. Amsterdam: Elsevier; 2021. p. 1184.
18. Kiazyk S, Ball T. Latent tuberculosis infection: an overview. Can Commun Dis Rep. 2017;43(3–4):62–6.
19. Ai JW, Ruan QL, Liu QH, Zhang WH. Updates on the risk factors for latent tuberculosis reactivation and their managements. Emerg Microbes Infect. 2016;5(2):e10.
20. Horsburgh CR, Rubin EJ. Clinical practice. Latent tuberculosis infection in the United States. N Engl J Med. 2011;364(15):1441–8.
21. Landry J, Menzies D. Preventive chemotherapy Where has it got us? Where to go next? Int J Tuberc Lung Dis. 2008;12:1352–64.
22. Aguado JM, Herrero JA, Gavaldá J, Torre-Cisneros J, Blanes M, Rufí G, et al. Clinical presentation and outcome of tuberculosis in kidney, liver, and heart transplant recipients in Spain. Spanish transplantation infection study group, GESITRA. Transplantation. 1997;63(9):1278–86.
23. Sakhuja V, Jha V, Varma PP, Joshi K, Chugh KS. The high incidence of tuberculosis among renal transplant recipients in India. Transplantation. 1996;61(2):211–5.
24. A double-blind placebo-controlled clinical trial of three antituberculosis chemoprophylaxis regimens in patients with silicosis in Hong Kong. Hong Kong chest service/tuberculosis research Centre, Madras/British Medical Research Council. Am Rev Respir Dis. 1992;145(1):36–41.
25. Cowie RL. The epidemiology of tuberculosis in gold miners with silicosis. Am J Respir Crit Care Med. 1994;150(5 Pt 1):1460–2.
26. Sutherland I. Recent studies in the epidemiology of tuberculosis, based on the risk of being infected with tubercle bacilli. Adv Tuberc Res. 1976;19:1–63.
27. Maini R, St Clair EW, Breedveld F, Furst D, Kalden J, Weisman M, et al. Infliximab (chimeric anti-tumour necrosis factor alpha monoclonal antibody) versus placebo in rheumatoid arthritis patients receiving concomitant methotrexate: a randomised phase III trial. ATTRACT Study Group. Lancet. 1999;354(9194):1932–9.
28. Keane J, Gershon S, Wise RP, Mirabile-Levens E, Kasznica J, Schwieterman WD, et al. Tuberculosis associated with infliximab, a tumor necrosis factor alpha-neutralizing agent. N Engl J Med. 2001;345(15):1098–104.
29. Solovic I, Sester M, Gomez-Reino JJ, Rieder HL, Ehlers S, Milburn HJ, et al. The risk of tuberculosis related to tumour necrosis factor antagonist therapies: a TBNET consensus statement. Eur Respir J. 2010;36(5):1185–206.

30. Hussein MM, Mooij JM, Roujouleh H. Tuberculosis and chronic renal disease. Semin Dial. 2003;16(1):38–44.
31. British Thoracic Society Standards of Care Committee and Joint Tuberculosis Committee, Milburn H, Ashman N, Davies P, Doffman S, Drobniewski F, et al. Guidelines for the prevention and management of mycobacterium tuberculosis infection and disease in adult patients with chronic kidney disease. Thorax. 2010;65(6):557–70.
32. Grzybowski S, Fishaut H, Rowe J, Brown A. Tuberculosis among patients with various radiologic abnormalities, followed by the chest clinic service. Am Rev Respir Dis. 1971;104(4):605–8.
33. International union against tuberculosis committee on prophylaxis. Efficacy of various durations of isoniazid preventive therapy for tuberculosis: five years of follow-up in the IUAT trial. International union against tuberculosis committee on prophylaxis. Bull World Health Organ. 1982;60(4):555–64.
34. Baussano I, Mercadante S, Pareek M, Lalvani A, Bugiani M. High rates of mycobacterium tuberculosis among socially marginalized immigrants in low-incidence area, 1991-2010, Italy. Emerg Infect Dis. 2013;19(9):1437–45.
35. Cain KP, Benoit SR, Winston CA, Mac Kenzie WR. Tuberculosis among foreign-born persons in the United States. JAMA. 2008;300(4):405–12.
36. Chu H, Shih CJ, Lee YJ, Kuo SC, Hsu YT, Ou SM, et al. Risk of tuberculosis among healthcare workers in an intermediate-burden country: a nationwide population study. J Infect. 2014;69(6):525–32.
37. Hernández M, Casar C, García P, Morales V, Mamani N, Gómez-Cofré N, et al. Latent tuberculosis infection screening in healthcare workers in four large hospitals in Santiago, Chile. Rev Chil Infectologia Organo Of Soc Chil Infectologia. 2014;31(3):254–60.
38. Catanzaro A. Nosocomial tuberculosis. Am Rev Respir Dis. 1982;125(5):559–62.
39. Nardell EA, Dodging droplet nuclei. Reducing the probability of nosocomial tuberculosis transmission in the AIDS era. Am Rev Respir Dis. 1990;142(3):501–3.
40. Graham NM, Nelson KE, Solomon L, Bonds M, Rizzo RT, Scavotto J, et al. Prevalence of tuberculin positivity and skin test anergy in HIV-1-seropositive and -seronegative intravenous drug users. JAMA. 1992;267(3):369–73.
41. Anderson C, Story A, Brown T, Drobniewski F, Abubakar I. Tuberculosis in UK prisoners: a challenge for control. J Epidemiol Community Health. 2010;64(4):373–6.
42. Harries AD, Lin Y, Satyanarayana S, Lönnroth K, Li L, Wilson N, et al. The looming epidemic of diabetes-associated tuberculosis: learning lessons from HIV-associated tuberculosis. Int J Tuberc Lung Dis. 2011;15(11):1436–44, i.
43. Dobler CC, Flack JR, Marks GB. Risk of tuberculosis among people with diabetes mellitus: an Australian nationwide cohort study. BMJ Open. 2012;2(1):e000666.
44. Leung CC, Lam TH, Chan WM, Yew WW, Ho KS, Leung GM, et al. Diabetic control and risk of tuberculosis: a cohort study. Am J Epidemiol. 2008;167(12):1486–94.
45. Chan ED, Keane J, Iseman MD. Should cigarette smoke exposure be a criterion to treat latent tuberculous infection? Am J Respir Crit Care Med. 2010;182(8):990–2.
46. Sia JK, Rengarajan J. Immunology of mycobacterium tuberculosis infections. Microbiol Spectr. 2019;7(4):10.
47. Bucşan AN, Chatterjee A, Singh DK, Foreman TW, Lee TH, Threeton B, et al. Mechanisms of reactivation of latent tuberculosis infection due to SIV coinfection. J Clin Invest. 2019;129(12):5254–60.
48. Foreman TW, Mehra S, LoBato DN, Malek A, Alvarez X, Golden NA, et al. CD4+ T-cell-independent mechanisms suppress reactivation of latent tuberculosis in a macaque model of HIV coinfection. Proc Natl Acad Sci U S A. 2016;113(38):E5636–44.
49. Lin PL, Flynn JL. CD8 T cells and mycobacterium tuberculosis infection. Semin Immunopathol. 2015;37(3):239–49.
50. Jasenosky LD, Scriba TJ, Hanekom WA, Goldfeld AE. T cells and adaptive immunity to mycobacterium tuberculosis in humans. Immunol Rev. 2015;264(1):74–87.

51. Lin PL, Rutledge T, Green AM, Bigbee M, Fuhrman C, Klein E, et al. CD4 T cell depletion exacerbates acute mycobacterium tuberculosis while reactivation of latent infection is dependent on severity of tissue depletion in Cynomolgus macaques. AIDS Res Hum Retroviruses. 2012;28(12):1693–702.
52. Diedrich CR, Rutledge T, Maiello P, Baranowski TM, White AG, Borish HJ, et al. SIV and mycobacterium tuberculosis synergy within the granuloma accelerates the reactivation pattern of latent tuberculosis. PLoS Pathog. 2020;16(7):e1008413.
53. Elkington P, Lerm M, Kapoor N, Mahon R, Pienaar E, Huh D, et al. In vitro granuloma models of tuberculosis: potential and challenges. J Infect Dis. 2019;219(12):1858–66.
54. Paige C, Bishai WR. Penitentiary or penthouse condo: the tuberculous granuloma from the microbe's point of view. Cell Microbiol. 2010;12(3):301–9.
55. Lin PL, Rodgers M, Smith L, Bigbee M, Myers A, Bigbee C, et al. Quantitative comparison of active and latent tuberculosis in the cynomolgus macaque model. Infect Immun. 2009;77(10):4631–42.
56. Gideon HP, Phuah J, Myers AJ, Bryson BD, Rodgers MA, Coleman MT, et al. Variability in tuberculosis granuloma T cell responses exists, but a balance of pro- and anti-inflammatory cytokines is associated with sterilization. PLoS Pathog. 2015;11(1):e1004603.
57. Fact Sheets | General | Latent TB Infection vs. TB Disease | TB | CDC [Internet]. 2022. https://www.cdc.gov/tb/publications/factsheets/general/ltbiandactivetb.htm. Accessed 6 Feb 2023.
58. CDCTB. Tuberculosis (TB) - Latent TB Infection and TB Disease [Internet]. Centers for Disease Control and Prevention. 2020. https://youtu.be/wA_fObLY6GE. Accessed 6 Feb 2023.
59. Guidelines for the Investigation of Contacts of Persons with Infectious Tuberculosis Recommendations from the National Tuberculosis Controllers Association and CDC. https://www.cdc.gov/mmwr/pdf/rr/rr5415.pdf. Accessed 25 Mar 2023.
60. Targeted Tuberculin Testing and Treatment of Latent Tuberculosis Infection. https://www.cdc.gov/mmwr/preview/mmwrhtml/rr4906a1.htm. Accessed 25 Mar 2023.
61. Updated Guidelines for Using Interferon Gamma Release Assays to Detect Mycobacterium tuberculosis Infection—United States. 2010. https://www.cdc.gov/mmwr/preview/mmwrhtml/rr5905a1.htm. Accessed 25 Mar 2023.
62. Sosa LE, Njie GJ, Lobato MN, Bamrah Morris S, Buchta W, Casey ML, et al. Tuberculosis screening, Testing, and treatment of U.S. health care personnel: recommendations from the National Tuberculosis Controllers Association and CDC, 2019. MMWR Morb Mortal Wkly Rep. 2019;68(19):439–43.
63. van Pinxteren LA, Ravn P, Agger EM, Pollock J, Andersen P. Diagnosis of tuberculosis based on the two specific antigens ESAT-6 and CFP10. Clin Diagn Lab Immunol. 2000;7(2):155–60.
64. US Preventive Services Task Force, Bibbins-Domingo K, Grossman DC, Curry SJ, Bauman L, Davidson KW, et al. Screening for latent tuberculosis infection in adults: US preventive services task force recommendation statement. JAMA. 2016;316:962.
65. Wallis RS, Pai M, Menzies D, Doherty TM, Walzl G, Perkins MD, et al. Biomarkers and diagnostics for tuberculosis: progress, needs, and translation into practice. Lancet. 2010;375(9729):1920–37.
66. Walzl G, Ronacher K, Hanekom W, Scriba TJ, Zumla A. Immunological biomarkers of tuberculosis. Nat Rev Immunol. 2011;11(5):343–54.
67. Zhang S, Shao L, Mo L, Chen J, Wang F, Meng C, et al. Evaluation of gamma interferon release assays using mycobacterium tuberculosis antigens for diagnosis of latent and active tuberculosis in Mycobacterium bovis BCG-vaccinated populations. Clin Vaccine Immunol. 2010;17(12):1985–90.
68. LTBI: A Guide for Primary Health Care Providers| Guides & Toolkits | Publications & Products | TB | CDC [Internet]. 2021 https://www.cdc.gov/tb/publications/ltbi/default.htm. Accessed 14 Feb 2023.
69. Huaman MA, Sterling TR. Treatment of latent tuberculosis infection—an update. Clin Chest Med. 2019;40(4):839–48.

70. Balcells ME, Thomas SL, Godfrey-Faussett P, Grant AD. Isoniazid preventive therapy and risk for resistant tuberculosis. Emerg Infect Dis. 2006;12(5):744–51.
71. den Boon S, Matteelli A, Getahun H. Rifampicin resistance after treatment for latent tuberculous infection: a systematic review and meta-analysis. Int J Tuberc Lung Dis. 2016;20(8):1065–71.
72. Akolo C, Adetifa I, Shepperd S, Volmink J. Treatment of latent tuberculosis infection in HIV infected persons. Cochrane Database Syst Rev. 2010;2010(1):CD000171.
73. Comstock GW. How much isoniazid is needed for prevention of tuberculosis among immunocompetent adults? Int J Tuberc Lung Dis. 1999;3(10):847–50.
74. LTBI: A Guide for Primary Health Care Providers| Guides & Toolkits | Publications & Products | TB | CDC. 2021. https://www.cdc.gov/tb/publications/ltbi/default.htm. Accessed 13 Feb 2023.
75. Clinically significant interactions with drugs used in the treatment of tuberculosis - PubMed [Internet]. https://pubmed.ncbi.nlm.nih.gov/11888353/. Accessed 13 Feb 2023.
76. Menzies D, Adjobimey M, Ruslami R, Trajman A, Sow O, Kim H, et al. Four months of rifampin or nine months of isoniazid for latent tuberculosis in adults. N Engl J Med. 2018;379(5):440–53.
77. Stagg HR, Zenner D, Harris RJ, Muñoz L, Lipman MC, Abubakar I. Treatment of latent tuberculosis infection: a network meta-analysis. Ann Intern Med. 2014;161(6):419–28.
78. Martínez Alfaro EM, Cuadra F, Solera J, Maciá MA, Geijo P, Sánchez Martínez PA, et al. Evaluation of 2 tuberculosis chemoprophylaxis regimens in patients infected with human immunodeficiency virus. The GECMEI Group. Med Clin. 2000;115(5):161–5.
79. Geijo MP, Herranz CR, Vaño D, García AJ, García M, Dimas JF. Short-course isoniazid and rifampin compared with isoniazid for latent tuberculosis infection: a randomized clinical trial. Enferm Infecc Microbiol Clin. 2007;25(5):300–4.
80. Sterling TR, Villarino ME, Borisov AS, Shang N, Gordin F, Bliven-Sizemore E, et al. Three months of rifapentine and isoniazid for latent tuberculosis infection. N Engl J Med. 2011;365(23):2155–66.
81. Borisov AS. Update of recommendations for use of once-weekly Isoniazid-Rifapentine regimen to treat latent Mycobacterium tuberculosis infection. MMWR Morb Mortal Wkly Rep. 2018;67:723. https://www.cdc.gov/mmwr/volumes/67/wr/mm6725a5.htm.
82. Mitnick CD, McGee B, Peloquin CA. Tuberculosis pharmacotherapy: strategies to optimize patient care. Expert Opin Pharmacother. 2009;10(3):381–401.
83. Deffur A, Mulder NJ, Wilkinson RJ. Co-infection with mycobacterium tuberculosis and human immunodeficiency virus: an overview and motivation for systems approaches. Pathog Dis. 2013;69(2):101–13.
84. Swindells S, Ramchandani R, Gupta A, Benson CA, Leon-Cruz J, Mwelase N, et al. One month of Rifapentine plus isoniazid to prevent HIV-related tuberculosis. N Engl J Med. 2019;380(11):1001–11.

Chapter 9
The Future of Serology-Based Diagnosis for Tuberculosis in India

Anvita Gupta Malhotra, Anil Kumar Gupta, and Amit Singh

Abstract Tuberculosis (TB) is a major public health problem in India, with an estimated 2.42 million cases in 2022. Traditional diagnostic methods for tuberculosis (TB) in India are limited in their accuracy and specificity. Sputum smear microscopy is the most widely used method, but it has a sensitivity of around 30–50%. Culture-based methods are more sensitive, but they are expensive, time-consuming, and require specialized laboratory facilities. As a result, there is a need for alternative diagnostic methods that are more accurate, specific, and cost-effective.

Serology-based diagnosis has been proposed as an alternative approach for TB diagnosis. Serological tests detect antibodies against *Mycobacterium tuberculosis* (*M. tuberculosis*) antigens in the blood of patients. They are simple, rapid, and can be performed in a standard laboratory setting. However, the performance of serological tests in the diagnosis of pulmonary TB has been poor, with low sensitivity and specificity. In addition, the interpretation of serological markers remains a challenge, and the heterogeneity of *M. tuberculosis* strains can complicate the diagnosis of TB. Also, the economic implications of serological testing for TB are also substantial. Therefore, the use of serological tests for TB diagnosis in India needs to be carefully evaluated in terms of cost-effectiveness.

Despite the limitations, serology-based diagnosis has the potential to improve TB diagnosis in India. Serological tests can provide rapid results, allowing for early detection and timely initiation of treatment. This is crucial in a country like India, where TB is a major public health concern. Serological tests can also help improve the overall diagnostic accuracy and enhance the detection of TB cases. However,

A. G. Malhotra (✉)
Department of Microbiology, All India Institute of Medical Sciences, Bhopal, India

A. K. Gupta
IRCH, All India Institute of Medical Sciences, New Delhi, India

A. Singh (✉)
Department of Gastroenterology & HNU, All India Institute of Medical Sciences, New Delhi, India

Department of Microbiology, Central University of Punjab, Bathinda, India
e-mail: amit.singh@cup.edu.in

A. Singh, D. Sharma (eds.), *Diagnosis of Mycobacterium*,
https://doi.org/10.1007/978-981-99-5624-1_9

more research is needed to improve the accuracy and specificity of serological tests and to evaluate their cost-effectiveness. Overall, serology-based diagnosis is a promising approach for TB diagnosis in India. However, more research is needed to make it a more reliable and cost-effective diagnostic tool.

Keywords Tuberculosis · *Mycobacterium tuberculosis* · Diagnosis · Serology · TB antigens · Point-of-care devices · ELISA · ESAT-6 · CFP-10

9.1 Introduction

Tuberculosis (TB) is a contagious bacterial infection that affects millions of people globally, with about two million deaths per year. It is a significant public health challenge in India, with the country accounting for approximately one-quarter of the global TB burden [1, 2]. Despite efforts to control the spread of the disease, delayed or missed diagnosis remains a significant contributor to the TB epidemic in India. Currently, the diagnosis of TB mostly relies on sputum smear microscopy, which limited sensitivity and specificity, especially in people with HIV. Culture-based tests and nucleic acid amplification tests are expensive and complex, making them difficult to use in low-income settings. However, missed diagnoses continue to fuel the global epidemic, especially in resource-limited countries where diagnostic tests are either too expensive or perform poorly.

Serological tests that detect antibodies to *M. tuberculosis* antigens in the blood have the potential to provide a rapid diagnosis with quick results, simplicity, and modest training requirements. But doubts about their accuracy have limited their use. Numerous commercial serological tests for TB diagnosis are sold throughout the world despite not being advised by international recommendations. These tests can be revised in point-of-care format and employed in peripheral health facilities which lack microscopy services. Thus implying their usage in high-burden countries [3]. However, its diagnostic value has been questioned by several studies [4]. Serology-based tests detect antibodies to *M. tuberculosis* antigens in patient's serum or plasma and are often used as a rapid and non-invasive diagnostic tool for TB. However, their accuracy in diagnosing active TB has been found to be limited, and they are often unhelpful in establishing the diagnosis of extrapulmonary TB (EPTB). Furthermore, serological tests are more expensive than other diagnostic methods and are primarily performed in the private medical sector, contributing to inequalities in access to TB diagnosis and treatment.

A meta-analysis of purified antigens for serodiagnosis of pulmonary TB found that antibody detection tests have limitations, and should not be relied upon as a sole diagnostic method for TB [5]. Another study conducted in India found that serology-based tests had low sensitivity and specificity, and concluded that these tests should not be used for TB diagnosis. A cost-effectiveness analysis comparing various diagnostic methods for active TB in India found that sputum smear microscopy is the most cost-effective test, and that liquid culture plus microscopy is more cost-effective than serological tests [6].

Although serology has diagnostic potential, key antigens for serodiagnosis of TB have not yet been identified. Despite their limitations, at least 13 different TB serological kits are on the market in India, and an estimated 1.5 million serological tests for active TB are performed every year, primarily in the private medical sector, at a cost of over US$15 million for testing alone. A root-cause analysis identified technical/medical, economic, and regulatory factors for the widespread use of inaccurate TB serological tests in the Indian private healthcare sector [6].

While serology-based tests have been widely used for TB diagnosis in India, their limitations and cost-effectiveness issues mean that alternative diagnostic methods are likely to be developed and adopted in the future. For example, nucleic acid amplification tests (NAATs) such as the GeneXpert system have been shown to be highly accurate and sensitive in detecting TB and are now being used in many countries, including India. Additionally, the use of artificial intelligence (AI) and machine learning (ML) algorithms in TB diagnosis has shown promising results and may become more widespread in the future.

Improving the availability and accessibility of accurate and cost-effective diagnostic methods for TB in India is crucial to improving TB diagnosis and control. This can be achieved through investments in research and development, as well as by increasing the availability of diagnostic tools in the public healthcare sector. As the burden of TB in India remains high, with a significant impact on morbidity and mortality, it is essential to continue exploring new and innovative approaches to TB diagnosis, including the development of new diagnostic tools and the expansion of existing diagnostic methods in the public healthcare sector.

It is important to note that while the serology-based diagnosis of TB has diagnostic potential, it has limitations and is not the most cost-effective option [7, 8]. Hence, while serology-based tests have been widely used for TB diagnosis in India, their limitations and cost-effectiveness issues mean that alternative diagnostic methods are likely to be developed and adopted in the future. As the burden of TB in India remains high, continued efforts are needed to explore new and innovative approaches to TB diagnosis, including the development of new diagnostic tools and the expansion of existing diagnostic methods in the public healthcare sector.

9.2 Serology-Based Diagnosis of Tuberculosis

Serology-based tests for TB diagnosis detect antibodies in the blood that are specific to the *M. tuberculosis* bacterium. These tests use purified antigens derived from *M. tuberculosis*, which are proteins or other molecules that are recognized by the immune system and induce the production of specific antibodies [9].

9.2.1 History

Serological tests for TB were first developed in the 1940s and 1950s, based on the detection of antibodies against *M. tuberculosis* antigens in patient's serum or plasma. These tests were initially used as screening tools for TB, but their clinical utility was limited due to low sensitivity and specificity. In the 1970s and 1980s, the development of new serological techniques, such as enzyme-linked immunosorbent assay (ELISA) and immunoblotting, led to the identification of new *M. tuberculosis*-specific antigens and improved the sensitivity and specificity of serological tests.

9.2.2 Principle and Examples

The principle of serology-based diagnosis tests is based on the fact that when a person is infected with *M. tuberculosis*, the body's immune system produces antibodies against the bacteria. These antibodies can be detected in the blood using serological tests, which involve the use of antigen-antibody reactions. These tests have contributed to a significant role in the diagnosis of various infectious diseases such as HIV, hepatitis, malaria, etc. Various immunological methods, i.e., Enzyme-linked immunosorbent assay (ELISA), tuberculin skin test (TST), and interferon-gamma release assay (IGRA) have been utilized enormously for TB diagnosis.

One such antigen is the 38-kDa antigen, which has been shown to be highly immunogenic in human. Several serological tests, such as the TB ELISA and immunochromatography tests (ICTs) have been developed based on the detection of antibodies against this antigen. Another promising antigen is the ESAT-6 (early secretory antigenic target) protein, which is encoded by a region of difference (RD1) that is present in *M. tuberculosis* but absent in most other mycobacterial species. The use of ESAT-6 in serological tests has been shown to improve the sensitivity and specificity of TB diagnosis. In addition to these antigens, several other *M. tuberculosis*-specific antigens, such as CFP-10, TB10.4, and Rv2626c, have been identified and are being evaluated for their potential use in serological tests.

Although serological tests that detect humoral and antibody-mediated responses are easily accessible, there are currently no guidelines recommending their use for diagnosing TB. In contrast, blood-based tests that measure cellular immune responses, such as T-cell-based IFN-γ release assays, have been recommended for use in some countries [10].

Despite the several decades of research and substantial funds investment by national and international agencies, only a few biomarkers (CFP-10, 19 kDa, 38 kDa Lipo-glycoprotein, Hsp16.3, Ag85B, RpfE, Hsp65, MTC-28, MPT-32, MTB-81, MTB-48, ESAT-6, Ag85A, Ag85B and 14 kDa antigens) have been validated and translated into new diagnostic test (Table 9.1). However, a meta-analysis of commercial TB serological test suggests zero value of this test due to variable sufficient sensitivity and specificity [6, 8]. In July 2011, WHO released a policy statement, concluding that, since "the harms/risks (Currently available commercial serodiagnostic tests) far outweigh any potential benefits (strong recommendation) and these tests should not be used in individuals suspected to have active PTB or EPTB, irrespective of their HIV status". After this WHO policy, the Revised National Tuberculosis Control Programme (RNTCP) published an advisory statement against the use of serological TB tests in India. The negative recommendation was clear that it only applied to existing commercial test and encouraged the development of the new test.

Few antigens have been identified and validated even after serology ban by WHO that shows better activity than earlier used antigen or combinations. However, these newly identified antigens do not fit well with the new guideline on serology test by WHO (Table 9.1).

Alternatively, detection of circulating *M. tuberculosis* antigens directly from clinical samples such as serum, sputum, urine, cerebrospinal spinal fluid (CSF), and pleural fluid (PF), provide higher specificities than antibody detection method. The sandwich ELISA or "antigen-capture" ELISA is commonly used for detection of specific antigen from clinical samples.

In comparison with conventional diagnostic methods, antigen detection tests seem to offer several advantages over antibody detection test: (1) provide direct evidence of active disease, therefore permitting to initiate TB treatment immediately; (2) TB detection using urine samples would provide an attractive diagnostic specimen in children, who may have difficulty in providing sputum; and finally (3) in suspected EPTB patients, might prevent the use of more invasive tests.

Lipoarabinomannan (LAM) was used enormously for TB detection using clinical samples. The pooled sensitivity and specificity of LAM in a sputum sample (both smear-positive and negative patients) were 87 and 70% respectively, In urine, pooled sensitivity in smear-positive was 54% (18 to 86%) and specificity 90% (83 to 95%), while pooled sensitivity in smear-negative was 51% (18 to 83%) and specificity 90% (79 to 96%) [23, 24]. In HIV-associated TB cases, sensitivity was significantly higher in HIV-infected (47%), than HIV-uninfected cases (14%), while similar specificity was detected in 96% and 97%, respectively. Some other studies, targeting LAM, ESAT-6, Ag85 complex, and 65 kDa in CSF show higher sensitivity 87% (61 to 98%) but low specificity 84% (47 to 95%) [25–27].

Detection of *M. tuberculosis*-specific antigen directly from blood samples of suspected TB patients has numerous advantages (rapid, cost-effective, etc.) compared to traditional culturing of bacteria [28, 29]. The detection of *M. tuberculosis* antigens in the clinical samples using specific mAbs has been shown to be a promising approach for the detection of active TB infection [30]. A number of mAbs were produced against the Antigen-85 complex, ManLAM, 38 kDa antigen, CFP-10, ESAT-6, hspX, α-crystallin protein and used for serodiagnosis of TB, and are listed in (Table 9.2). Although, tests based on mAbs have been successfully differentiating *M. tuberculosis* from other mycobacterial strains from culture or using other clinical samples, but failed to fulfil the serodiagnostic test guideline of WHO published in 2014.

Table 9.1 Newly identified biomarkers studied for the serodiagnosis of *M. tuberculosis*

S. no	Protein/antigen	Technique	Sensitivity (95% CI)	References
1	Mtb81, MPT32, MPT-64, Ag85A, Ag85C, Ag85, HspX,	ELISA	81	Tucci et al. (2014) [11]
2	PstS1, Rv0831c, FbpA, EspB, bfrB, HspX and ssb	LIPS	73.5	Burbelo et al. (2015) [12]
3	Rv0054, Rv0831c, Rv2031c, Rv0222, Rv0948c, Rv2853, Rv3405c, Rv3544c	HD-NAPPA	80%	Song et al. (2017) [13]
4	Mtb11, Mtb8, Mtb48	ELISA	54–85	She and Litwin (2015) [14]
5	Fusion protein (CFP-10, ESAT-6, PPE68)	ELISA	73.3%	Xu et al. (2012) [15]
6	CMX fusion protein (Ag85C, MPT51 and HspX)	ELISA	80.1%	Zagmignan et al. (2017) [16]
7	Rv3871, Rv3874, Rv3875, Rv3876, and Rv3879	ELISA	32.21–83.56	Liu et al. (2016) [17]
8	MTB-48 kDa, 8 kDa, 38 kDa, LAM, MPT-64, 16 kDa	Rapid test	83.9–48.4%	Bai et al. (2018) [18]
9	(P38 or PstS1), HspX, Ag85b, MPT32, CFP-10, ESAT-6, Ag85a, GroES, Rv3507, Rv1926c, Rv2878c, (CFP10-ESAT) fusion, Rv1099, Rv3619, Rv1677, Rv2220, Rv2032, Rv1984c Rv3873, Rv0054, Rv3841, MPT64, Ag85c, Rv1566c, Rv2875, Rv1009, Rv0831c	Multiplex bead Luminex-based assay	88–95%	Khaliq et al. (2017) [19]
10	Rv0054, Rv0831c, Rv2031c, Rv0222, Rv0948c, Rv2853, Rv3405c, Rv3544c	ELISA	42–76%	Song et al. (2017) [20]
11	Rv3881c, Rv0934, HspX, MPT32, Rv3804c, Ag85a, Rv1886c, Ag85b, Rv0129c Ag85c, ESAT-6, CFP-10, Rv3841, Rv3418c, MPT70, CFP21, MPT64, Rv0054, CFP10-ESAT fusion, Rv3873, Rv3619, Rv2220, Rv0831c, Rv1009, Rv1099, Rv2032, Rv1926c, Rv2878c, Rv1677, Rv1566c, Rv3507	In-house multiplex microbead assay	>90%	Shete et al. (2017) [21]
12	Rv2145c, Rv0164, Rv1437, Rv1827, Rv2970c	Dot blot and ELISA	86.2 to 99.1%	Singh et al. (2017) [22]

9.2.3 Types of Serological Tests

There are several types of serological tests used for TB diagnosis: the enzyme-linked immunosorbent assay (ELISA), the lateral flow assay (immunochromatographic assays), and multiplex bead assays [45].

Table 9.2 Antibodies produced against *M. tuberculosis* antigens and evaluated for the diagnosis of TB

S. no	Antigens used for mAbs	Techniques	Specificity	Sensitivity	References
1	Antigen 85 complex)	ELISA	82%	86%	Kashyap et al. (2007) [31]
	19 kDa, 38 kDa and Tb72 Ab	ELISA	50–82%	–	Bothamley and Rudd (1994) [32]
2	38 kDa	Competitive ELISA	94%	66%	Kumar Verma and Jain (2007) [33]
4	CFP-10 ESAT-6	ELISA, ICT	80–100% (ESAT-6) 92.2% CFP-10	50–95.4% (ESAT-6) 81.6% (CFP-10)	Feng et al. (2014) [34] and Leng et al. (2014) [35]
5	LAM	ELISA	78.3–95%	50.2–93%	Arias-Bouda et al., (2000) [29], Lawn and Gupta-Wright (2016) [23] and Shah et al. (2016) [27]
6	Hsp antigen (16.3 and 85 kDa)	ELISA	81.8–100%	57.1–74%	Trilling et al. (2011) [36]
7	MPT-64	ICT	100%	96.5–100%	Jørstad et al. (2018) [37]
8	ManLAM	Sandwich ELISA	92.3% Urine, 93.3% Serum	26.6–46.7% (urine) 26.6–33.3% (serum)	Chan et al. (2015) [38]
9	Crude antigen (detection from Mtb culture)	*Patho-TB*	72.7–92.9	63 and 97.4%	Nour-Neamatollahi et al. (2018) [39]
10	TB-LAM	ELISA	15.7–89.2	98	Shah et al. (2016) [27] and Swaminathan and Rekha (2012) [24]
11	HSP 65 kDa	ELISA	84% (CSF)	90	Mudaliar et al. (2006) [40]
12	CFP-10	ICT	97.48	95.79	Tiwari et al. (2017) [41]
13	Ag38 kDa	Biosensor	74	–	McNerney et al. (2010) [42]
14	Cocktail of Ab (TB-ES-31, ES-43, ESAT-6)	ELISA	70–77	90	Harinath et al. (2006) [43]
15	Fusion protein (38 kDa, ESAT-6 and CFP-10)	ELISA	68%	–	Dai et al. (2017) [44]

9.2.3.1 Elisa

ELISA is a laboratory-based test that uses specific antigens from *M. tuberculosis* to detect antibodies in the patient's serum or plasma. The test involves coating a plate with *M. tuberculosis* antigens and then adding the patient's serum or plasma to the plate. If antibodies are present, they will bind to the antigens on the plate. The plate is then washed to remove any unbound materials, and an enzyme-linked secondary antibody is added to the plate. If the secondary antibody binds to the bound antibodies, a color change occurs, indicating a positive result.

Examples

1. TB-ELISA/TB IgG/IgM Rapid Test: This type of ELISA test uses specific antigens derived from *M. tuberculosis* to detect antibodies in patient samples. For example, the ELISA may utilize recombinant proteins such as ESAT-6 (early secretory antigenic target-6), CFP-10 (culture filtrate protein-10), or Ag85 (Antigen 85 complex). The patient's serum or blood sample is added to a plate coated with these antigens, and if TB-specific antibodies are present, they bind to the antigens.

9.2.3.2 Lateral Flow Assays

Lateral flow assays, also known as rapid diagnostic tests (RDTs), are point-of-care tests that use antigens from *M. tuberculosis* to detect antibodies in a patient's blood. These tests work by applying a patient's blood sample to the test strip, which contains the *M. tuberculosis* antigens. If antibodies are present, they bind to the antigens, causing a color change that can be visually interpreted as a positive result. They are simpler and faster than ELISA but have lower sensitivity and specificity, particularly in the diagnosis of extrapulmonary TB. Lateral flow assays are also more prone to user error and require careful interpretation of results, which can be challenging in resource-limited settings where trained personnel may be scarce.

Examples:

1. Determine TB LAM Ag test: targets LAM antigen.
2. SD BIOLINE TB Ag 38 rapid: targets 38 kDa antigen.
3. SD BIOLINE TB IgG/IgM Rapid Test: targets multiple antigens including ESAT-6, CFP-10, and Ag85B.

9.2.3.3 Multiplex Bead Assays

Multiplex bead assays are based on the use of fluorescently labeled microspheres that can detect multiple antibodies simultaneously in a single sample. These assays can detect antibodies to several different *M. tuberculosis* antigens, potentially increasing sensitivity and specificity compared to single-antigen ELISA tests. Multiplex bead assays have several advantages over traditional serological tests. Firstly, they can detect multiple antibodies in a single sample, reducing the amount of blood required for testing and increasing the speed and efficiency of diagnosis. Secondly, they can detect antibodies to multiple *M. tuberculosis* antigens, potentially improving diagnostic accuracy compared to single-antigen tests. Thirdly, the use of fluorescently labeled microspheres allows for the quantitative measurement of antibody levels, providing additional information about the stage and severity of TB infection.

However, there are also some limitations to the use of multiplex bead assays for TB diagnosis. Firstly, these assays are more complex and expensive than traditional serological tests, limiting their widespread adoption in resource-limited settings. Secondly, while multiplex assays may improve diagnostic accuracy compared to single-antigen tests, the sensitivity and specificity of these tests can still be affected by factors such as host variability and cross-reactivity with other bacterial or viral infections. Finally, while multiplex bead assays can detect antibodies to several different *M. tuberculosis* antigens, there is still no consensus on which antigens are most reliable for TB diagnosis. Overall, multiplex bead assays show promise as a tool for TB diagnosis, particularly in research settings where the availability of resources and expertise can support their use. However, further research is needed to validate their accuracy and clinical utility in different populations and settings, and to determine the optimal combination of *M. tuberculosis* antigens for detection of active TB.

Examples:

1. QuantiFERON-TB Gold Plus: targets ESAT-6, CFP-10, and TB7.7 antigens.
2. BioPlex 2200 *M. tuberculosis* IgG: targets multiple antigens including ESAT-6, CFP-10, Ag85A, Ag85B, and Ag85C.

9.3 Implementation of Serology-Based Diagnosis in Pediatric TB and HIV-TB Patients

Serological TB diagnosis in pediatric and HIV-positive cases is challenging due to the limitations of serological tests especially in the diagnosis of pulmonary TB [46]. A study evaluated the performance of a multiplex bead assay for the diagnosis of childhood pulmonary TB [47]. The assay measured IgG responses against 10 *M. tuberculosis* antigens, including Ag85B, Ag85A, Rv2626c, Rv1813, Rv2875, Rv2958c, Rv1808, Rv1886c, Rv3804c, and Rv1884c. The sensitivity of the assay ranged from 0 to 22% for individual antigens, and the specificity ranged from 70 to 100% [49]. The combination of Ag85B, Ag85A, and Rv2626c had the highest sensitivity (22%) and specificity (100%) for the diagnosis of childhood pulmonary TB [47].

In HIV-positive cases, the QuantiFERON-TB Gold test has been shown to have higher sensitivity and specificity than the tuberculin skin test (TST) [48]. However, the accuracy of the QuantiFERON-TB Gold test in HIV-positive pediatric cases is still under investigation [48]. Therefore, further research is needed to evaluate the accuracy of serological tests, including the multiplex bead assay and QuantiFERON-TB Gold test, in the diagnosis of TB in pediatric and HIV-positive cases.

9.4 Serological Diagnosis of Extrapulmonary Tuberculosis (EPTB)

EPTB is a form of tuberculosis that affects parts of the body other than the lungs. EPTB is a significant public health problem in many countries, and the diagnosis of EPTB can be challenging due to the lack of reliable diagnostic methods. Serological tests have been evaluated for their potential use in the diagnosis of EPTB, but their diagnostic value remains limited.

Studies have shown that serological tests have lower sensitivity and specificity for EPTB compared to pulmonary TB [25]. The diagnostic accuracy of serological tests for EPTB can also vary depending on the type of antigen used and the location of the affected site. For example, some antigens may be more specific for certain types of EPTB, such as sarabinomannan (LAM) for childhood TB meningitis, while others may be more sensitive for certain types of EPTB, such as the 38-kDa antigen for TB pleural effusion [49, 50].

Despite their limitations, serological tests may have some potential use in the diagnosis of EPTB, particularly in resource-limited settings where other diagnostic methods may not be available. Serological tests may also have a role in ruling out TB in individuals suspected of having EPTB, particularly if the test result is negative since these tests have better specificity than sensitivity. However, it is important to note that a negative serological test does not exclude the diagnosis of TB, and other diagnostic methods should be used in combination for accurate diagnosis.

Thus, serological tests have limited diagnostic value for EPTB, and their use should be considered in combination with other diagnostic methods for accurate diagnosis. Further research is needed to identify specific antigens for EPTB diagnosis and to evaluate the performance of serological tests in different types of EPTB.

The usefulness of serology diagnosis for extrapulmonary tuberculosis (EPTB) diagnosis is limited. A systematic review of commercial serological antibody detection tests for the diagnosis of EPTB found that their accuracy is unknown [8]. The review included 21 studies that reported on the accuracy of commercial serological antibody detection tests for the diagnosis of EPTB, and the results were inconsistent. Therefore, a combination of clinical, radiological, and microbiological investigations is necessary for the accurate diagnosis and management of EPTB [51].

9.5 Major Limitation of Serology Diagnosis of Tuberculosis

The diagnosis of tuberculosis (TB) has been a major challenge in India, where the disease burden is high. Traditional diagnostic methods, such as sputum smear microscopy and culture, have limitations in terms of sensitivity and specificity. Serology-based diagnosis has been proposed as an alternative approach for the diagnosis of TB. However, the performance of serological tests in the diagnosis of pulmonary TB has been poor, and the economic implications of serological testing for

TB are substantial. Therefore, the future of serology-based TB diagnosis in India remains uncertain.

While serology-based tests have been widely used for TB diagnosis in the past, their accuracy has been questioned, and they are not currently recommended as a primary diagnostic method for TB. This is because the presence of antibodies does not necessarily indicate active TB disease, as individuals who have been previously vaccinated with the Bacillus Calmette-Guérin (BCG) vaccine or who have been exposed to non-tuberculous mycobacteria (NTM) can also have antibodies to *M. tuberculosis* antigens. Furthermore, antibodies can persist in the bloodstream even after successful treatment of TB, leading to false positive results.

Some formats of serological tests for TB may be suitable for resource-limited areas due to their speed and potential simplicity compared to microscopy, whereas, other tests such as the interferon gamma release assay (IGRA) do require specialized laboratory facilities and trained laboratory staff, costly assay reagents/kits which can be a challenge in resource limiting countries [52].

Therefore, the feasibility and appropriateness of using specific TB diagnostic tests, including serological tests or IGRAs, in different settings should be carefully evaluated based on factors such as local resources, infrastructure, and epidemiology of TB.

9.6 WHO's (World Health Organization) Take on Serological Diagnosis for Tuberculosis

WHO does not recommend the use of serological tests for the diagnosis of active pulmonary or extrapulmonary tuberculosis (TB) due to their low sensitivity and specificity. According to a systematic review and meta-analysis of commercial serological tests for the diagnosis of active pulmonary and extrapulmonary TB, the accuracy of these tests is unknown, and their use is not recommended for TB diagnosis [53]. The WHO recommends the use of microbiological and molecular tests, such as sputum smear microscopy, culture, and the GeneXpert MTB/RIF assay, for the diagnosis of active TB. Additionally, imaging techniques, such as chest radiography and computed tomography (CT), can aid in the diagnosis of TB. The WHO also recommends the use of the tuberculin skin test (TST) or interferon-gamma release assays (IGRAs) for the diagnosis of latent TB infection [54].

Despite the poor performance of commercial serological tests, WHO stresses the importance of continued research on these and other tests that could provide quick and accurate diagnosis of TB. Until better diagnostic tools become available, the WHO recommends the use of existing diagnostic methods, including microscopic examination of sputum, chest radiography, mycobacterial culture, and nucleic acid amplification tests. These methods, though imperfect, are currently the most reliable methods for diagnosing tuberculosis and are recommended for use in routine practice. However, there are some conditions where serological tests may be useful in the diagnosis of tuberculosis. For example, in cases where a patient has symptoms of tuberculosis but sputum microscopy, culture, and chest radiography are negative, a serological test can be used as an additional diagnostic tool. Similarly, in cases where a patient has extrapulmonary tuberculosis, serological tests can be used as a complementary diagnostic tool in conjunction with other tests. In resource-limited settings where other diagnostic tools are not available, serological tests may also be considered as a last resort for the diagnosis of tuberculosis.

It is important to note that serological tests should not be used as the sole diagnostic tool for tuberculosis, and their results should be interpreted in conjunction with other radiological investigations, diagnostic tests, and clinical findings. The use of serological tests for the diagnosis of tuberculosis should be guided by the recommendations of national and international guidelines and should be based on careful consideration of the local epidemiology, diagnostic infrastructure, and patient population.

WHO advises against using LF-LAM to diagnose active TB in outpatient settings in HIV-positive adults, adolescents, and children without first assessing TB symptoms, as well as HIV-positive patients without TB symptoms and an unknown CD4 cell count or those with a CD4 cell count greater than or equal to 100 cells/mm^3, HIV-negative people, and people who are not infected with HIV. Additionally, WHO suggests that interferon-gamma release assays may be used to detect latent TB infection in low- and middle-income countries, but these tests should not be used to diagnose pulmonary or extrapulmonary TB or to diagnose active TB in adults (including HIV-positive individuals) suspected of having TB [55].

9.7 Role of Serology-Based Diagnosis for Latent Tuberculosis

IGRA, which stands for interferon gamma release assay, is a blood test used for the diagnosis of latent tuberculosis infection (LTBI) in India. The IGRA test is performed on a blood sample, and it measures the response of T-cells to TB-specific antigens. The two main types of IGRA tests are QuantiFERON-TB Gold In-Tube (QFT-GIT) and T-SPOT.TB.

In India, the QFT-GIT test is the most commonly used IGRA for the diagnosis of LTBI. The test involves collecting a blood sample from the patient, which is then incubated with TB-specific antigens. After incubation, the levels of interferon-gamma (IFN-γ) released by the patient's T-cells are measured. A positive test result indicates the presence of LTBI, while a negative result indicates the absence of infection.

The IGRA test is a useful tool for the diagnosis of LTBI, especially in individuals who have been vaccinated with the Bacille Calmette-Guérin (BCG) vaccine, which is commonly used in India. The BCG vaccine can cause false-positive results in the tuberculin skin test (TST), another test used for the diagnosis of LTBI. However, the IGRA test is not affected by the BCG vaccine and can provide more accurate results in BCG-vaccinated individuals.

It is important to note that the IGRA test is not a diagnostic test for active tuberculosis disease, and further testing is required to confirm the diagnosis of TB. Additionally, the IGRA test may not be suitable for all individuals, and medical professionals should consider the patient's medical history and other risk factors before deciding to perform the test.

The current tests for LTBI are the TST and IGRAs, which detect memory T-cell responses to *M. tuberculosis*. However, the term "latent tuberculosis infection" may not accurately reflect the nature of these responses, which may represent lasting immune responses to *M. tuberculosis* rather than a true latent infection.

9.8 Moving Forward with Serological Tests for Tuberculosis Diagnosis

India faces significant challenges in addressing the burden of TB, with over two million cases reported annually, indicating the need for improved diagnosis and management. For this, the government has an ambitious objective of giving all TB patients in the nation access to proper diagnosis and treatment. This necessitates the adoption of precise, WHO-approved tools as well as the replacement of ineffective tests with ones that can enhance patient outcomes and lower the spread of TB. To do this, new distribution methods and tools integrating both the public and commercial sectors are required. Medical associations and regulatory bodies must discourage the use of unreliable serological tests and tighten the regulation of all in vitro diagnostics. Clear specifications for TB diagnostics should be set, and quality assurance in laboratories must be improved. Greater private sector engagement is needed to reduce misdiagnosis and promote the use of validated technologies.

The One Health approach [56] recognizes the interconnectedness of human, animal, and environmental health and emphasizes the importance of collaboration between different sectors to tackle infectious diseases such as tuberculosis (TB). In this context, a new study [57] has shown promising results in the development of a universal serological test for detecting TB infection in multiple animal species, which could improve surveillance efforts and advance the One Health approach.

The study established an indirect ELISA for detecting *M. tuberculosis* complex infection using a fusion protein of common antigens (MPB70, MPB83, ESAT-6, and CFP-10 as the coating antigen and HRP-labeled fusion protein A/G as a secondary antibody to detect *Mycobacterium tuberculosis* complex infection) present in *M. tuberculosis*, *M. bovis*, and *M. caprae*. The diagnostic sensitivity and specificity of the test were determined to be high for *M. bovis* infection in cattle, sheep, cervids, monkeys, and for *M. tuberculosis* infection in humans. Furthermore, the test showed potential for detecting *M. caprae* infection in roe deer.

These findings have significant implications for TB control and surveillance, particularly in the context of the One Health approach. The development of a universal serological test that can detect TB infection in multiple animal species could improve the accuracy and efficiency of TB surveillance, allowing for better tracking and control of the disease. Moreover, the use of a common test across different animal species could facilitate cross-sector collaboration and data sharing, advancing the One Health approach to TB control.

In a systematic review and meta-analysis [53], researchers evaluated the diagnostic accuracy of commercial serological tests for pulmonary and extrapulmonary tuberculosis. They discovered that the commercial serological tests do not provide an accurate diagnosis of tuberculosis and that the overall quality of the data from the studies of the serological tests was very low. The results point to the need for more study/research on current serological assays and novel methods for tuberculosis serological diagnosis. The World Health Organization has recommended against the use of currently available serological tests for the diagnosis of TB, while emphasizing the importance of continued research on existing and novel tests that could provide quick, accurate, and affordable diagnosis of TB to reduce missed diagnosis and control the spread of tuberculosis, especially in resource-limited countries.

The need for rapid and accurate tests for tuberculosis (TB) diagnosis is urgent. This review [58] discusses key technologies and programmatic and resource issues that could affect the impact of TB diagnostics. Mid-to-early-stage technologies such as automated digital chest X-radiography and point-of-care assays using capillary blood are particularly promising. The diagnostic pipeline has pitfalls, such as a lack of community-based tools. These technologies can complement each other within the TB care cascade, help diagnose subclinical TB, and expand options for extrapulmonary TB. However, challenges such as detecting paucibacillary TB and limitations of current reference standards exist. The review also discusses how researchers and developers can design and evaluate assays to optimize programmatic uptake. Finally, leveraging the urgency and innovation applied to COVID-19 is critical to improving TB patients' diagnostic quality of care.

The current diagnostic pipeline has its downsides like lack of community-based diagnostic tools. However, mid- to early-stage technologies are particularly intruding such automated digital chest X-rays and point-of-care tests using capillary blood. Therefore, these technologies can support one another in the TB care cascade and aid in the detection of subclinical TB, and increase the possibilities for treating extrapulmonary TB. Though, detecting paucibacillary TB is still a challenge. The urgency and innovation applied to COVID-19 can be used to improve the diagnostic quality and care for TB patients by utilizing similar strategies and technologies.

There is an urgent need for improved and decentralized testing for tuberculosis (TB), particularly for sputum-free tests that can be implemented as point-of-care diagnosis with minimal training. Despite the availability of recommended rapid molecular tests, the majority of the globe still uses sputum smear microscopy, which has poor sensitivity. The lack of efficient utilization of available tests and the loss-to-follow-up of diagnosed patients further underscore the need for better testing strategies. Additionally, the prevalence surveys highlight the high proportion of identified cases who are pre-symptomatic, indicating urgent need for earlier diagnosis through non-invasive and sputum-free tests, which can prevent disease progression, morbidity, mortality, and transmission. These tests can also play a critical public health role by identifying which patients are infectious and for how long. The World Health Organization has identified the development of non-invasive, rapid, accurate, and sputum-free tests as priorities for new TB diagnostics.

The ease of use and the high demand for serological tests in countries with a significant TB burden have made them appealing to smaller laboratories seeking quick profits. Regrettably, the disregard for test quality, the risk of inaccurate outcomes, and unethical medical approaches have all played a role in the widespread adoption of these tests. Despite the endorsement of WHO guidelines and a ban by the TB Division of the Government of India, the lack of control over the import and manufacturing of these kits means that they continue to confuse the Indian market and profit-seeking parties. It is hoped that without adequate re-validation on well-characterized samples, these tests won't be recommended or used in India or other TB-endemic nations. However, this does not mean the end for serology but as a potential avenue for creating an immuno-diagnosis for tuberculosis [59].

9.9 Conclusion

Antibody (Ab) detection assays can be adapted for the development of rapid, inexpensive, easy-to-use tests that require neither laboratory infrastructure nor specific training. Importantly, the Ab profiles of TB patients are heterogeneous, and tests that are based on a limited number of antigens, often only one or two [2, 3], are insufficient to capture the diversity of TB cases. Because of the potential to turn Ab detection assays into simple dipstick formats, TB serology, despite its known limitations, remains a field of study that is worthwhile pursuing further and new biomarker targets need to be identified.

The World Health Organization strongly promotes more research despite acknowledging the shortcomings of the currently existing serologic tests and even advising against using them [60], in order to address the demand for accurate, straightforward testing for TB in endemic countries. Finding Ab targets that are reliable indicators of TB is worthwhile since Ab detection is easy to use in dipstick format and incorporates a variety of antigens.

Moreover, there is a need to improve the availability and accessibility of accurate and cost-effective diagnostic methods for TB in India. This can be achieved through investments in research and development, as well as by increasing the availability of diagnostic tools in the public healthcare sector.

In conclusion, there are both opportunities and difficulties for serodiagnosis of tuberculosis (TB) in India. Serological assays may be able to provide quick and additional TB diagnostic techniques, especially in cases where conventional approaches may be constrained. To satisfy the exacting diagnostic standards needed for TB diagnosis, these tests' dependability and accuracy must yet be improved. The standardization of serological testing, the identification of precise and sensitive antigen targets, and the problem of cross-reactivity with other mycobacterial species should be the main areas of future study and development. With advancements in technology and a concerted effort to overcome existing limitations, serodiagnosis of TB could play a valuable role in India's fight against this persistent and burdensome disease.

References

1. 5646719104TB_AR_2023_04-04-2023_LRP_final.pdf [Internet]. https://tbcindia.gov.in/WriteReadData/l892s/5646719104TB_AR_2023_04-04-2023_LRP_final.pdf. Accessed 23 May 2023.
2. Kanabus A. TB facts—tests, drugs, statistics & lots more about "information about tuberculosis". GHE. 2022 TBFacts. https://tbfacts.org/. Accessed 23 May 2023.
3. Steingart KR, Ramsay A, Dowdy DW, Pai M. Serological tests for the diagnosis of active tuberculosis: relevance for India. Indian J Med Res. 2012;135(5):695–702.
4. Singh S, Singh J, Kumar S, Gopinath K, Balooni V, Singh N, et al. Poor performance of serological tests in the diagnosis of pulmonary tuberculosis: evidence from a contact tracing field study. PloS One. 2012;7(7):e40213.
5. Steingart KR, Dendukuri N, Henry M, Schiller I, Nahid P, Hopewell PC, et al. Performance of purified antigens for serodiagnosis of pulmonary tuberculosis: a meta-analysis. Clin Vaccine Immunol. 2009;16(2):260–76.
6. Dowdy DW, Steingart KR, Pai M. Serological testing versus other strategies for diagnosis of active tuberculosis in India: a cost-effectiveness analysis. PLoS Med. 2011;8(8):e1001074.
7. Jarosławski S, Pai M. Why are inaccurate tuberculosis serological tests widely used in the Indian private healthcare sector? A root-cause analysis. J Epidemiol Glob Health. 2012;2(1):39–50.
8. Steingart KR, Henry M, Laal S, Hopewell PC, Ramsay A, Menzies D, et al. A systematic review of commercial serological antibody detection tests for the diagnosis of extrapulmonary tuberculosis. Postgrad Med J. 2007;83(985):705–12.
9. Weldingh K, Rosenkrands I, Okkels LM, Doherty TM, Andersen P. Assessing the Serodiagnostic potential of 35 mycobacterium tuberculosis proteins and identification of four novel serological antigens. J Clin Microbiol. 2005;43(1):57–65.
10. Menzies D, Pai M, Comstock G. Meta-analysis: new tests for the diagnosis of latent tuberculosis infection: areas of uncertainty and recommendations for research. Ann Intern Med. 2007;146(5):340–54.
11. Tucci P, González-Sapienza G, Marin M. Pathogen-derived biomarkers for active tuberculosis diagnosis. Front Microbiol. 2014;5:549.
12. Burbelo PD, Keller J, Wagner J, Klimavicz JS, Bayat A, Rhodes CS, et al. Serological diagnosis of pulmonary mycobacterium tuberculosis infection by LIPS using a multiple antigen mixture. BMC Microbiol. 2015;15:205.
13. Yu X, Song L, Petritis B, Bian X, Wang H, Viloria J, et al. Multiplexed nucleic acid programmable protein arrays. Theranostics. 2017;7(16):4057–70.
14. She RC, Litwin CM. Performance of a tuberculosis serologic assay in various patient populations. Am J Clin Pathol. 2015;144(2):240–6.
15. Xu JN, Chen JP, Chen DL. Serodiagnosis efficacy and immunogenicity of the fusion protein of mycobacterium tuberculosis composed of the 10-kilodalton culture filtrate protein, ESAT-6, and the extracellular domain fragment of PPE68. Clin Vaccine Immunol. 2012;19(4):536–44.
16. Zagmignan A, da Costa AC, Viana JL, Lima Neto LG, de Monteiro C, Gaioso Neto AG, et al. Identification of specific antibodies against the Ag85C-MPT51-HspX fusion protein (CMX) for serological screening of tuberculosis in endemic area. Expert Rev Clin Immunol. 2017;13(8):837–43.

17. Liu Z, Qie S, Li L, Xiu B, Yang X, Dai Z, et al. Identification of novel RD1 antigens and their combinations for diagnosis of sputum smear–/culture+ TB patients. Biomed Res Int. 2016;2016:7486425.
18. Bai XJ, Yang YR, Liang JQ, An HR, Wang J, Ling YB, et al. Diagnostic performance and problem analysis of commercial tuberculosis antibody detection kits in China. Mil Med Res. 2018;5(1):10.
19. Khaliq A, Ravindran R, Hussainy SF, Krishnan VV, Ambreen A, Yusuf NW, et al. Field evaluation of a blood based test for active tuberculosis in endemic settings. PloS One. 2017;12(4):e0173359.
20. Song L, Wallstrom G, Yu X, Hopper M, Van Duine J, Steel J, et al. Identification of antibody targets for tuberculosis serology using high-density nucleic acid programmable protein arrays. Mol Cell Proteomics. 2017;16(4 Suppl 1):S277–89.
21. Shete PB, Ravindran R, Chang E, Worodria W, Chaisson LH, Andama A, et al. Evaluation of antibody responses to panels of M. tuberculosis antigens as a screening tool for active tuberculosis in Uganda. PloS One. 2017;12(8):e0180122.
22. Singh A, Kumar Gupta A, Gopinath K, Sharma P, Singh S. Evaluation of 5 novel protein biomarkers for the rapid diagnosis of pulmonary and extra-pulmonary tuberculosis: preliminary results. Sci Rep. 2017;7:44121.
23. Lawn SD, Gupta-Wright A. Detection of lipoarabinomannan (LAM) in urine is indicative of disseminated TB with renal involvement in patients living with HIV and advanced immunodeficiency: evidence and implications. Trans R Soc Trop Med Hyg. 2016;110(3):180–5.
24. Swaminathan S, Rekha VVB. Antigen detection as a point-of-care test for TB: the case of lipoarabinomannan. Future Microbiol. 2012;7(5):559–64.
25. Flores LL, Steingart KR, Dendukuri N, Schiller I, Minion J, Pai M, et al. Systematic review and meta-analysis of antigen detection tests for the diagnosis of tuberculosis. Clin Vaccine Immunol. 2011;18(10):1616–27.
26. Flores J, Cancino JC, Chavez-Galan L. Lipoarabinomannan as a point-of-care assay for diagnosis of tuberculosis: how far are we to use it? Front Microbiol. 2021;12:638047.
27. Shah M, Hanrahan C, Wang ZY, Dendukuri N, Lawn SD, Denkinger CM, et al. Lateral flow urine lipoarabinomannan assay for detecting active tuberculosis in HIV-positive adults. Cochrane Database Syst Rev. 2016;5:CD011420.
28. Achkar JM, Ziegenbalg A. Antibody responses to mycobacterial antigens in children with tuberculosis: challenges and potential diagnostic value. Clin Vaccine Immunol. 2012;19(12):1898–906.
29. Pereira Arias-Bouda LM, Nguyen LN, Ho LM, Kuijper S, Jansen HM, Kolk AHJ. Development of antigen detection assay for diagnosis of tuberculosis using sputum samples. J Clin Microbiol. 2000;38(6):2278–83.
30. Anochie PI, Onyeneke EC, Ogu AC, Onyeozirila AC, Aluru S, Onyejepu N, et al. Recent advances in the diagnosis of mycobacterium tuberculosis. Germs. 2012;2(3):110–20.
31. Kashyap RS, Rajan AN, Ramteke SS, Agrawal VS, Kelkar SS, Purohit HJ, et al. Diagnosis of tuberculosis in an Indian population by an indirect ELISA protocol based on detection of antigen 85 complex: a prospective cohort study. BMC Infect Dis. 2007;7:74.
32. Bothamley GH, Rudd RM. Clinical evaluation of a serological assay using a monoclonal antibody (TB72) to the 38 kDa antigen of mycobacterium tuberculosis. Eur Respir J. 1994;7(2):240–6.
33. Verma RK, Jain A. Antibodies to mycobacterial antigens for diagnosis of tuberculosis. FEMS Immunol Med Microbiol. 2007;51(3):453–61.
34. Feng F, Zhang H, Zhu Z, Li C, Shi Y, Zhang Z. The application of anti-ESAT-6 monoclonal antibody fluorescent probe in ex vivo near-infrared fluorescence imaging in mice with pulmonary tuberculosis. Lumin J Biol Chem Lumin. 2014;29(6):614–20.
35. Leng J, Ding Y, Shou C, Wu Z, Zhuo G, Wang K, et al. Development of a novel anti ESAT-6 monoclonal antibody for screening of mycobacterium tuberculosis. Int J Clin Exp Med. 2014;7(11):4238–43.

36. Trilling AK, de Ronde H, Noteboom L, van Houwelingen A, Roelse M, Srivastava SK, et al. A broad set of different llama antibodies specific for a 16 kDa heat shock protein of mycobacterium tuberculosis. PloS One. 2011;6(10):e26754.
37. Jørstad MD, Marijani M, Dyrhol-Riise AM, Sviland L, Mustafa T. MPT64 antigen detection test improves routine diagnosis of extrapulmonary tuberculosis in a low-resource setting: a study from the tertiary care hospital in Zanzibar. PloS One. 2018;13(5):e0196723.
38. Chan CE, Götze S, Seah GT, Seeberger PH, Tukvadze N, Wenk MR, et al. The diagnostic targeting of a carbohydrate virulence factor from M. Tuberculosis. Sci Rep. 2015;5:10281.
39. Alavi-Naini R, Metanat M, Alijani E, Mozaffar H. Patho-TB test for the rapid diagnosis of pulmonary tuberculosis. J Res Med Sci. 2009;14(5):301–7.
40. Mudaliar AV, Kashyap RS, Purohit HJ, Taori GM, Daginawala HF. Detection of 65 kD heat shock protein in cerebrospinal fluid of tuberculous meningitis patients. BMC Neurol. 2006;6:34.
41. Tiwari D, Haque S, Tiwari RP, Jawed A, Govender T, Kruger HG. Fast and efficient detection of tuberculosis antigens using liposome encapsulated secretory proteins of mycobacterium tuberculosis. J Microbiol Immunol Infect. 2017;50(2):189–98.
42. McNerney R, Wondafrash BA, Amena K, Tesfaye A, McCash EM, Murray NJ. Field test of a novel detection device for mycobacterium tuberculosis antigen in cough. BMC Infect Dis. 2010;10:161.
43. Harinath BC, Kumar S, Roy SS, Hirudkar S, Upadhye V, Shende N. A cocktail of affinity-purified antibodies reactive with diagnostically useful mycobacterial antigens ES-31, ES-43, and EST-6 for detecting the presence of mycobacterium tuberculosis. Diagn Microbiol Infect Dis. 2006;55(1):65–8.
44. Dai Z, Liu Z, Xiu B, Yang X, Zhao P, Zhang X, et al. A multiple-antigen detection assay for tuberculosis diagnosis based on broadly reactive polyclonal antibodies. Iran J Basic Med Sci. 2017;20(4):360–7.
45. Houghton RL, Lodes MJ, Dillon DC, Reynolds LD, Day CH, McNeill PD, et al. Use of multiepitope polyproteins in Serodiagnosis of active tuberculosis. Clin Diagn Lab Immunol. 2002;9(4):883–91.
46. Mandal N, Anand PK, Gautam S, Das S, Hussain T. Diagnosis and treatment of paediatric tuberculosis: an insight review. Crit Rev Microbiol. 2017;43(4):466–80.
47. Raqib R, Mondal D, Karim MA, Chowdhury F, Ahmed S, Luby S, et al. Detection of antibodies secreted from circulating mycobacterium tuberculosis-specific plasma cells in the diagnosis of pediatric tuberculosis. Clin Vaccine Immunol. 2009;16(4):521–7.
48. Lee CM, Lee E, Bang JH, Park SW, Park WB, Oh MD, et al. Prevalence of multidrug-resistant tuberculosis in HIV/tuberculosis co-infected patients. Infect Chemother. 2021;53(4):792–5.
49. Kwizera R, Cresswell FV, Mugumya G, Okirwoth M, Kagimu E, Bangdiwala AS, et al. Performance of Lipoarabinomannan assay using cerebrospinal fluid for the diagnosis of tuberculous meningitis among HIV patients. Wellcome Open Res. 2019;4:123.
50. Blok N, Visser DH, Solomons R, Van Elsland SL, den Hertog AL, van Furth AM. Lipoarabinomannan enzyme-linked immunosorbent assay for early diagnosis of childhood tuberculous meningitis. Int J Tuberc Lung Dis. 2014;18(2):205–10.
51. Fahmi MN, Harti AP. A diagnostic approach for differentiating abdominal tuberculosis from ovarian malignancy: a case series and literature review. BMC Proc. 2019;13(Suppl 11):13.
52. Nurwidya F, Handayani D, Burhan E, Yunus F. Molecular diagnosis of tuberculosis. Chonnam Med J. 2018;54(1):1–9.
53. Steingart KR, Flores LL, Dendukuri N, Schiller I, Laal S, Ramsay A, et al. Commercial serological tests for the diagnosis of active pulmonary and Extrapulmonary tuberculosis: An updated systematic review and meta-analysis. PLoS Med. 2011;8(8):e1001062.
54. Commercial Serodiagnostic Tests for Diagnosis of Tuberculosis - NCBI Bookshelf. https://www.ncbi.nlm.nih.gov/books/NBK304212/. Accessed 23 May 2023.

55. WHO operational handbook on tuberculosis: module 3: diagnosis: rapid diagnostics for tuberculosis detection, 2021 update [Internet]. https://www.who.int/publications/i/item/9789240030589. Accessed 23 May 2023.
56. Aggarwal D, Ramachandran A. One health approach to address zoonotic diseases. Indian J Commun Med. 2020;45(Suppl 1):S6–8.
57. Sun L, Chen Y, Yi P, Yang L, Yan Y, Zhang K, et al. Serological detection of mycobacterium tuberculosis complex infection in multiple hosts by one universal ELISA. PloS One. 2021;16(10):e0257920.
58. Nathavitharana RR, Garcia-Basteiro AL, Ruhwald M, Cobelens F, Theron G. Reimagining the status quo: how close are we to rapid sputum-free tuberculosis diagnostics for all? EBioMedicine. 2022;78:103939.
59. Singh S, Katoch VM. Commercial serological tests for the diagnosis of active tuberculosis in India: time for introspection. Indian J Med Res. 2011;134(5):583–7.
60. Morris K. WHO recommends against inaccurate tuberculosis tests. Lancet. 2011;377(9760):113–4.

Chapter 10
Point-of-Care (POC) Detection Technique for Mycobacterium

Saumya Srivastava, Anil Kumar Gupta, Pawan Sharma, and Amit Singh

Abstract Globally tuberculosis (TB) has the highest mortality rate among all infectious diseases, and is a major public health issue. The main aim of WHO End TB strategy is to end the global TB epidemic by achieving 90% cut in new cases of TB and 95% reduction in TB mortality by the year 2035. It can only be achieved by a proper management of TB treatment. For effective treatment of TB, the accurate and early detection of the disease is crucial and essential. The management of this disease needs a point-of-care (POC) diagnostic test. This includes the diagnostic testing of patients at hospitals, peripheral clinics, community health care centres or at home. POC tests enable the early detection and timely treatment of TB which reduces the disease transmission. Currently, several molecular tests have been developed and evaluated for their use as POC tests.

This chapter includes current diagnostic assays endorsed by WHO like original Xpert MTB/RIF, GeneXpert Omni platform, Loop mediated isothermal amplification (LAMP), Lateral flow lipoarabinomannan commercial antigen tests (LF-LAM)

The original version of this chapter has been revised. The affiliation of author Anil Kumar Gupta has been updated. The correction to this chapter can be found at
https://doi.org/10.1007/978-981-99-5624-1_15

S. Srivastava (✉)
Department of Ocular Pharmacology, All India Institute of Medical Sciences (AIIMS) New Delhi, New Delhi, India

A. K. Gupta
IRCH, All India Institute of Medical Sciences, New Delhi, India

P. Sharma
International Centre for Genetic Engineering and Biotechnology (ICGEB), New Delhi, India

A. Singh (✉)
Department of Gastroenterology & HNU, All India Institute of Medical Sciences, New Delhi, India

Department of Microbiology, Central University of Punjab, Bathinda, India
e-mail: amit.singh@cup.edu.in

A. Singh, D. Sharma (eds.), *Diagnosis of Mycobacterium*,
https://doi.org/10.1007/978-981-99-5624-1_10

and their use in POC testing programme. Here we also discuss the gaps in existing conventional tests that can be addressed by POC tests. We discuss the new research studies for novel biomarkers and new rapid tests. Last, we highlight the advancement of microfluidics technology that may aid the development of new POC tests.

Keywords Tuberculosis · Point-of-care · Diagnosis · Xpert · NAAT LAM · LAMP · Microfluidics

10.1 Introduction

Tuberculosis (TB) is an important public health problem at the global level. As per the recent reports around ten million population had TB every year, of which almost one-third were never diagnosed or reported [1]. Currently smear microscopy, which has a sensitivity of 34–60%, is the only point-of-care TB diagnostic test available for the diagnosis of TB. Due to this around 3.6 million of the TB cases are not diagnosed by health systems and left untreated. World Health Organization (WHO) and the End TB strategy goals aim to lower TB incidence and deaths by 90 and 95% respectively, by 2035, which can be achieved only if the patients will be diagnosed accurately and timely. Rapid diagnosis is a key pillar to end TB. It will not be achieved without a new point-of-care tests (POCTs) with high sensitivity that could be applied at resource-limited settings [2]. POCT provides early diagnosis, immediate treatment and reduces transmission of the disease. These POC test should able to give results in non-sputum samples also and provide results within hours. So that clinician can take the decision and start the treatment without losing the time. Despite major developments in recent past, a simple, accurate, reliable and effective POC test is still not available for the diagnosis of TB.

In this chapter we are going to discuss the current and future pipeline rapid diagnostic assay for TB that may provide the TB community a tool to identify and treat missed cases, and thus would contribute towards achieving the WHO goals and the gap between existing conventional tests and the ideal POCT.

10.2 Current Point-of-Care Tests

Point-of-care test is one of the key pillars to end tuberculosis (TB), and here we discuss the different kind of POCT available for TB based on molecular technique, biomarkers and AI (Fig. 10.1), (Table 10.1).

10.2.1 Molecular Technique as a Point-of-Care Test

The traditional molecular techniques are based on detecting the target genetic material of microbes. DNA or RNA, extracted directly from the patient sample can be analysed using molecular techniques like polymerase chain reactions (PCR).

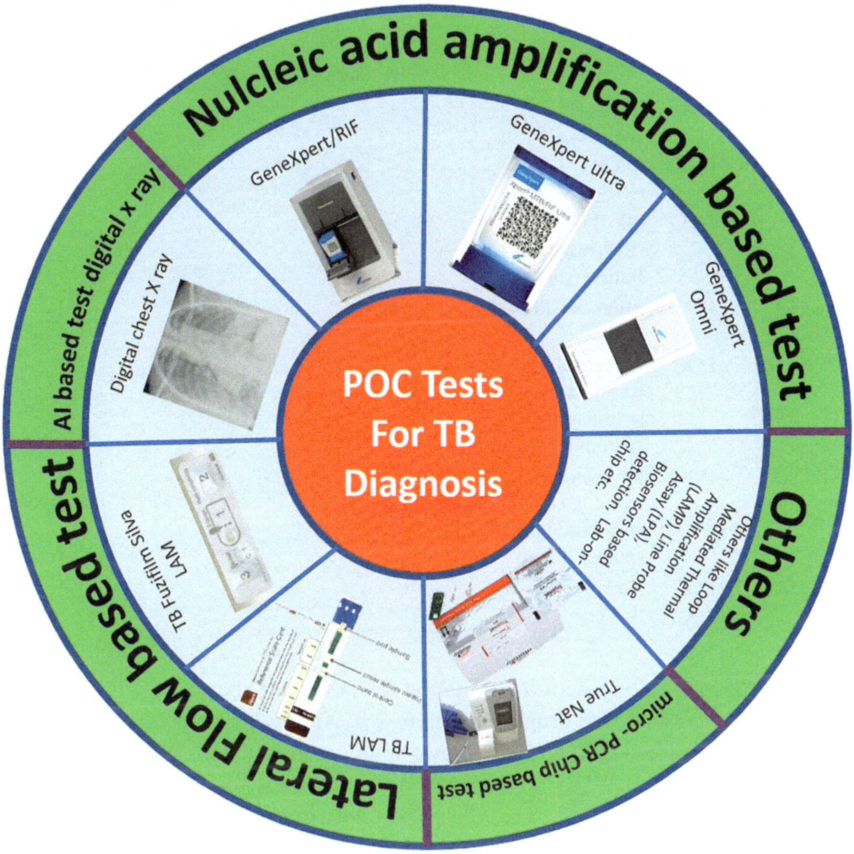

Fig. 10.1 Different type of POCTs available for the diagnosis of tuberculosis

Specific primers bind to the DNA or RNA segment, and amplify the targeted gene. This amplified product can be detected by the instrument, even if small amounts of genetic material are present in the clinical samples. Most PCR assays require long hours, specific laboratories and experts to get the results, but newer nucleic acid amplification-based point-of-care tests have taken this testing from molecular diagnostics laboratories to general laboratories and even clinics and fields. Here we are discussing the POCT of TB based on molecular technique.

10.2.1.1 Xpert MTB/RIF

The Xpert MTB/RIF is a cartridge-based nucleic acid amplification test (NAAT) which has revolutionised tuberculosis diagnosis by contributing to the rapid diagnosis of TB disease and drug resistance. The cartridge's microfluidics regulate all aspects of the testing process: sample prep, nucleic acid extraction, amplification and detection. The test simultaneously detects the *Mycobacterium tuberculosis*

Table 10.1 Summary of different POCT available for the diagnosis of TB

Test	Type of test	Principle	Time duration	Sensitivity	Specificity	Reference
Xpert MTB/RIF	POC	qPCR (RIF-resistance)	<2 h	85%	99%	[3]
Xpert MTB/RIF ultra	POC	qPCR (RIF-resistance)	<2 h	90%	96%	[4]
GeneXpert OMNI	POC	qPCR (RIF-resistance) Battery support	>1 h			[5])
TrueNat MTB	POC	Micro RT-PCR	<2 h	80%	95%	[6]
LF-LAM	POC	Polyclonal antibody-based, lateral flow strip-based test (non-sputum based)	~30 min	42%	92%	[7]
FujiLAM	POC	Monoclonal antibody based lateral flow strip-based test (non-sputum based)	~30 min	71%		[8]
DIGITAL CXR	POC	X-ray		93%	65%	[9]

complex (MTBC) along with the mutation in rpoB gene that causes resistance to rifampin (RIF), a first line drug in less than 2 h. Xpert was introduced in 2011 by Cepheid, Sunnyvale, USA, and has shown high sensitivity and specificity to diagnose TB [3]. Xpert is WHO-approved as a frontline test for both pulmonary and extrapulmonary tuberculosis. The GeneXpert Instrument System is a fully automated system that uses a disposable cartridge containing sputum sample of suspected TB patient mixed with the specific reagents (Fig. 10.1). The advantages of this system are a shorter turn-around time and the requirement of a minimal technical training. The quick results given by the Xpert MTB/RIF assay helps in selecting treatment regimens and making quick decisions for infection control and treatment. It is used for the diagnosis of tuberculosis (TB) and RIF-resistant tuberculosis (RR-TB) in National TB Elimination Program (NTEP) of India. However Xpert is not without limitations. Despite its excellent sensitivity in tests of smear-positive sputum samples, Xpert is somewhat less sensitive particularly in smear negative and HIV-associated TB patient's sample [3].

10.2.1.2 Xpert MTB/RIF Ultra

Xpert MTB/RIF Ultra is an improved version of Xpert MTB/RIF to overcome its limitations. It has higher sensitivity especially in smear-negative TB cases and improved detection of mixed infections like HIV. Xpert MTB/RIF Ultra is same in use as Xpert MTB/RIF as it uses the same test hardware and easy-to-use process. Xpert MTB/RIF Ultra gives results in <80 min with increased accuracy of Rifampicin results [10]. In Xpert MTB/RIF Ultra two different multicopy amplification targets (IS6110 and IS1081) along with improved assay chemistry and cartridge design have been used [4] (Fig. 10.1). These revisions resulted in an approximately 1-log improvement in the lower limit of detection compared with Xpert [4]. Overall, sensitivity of the Xpert MTB/RIF Ultra is 5% higher than that of Xpert MTB/RIF (95% CI +2.7, +7.8) but specificity is 3.2% lower (−2.1, −4.7) [11]. This higher sensitivity but lower specificity of Xpert MTB/RIF Ultra than the Xpert MTB/RIF assay is due to its inability to accurately distinguish between dormant and active TB DNA [12].

10.2.1.3 GeneXpert Omni

Since both the traditional GeneXpert platforms use continuous power supply, Cepheid has developed a new platform GeneXpert Omni. It has a battery that gives a 4-h battery life. Like other Xpert, the GeneXpert Omni also provides a portable clinical molecular diagnostic testing using cartridge technology. The microfluidics of cartridge regulate all the process including sample preparation, nucleic acid extraction, amplification and detection. GeneXpert Omni is a single standalone instrument that has 4 h of battery life and is capable of processing Xpert cartridges in even more extreme settings like elevated temperatures and humidity and challenging environment condition [11]. Omni is very handy, small and portable, weighing only 1.0 kg. Due to its high cost and limited accessibility in high-endemic areas, it was neither endorsed by WHO, nor there is any available evidence to support its use [13].

10.2.1.4 TrueNat TB Test

TrueNat was originally developed by Molbio Diagnostics Pvt. Ltd. to detect tuberculosis in 1 h. It is a rapid molecular test used for diagnosis of TB and RIF-resistance [6]. It was developed by utilising real-time polymerase chain reaction (RT-PCR) technology built into micro-PCR chips. So basically, TrueNat is a chip-based, battery-operated RT-PCR kit, which integrates all operations (heating, fluid mixing, magnetic control, amplification steps and timing) using a programmed micro-controller with easy-to-follow screen instructions. It enables the nucleic acid isolation without the need for any additional equipment [14]. A battery-operated sample preparation device is used to process the samples, and a nanoparticle-based protocol (optimized for sputum samples) is used to extract nucleic acids by a user-friendly, menu-driven process. TrueNat simplifies the RT-PCR process, thus facilitating laboratories with minimal infrastructure to perform these tests routinely, with a shorter turn-around time.

On the basis of its low cost and miniature version of the PCR test the WHO has approved TrueNat for detecting TB as a POCT [15].

10.2.2 Biomarker-Based Point-of-Care Test

A biomarker or a diagnostic marker is a substance that can estimate the occurrence of a particular disease and a point-of-care test needs a minimum amount of biological samples like blood, urine, saliva, etc. to determine the amount of targeted biomarker present in the sample in less duration, even in low concentrations. For their use in a POCT, biomarkers need to be evaluated in a low-cost, limited laboratory setting, and should be easy to use. In the last few years, multiple biomarkers like Mtb Ag85, C-reactive protein, pleural fluid adenosine deaminase, serum amyloid protein A, CA-125, ESAT-6 and lypoarabinomannan (LAM) have been used as a biomarker for TB diagnosis [16–21]. Although many of these markers are promising, none except LAM are available in POC format [22].

10.2.2.1 Lateral Flow Lypoarabinomannan Commercial Tests (LF-LAM)

Lipoarabinomannan (LAM) is a kind of glycolipid present on the outer cell wall of mycobacterium and released in urine from the degenerating or metabolically active mycobacterial cells [23]. It was first developed and demonstrated as a polyclonal antibody-based lateral flow rapid diagnostic test in 2015. The first commercially available kit using lateral flow for LAM detection in TB patient was the Alere Determine™ TB LAM Ag (AlereLAM) [24]. This test is non-sputum-based and facilitates the early initiation of anti-TB treatment. AlereLAM is performed manually by applying 60 μL of urine to the test strip followed by the incubation at room temperature for 25 min. The strip is then examined by visual observation of bands. The intensity of any visible band on the test strip is graded by comparing it with the intensities of the bands on a manufacturer-supplied reference scale card [25]. LAM is a potential marker of active tuberculosis but due to the use of the reading scale card when applying the test and its suboptimal sensitivity it was not found suitable as general diagnostic tests for TB [25]. In a meta-analysis, AlereLAM sensitivity was estimated at 35–42% in HIV-positive patients with symptoms of tuberculosis (29% in ambulatory settings) with 92% specificity [7]. WHO has recommended its use mostly in inpatient settings for TB disease in individuals living with HIV infection, including adults, adolescents, and children who are seriously ill, particularly those who are seriously ill and cannot produce sputum samples, defined as having fever above 39 °C, being tachypneic, tachycardic and unable to walk without help, providing results in just few minutes [25].

10.2.2.2 SILVAMP TB LAM (FujiLAM) Assay

In recent years, a new LAM detection assay, Fujifilm SILVAMP TB LAM (FujiLAM, Fujifilm, Tokyo, Japan) has been introduced. It is a next-generation, rapid, urine-based LAM test that has been demonstrated to have twofold higher sensitivity than

LF-LAM in patient with HIV-TB coinfection (71 vs. 35%) [25]. This sensitivity could be attributed to the use of a combination of high-affinity monoclonal antibodies against LAM epitopes specific for *M. tuberculosis*, as well as the technique of silver amplification immunochromatography, which improves the visibility of the control and test lines [24, 26, 27]. A previously published study has reported the increased sensitivity of FujiLAM with respect to AlereLAM in advanced cases of paediatric TB [28]. Hence while FujiLAM assay shows promise in terms of an effective diagnostic technique in children, further improvements need to be made in the same for its utilization as a rule-out test.

10.3 Portable Digital Chest X-Ray (CXR)

Chest X-ray (CXR) is indubitably a very sensitive test. It gives the images of the heart, lungs, blood vessels and airways. Chest X-rays can also reveal fluid and air in or around the lungs. In addition, the use of artificial intelligence (AI) in the medical diagnostics and AI-powered deep learning have rapidly increased in the past decade and are increasingly being used to analyse medical images, such as chest radiographs or X-rays [29]. The development of digital and portable X-ray machines along with the automated software is obviating the need of an experienced radiologists, which allows CXR to be considered as POC test. A simple van can carry a portable digital X-ray device to any location. It's all making them a very attractive option in low-resourced settings and remote areas [11].

10.4 Future Assays with Potential for POC for TB Diagnosis

Although there are some molecular technologies like GeneXpert, which are available as a recent advancement, these still have a scope for improvement to enable clinicians to diagnose drug-resistant and sensitive TB cases in the shortest time possible. These potential POC diagnostic tests may come from isothermal technology like the loop-mediated isothermal amplification (LAMP) [30, 31] which does not need thermocycler and lateral-flow-based innovations [32]. Additionally, analytical devices such as biosensors, which utilise a biological signalling molecule combined with a physicochemical detector, have a high sensitivity and specificity [33]. Biosensing technology, when combined with nanotechnology has immense potential in the field of medical diagnostics. Lab-on-chip (LOC) platform which can perform label-free and rapid single-cell capture [34] or a colorimetric sensing strategy employing gold nanoparticles [35] are also very promising assays. These are the newly advanced technologies which is being revolutionised for the POCT for the diagnosis of multiple diseases. TB researchers are also working on these platforms and we hope very soon a promising POCT will be available for us.

10.5 Limitations of Conventional Tests and the POC Tests

A majority of conventional tests and current POC tests are based on sputum specimens. However large proportions of community and peripheral health centres are not equipped to induce sputum and certain populations (PLHIV, infants, elderly) are unable to expectorate.

The development of modified diagnostic tests based on minimally invasive specimen types such as urine, blood and stool which offer the promising diagnostic result including genotypic, drug-resistance detection and treatment monitoring is the need of the hour. Non-sputum-based POC tests for TB diagnosis are needed to narrow the diagnostic gap and ensure timely treatment. As discussed earlier, a lateral flow assay based on TB LAM Ag 'AlereLAM' is a simple test for diagnosis of active TB in people living with HIV and available commercially.

This test uses urine which is very easy to collect and also it takes a maximum of 30 min to get the results. However, the application of this test is limited due to its low sensitivity and specificity, due to which WHO has recommended its use mostly in in-patient settings. Hence, there is an urgent need to develop a non-sputum-based point-of-care test for TB diagnosis.

10.6 Challenges Ahead

The development of highly accurate diagnostic tests (including home-based testing kits) for SARS-CoV-2 in record time during the COVID-19 pandemic was a milestone for the rapid diagnosis of patients, as it played a major role in the control and treatment of the disease. Similarly, the TB researcher community needs to follow the same pathway to develop a rapid, accurate, highly sensitive and specific point-of-care diagnostic test for not only TB but Multi-Drug Resistance (MDR) TB, utilizing multi-molecular diagnostic platforms.

10.7 Conclusion

In this chapter we have discussed some of the TB diagnostic tests suitable for POC use that are currently in use or in the late stages of development. While tests such as Xpert, trueNt and FujiLAM have shown promising results in a POC clinical setting, their global use is limited as none of them meet the WHO criteria, due to their low sensitivity. Researchers working on TB diagnostic fields should improve the simplicity and efficacy of POC tests. They should also direct their approaches in a more streamlined way so that more sophisticated tests are possible. This will create a holistic approach to ending TB and will bring the global medical community a step closer to the goal of eliminating TB by 2035.

Key Points

There is an urgent need of accurate, quick, non-sputum-based tests for the diagnosis of TB. Recently developed technologies for such tests are in the early stages of development and further research is needed to improve their efficacy. Additionally, diagnostic methods using non-sputum specimens should be developed further to obtain POC kits.

References

1. Organization, World Health. Lateral flow urine lipoarabinomannan assay (LF-LAM) for the diagnosis of active tuberculosis in people living with HIV: policy update (2019): evidence to decision tables. World Health Organization; 2019.
2. Global Tuberculosis Report 2022. n.d.; Accessed 23 Mar 2023. https://www.who.int/publications/i/item/9789240061729.
3. Marlowe EM, Novak-Weekley SM, Cumpio J, Sharp SE, Momeny MA, Babst A, Carlson JS, Kawamura M, Pandori M. Evaluation of the cepheid Xpert MTB/RIF assay for direct detection of mycobacterium tuberculosis complex in respiratory specimens. J Clin Microbiol. 2011;49(4):1621–3.
4. Chakravorty S, Simmons AM, Rowneki M, Parmar H, Cao Y, Ryan J, Banada PP, Deshpande S, Shenai S, Gall A. The new Xpert MTB/RIF ultra: improving detection of mycobacterium tuberculosis and resistance to Rifampin in an assay suitable for point-of-care testing. MBio. 2017;8(4):e00812–7.
5. "Lessells2019." n.d.
6. Penn-Nicholson A, Gomathi SN, Ugarte-Gil C, Meaza A, Lavu E, Patel P, Choudhury B, Rodrigues C, Chadha S, Kazi M. A prospective multicentre diagnostic accuracy study for the truenat tuberculosis assays. Eur Respir J. 2021;58(5):2100526.
7. Bulterys MA, Wagner B, Redard-Jacot M, Suresh A, Pollock NR, Moreau E, Denkinger CM, Drain PK, Broger T. Point-of-care urine LAM tests for tuberculosis diagnosis: a status update. J Clin Med. 2019;9(1):111.
8. Huerga H, Bastard M, Lubega AV, Akinyi M, Antabak NT, Ohler L, Muyindike W, Taremwa IM, Stewart R, Bossard C. Novel FujiLAM assay to detect tuberculosis in HIV-positive ambulatory patients in four African countries: a diagnostic accuracy study. Lancet Glob Health. 2023;11(1):e126–35.
9. Nalunjogi J, Mugabe F, Najjingo I, Lusiba P, Olweny F, Mubiru J, Kayongo E, Sekibira R, Katamba A, Kirenga B. Accuracy and incremental yield of the chest X-Ray in screening for tuberculosis in Uganda: a cross-sectional study. Tuberc Res Treat. 2021;2021:6622809.
10. Jones M, Chakravorty S, Simmons M. MTB-RIF ultra–design and analytical performance of a second generation GeneXpert assay (Poster 0475). ECCMID; 2016.
11. García-Basteiro AL, DiNardo A, Saavedra B, Silva DR, Palmero D, Gegia M, Migliori GB, et al. Point of care diagnostics for tuberculosis. Pulmonology. 2018;24(2):73–85. https://doi.org/10.1016/j.rppnen.2017.12.002.
12. "View of MODERN EVOLUTIONS IN TUBERCULOSIS DIAGNOSIS.Pdf." n.d.
13. Unitaid. Tuberculosis diagnostics technology landscape. Geneva: World Health Organization; 2017.
14. Nikam C, Jagannath M, Narayanan MM, Ramanabhiraman V, Kazi M, Shetty A, Rodrigues C. Rapid diagnosis of mycobacterium tuberculosis with truenat MTB: a near-care approach. PLoS One. 2013;8(1):e51121.
15. Lee DJ, Kumarasamy N, Resch SC, Sivaramakrishnan GN, Mayer KH, Srikanth Tripathy A, Paltiel D, Freedberg KA, Reddy KP. Rapid, point-of-care diagnosis of tuberculosis

with novel truenat assay: cost-effectiveness analysis for India's public sector. PLoS One. 2019;14(7):e0218890.
16. Ali N, Nath NC, Parvin R, Rahman A, Bhuiyan TM, Rahman M, Mohsin MN. Role of ascitic fluid adenosine deaminase (ADA) and serum CA-125 in the diagnosis of tuberculous peritonitis. Bangladesh Med Res Counc Bull. 2014;40(3):89–91.
17. Babaki MK, Zadeh SS, Rezaee SA. Antigen 85 complex as a powerful mycobacterium tuberculosis immunogene: biology, immune-pathogenicity, applications in diagnosis, and vaccine design. Microb Pathog. 2017;112:20–9.
18. Broger T, Tsionksy M, Mathew A, Lowary TL, Pinter A, Plisova T, Bartlett D, Barbero S, Denkinger CM, Moreau E. Sensitive electrochemiluminescence (ECL) immunoassays for detecting lipoarabinomannan (LAM) and ESAT-6 in urine and serum from tuberculosis patients. PLoS One. 2019b;14(4):e0215443.
19. Porcel JM, Bielsa S, Esquerda A, Ruiz-González A, Falguera M. Pleural fluid C-reactive protein contributes to the diagnosis and assessment of severity of parapneumonic effusions. Eur J Intern Med. 2012;23(5):447–50.
20. Şahin F, Yildiz P. Serum CA-125: biomarker of pulmonary tuberculosis activity and evaluation of response to treatment. Clin Invest Med. 2012;1:E223–8.
21. Vietri L, Fui A, Bergantini L, d'Alessandro M, Cameli P, Sestini P, Rottoli P, Bargagli E. Serum amyloid a: a potential biomarker of lung disorders. Respir Investig. 2020;58(1):21–7.
22. Corrigan DT, Ishida E, Chatterjee D, Lowary TL, Achkar JM. Monoclonal antibodies to lipoarabinomannan/arabinomannan–characteristics and implications for tuberculosis research and diagnostics. Trends Microbiol. 2022;1:1.
23. Liu D, Ling G, Zhang R, Liu L, Shen Y, Shao Y, Wang J, Sun J, Qi T, Wang Z. Utility of urine lipoarabinomannan (LAM) in diagnosing mycobacteria infection among hospitalized HIV-positive patients. Int J Infect Dis. 2022;118:65–70.
24. Broger T, Sossen B, du Toit E, Kerkhoff AD, Schutz C, Reipold EI, Ward A, Barr DA, Macé A, Trollip A. Novel lipoarabinomannan point-of-care tuberculosis test for people with HIV: a diagnostic accuracy study. Lancet Infect Dis. 2019a;19(8):852–61.
25. Broger T, Nicol MP, Szekely R, Bjerrum S, Sossen B, Schutz C, Opintan JA, Johansen IS, Mitarai S, Chikamatsu K. Diagnostic accuracy of a novel tuberculosis point-of-care urine lipoarabinomannan assay for people living with HIV: a meta-analysis of individual in-and outpatient data. PLoS Med. 2020;17(5):e1003113.
26. Vaezipour N, Fritschi N, Brasier N, Bélard S, Domínguez J, Tebruegge M, Portevin D, Ritz N. Towards accurate point-of-care tests for tuberculosis in children. Pathogens. 2022;11(3):327. https://doi.org/10.3390/PATHOGENS11030327.
27. Mitamura K, Shimizu H, Yamazaki M, Ichikawa M, Nagai K, Katada J, Wada A, Kawakami C, Sugaya N. Clinical evaluation of highly sensitive silver amplification immunochromatography systems for rapid diagnosis of influenza. J Virol Methods. 2013;194(1–2):123–8. https://doi.org/10.1016/j.jviromet.2013.08.018.
28. Comella-del-Barrio P, Molina-Moya B, Gautier J, Villar-Hernández R, Doresca MJC, Sallés-Mingels B, Canales-Aliaga L, Narcisse M, Pérez-Porcuna TM, Creswell J. Diagnostic performance of the fujifilm SILVAMP TB-LAM in children with presumptive tuberculosis. J Clin Med. 2021;10(9):1914.
29. Vo LN, Quang AC, Ngo TD, Dao TP, Dong TTT, Mo HTL, Forse R, Nguyen TT, Van Cung C, Nguyen HB. Early evaluation of an ultra-portable x-Ray system for tuberculosis active case finding. Trop Med Infect Dis. 2021;6(3):163.
30. Alhakeem R, Arruda V, Steinberg K, Code C, Neves AF. Using saliva and LAMP for non-invasive detection of communicable pathogens in low medical access regions reham; 2023.
31. Yadav B, Sharma M, Singla N, Shree R, Goyal M, Modi T, Sharma A, Sharma A, Sharma N, Ray P. Molecular diagnosis of tuberculous meningitis: SdaA-based multi-targeted LAMP and GeneXpert ultra. Tuberculosis. 2023;140:102339.

32. Talreja J, Peng C, Nguyen T-M, Draghici S, Samavati L. Discovery of novel transketolase epitopes and the development of IgG-based tuberculosis serodiagnostics. Microbiol Spectr. 2023;11:e03377–22.
33. Zhou L, He X, He D, Wang K, Qin D. Biosensing technologies for mycobacterium tuberculosis detection: status and new developments. Clin Dev Immunol. 2011;2011:193963.
34. Esfandyarpour R, DiDonato MJ, Yang Y, Durmus NG, Harris JS, Davis RW. Multifunctional, inexpensive, and reusable nanoparticle-printed biochip for cell manipulation and diagnosis. Proc Natl Acad Sci. 2017;114(8):E1306–15.
35. Tsai T-T, Huang C-Y, Chen C-A, Shen S-W, Wang M-C, Cheng C-M, Chen C-F. Diagnosis of tuberculosis using colorimetric gold nanoparticles on a paper-based analytical device. ACS Sens. 2017;2(9):1345–54.

Chapter 11
Recent Developments in the Diagnosis and Treatment of Non-Tuberculous Mycobacterial Infection

Shipra Tomar and Anand Kumar Maurya

Abstract Non-tuberculous mycobacteria are known to have approximately 200 species that caused a wide array of infections and are present extensively in the environment. Patients suspected of non-tuberculous mycobacteria (NTM) infection must follow all clinical and microbiological norms for its diagnosis. Molecular techniques help describe the new types of NTM, as well as their subspecies. Treatment for such infection is complicated and species-specific. A careful analysis by a medical professional is necessary for deciding to initiate treatment for NTM infection. We have summarized the evolution of our diagnostic and treatment options for non-tuberculous mycobacteria infection based on this chapter.

Keywords Non-tuberculous mycobacteria (NTM) · Diagnosis · Treatment

11.1 Introduction

Non-tuberculous mycobacteria are a common microorganism that may be found in all habitats and are a member of the Mycobacteriaceae family. They are particularly prevalent in soil, water, biofilm, damp walls, and wild animals. They are also the cause of several significant human disorders. The group of more than 190 *M. mycobacteria*, known by several names such as environmental mycobacteria, anomalous mycobacteria, or mycobacteria other than tubercle bacilli (MOTT) is collectively called non-tuberculous mycobacteria [1] (Fig. 11.1).

NTM are distinguished by a thin peptidoglycan layer that is encircled by a thick, lipid-rich outer coating. This characteristic allows NTM to adhere to rough surfaces and provides resistance to antibiotics and disinfectants, allowing them to persist in environments with low oxygen and carbon concentrations [2]. NTM are classified into two categories based on the characteristics of the subculture's growth: (1)

S. Tomar · A. K. Maurya (✉)
Department of Microbiology, AIIMS Bhopal, Bhopal, Madhya Pradesh, India
e-mail: anand.microbiology@aiimsbhopal.edu.in

A. Singh, D. Sharma (eds.), *Diagnosis of Mycobacterium*,
https://doi.org/10.1007/978-981-99-5624-1_11

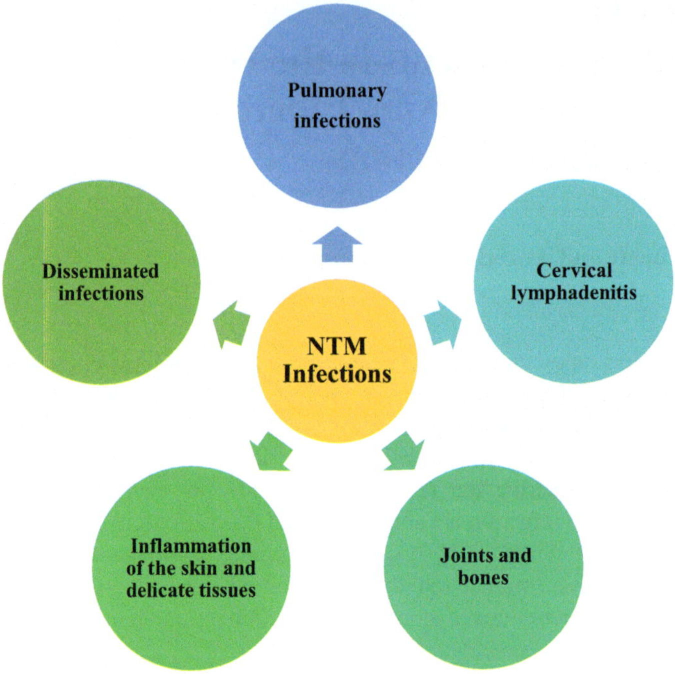

Fig. 11.1 Spectrum of infections caused by non-tubercular mycobacteria

Table 11.1 NTM that cause diseases in humans

Rapid growing NTM (<7 days)	Slow growing NTM (≥ 7 days)
M. fortuitum complex	**Photo-chromogens**
M. fortuitum	*M. kansasii*
M. peregrinum	*M. marinum*
M. porcinum	**Scoto-chromogens**
M. chelonae	*M. gordonae*
M. abscessus	*M. scrofulaceum*
M. abscessus subspecies abscessus	**Non-chromogens**
M. abscessus subspecies bolletii	*M. avium complex*
	M. avium
	M. intracellulare
	M. chimaera
	M. terrae complex
	M. ulcerans
	M. xenopi

mycobacteria that develop quickly known as **rapid growing NTM** (<7 days); (2) mycobacteria that grow slowly known as **slow growing NTM** (≥ 7 days) (Table 11.1).

11.2 Epidemiology

The frequency and prevalence of NTM conditions continue to increase worldwide, leading to an arising public health problem. Indeed, though the distribution of NTM species varies markedly grounded on terrain, *M. avium* complex (MAC) is the most common pathogen in numerous areas followed by *M. abscessus* complex (MABC) and *M. kansasii* [3]. Belgium (2.1), Italy (2.5), Denmark (5.3), United Kingdom (6.0), France (6.5), Finland (6.7), Spain (10.8), Germany (12.2), Portugal (16.5), Czech Republic (17.5), Switzerland (17.5), and Turkey (33.9) were the countries with the loftiest recorded frequentness of *M. fortuitum* (4). According to studies from England and Wales, *M. kansasii* was the most pathogenic species isolated, whereas *M. malmoense* and *M. xenopi* were the two species that were most current in Scotland and Southeast England, independently [4, 5]. NTM frequency was also set up in India, where *M. fortuitum* and *M. avium intracellulare* (MAI) were the two most constantly isolated species [4, 6].

Clinical and microbiological criteria are needed for diagnosing NTM diseases due to the difficulty in differentiating between NTM isolation and complaint. NTM infections can be extremely delicate to treat. According to the numerous NTM species, suggested treatment rules, medicine resistance patterns, and treatment issues presently vary, and operation is a time-consuming process with many remedial choices. The current suggestions, still, heavily calculate on professional judgment but several recommendations' specifics are still debatable. Also, there have been developments in molecular individual ways for the discovery of NTM species, the discovery of medicine resistance, and the treatment of NTM lung illness with antibiotics. As a result, streamlined substantiation-grounded recommendations, useful knowledge, and education are needed for working clinicians [3]. We summarize the pathogenesis and contemporary developments in this chapter.

Table 11.2 Runyon's classification of non-tuberculous mycobacteria

Runyon group	Characteristic feature	Species
Group-1 photo-chromogens	When light is present, the pigment is produced	*M. marinum, M. kansasii, M. asiaticum, M. simiae*
Group-2 Scoto-chromogens	In the presence or absence of light, the pigment is formed	*M. gordonae, M. scrofulaceum, M. szulgai, M. interjectum*
Group-3 non-photochromogens	No pigment is produced by them	*M. ulcerans, M. malmoense, M. xenopi,Mycobacterium avium* complex
Group-4 rapid growers	Growth happens after a week	*M. fortuitum, M. smegmatis, M. abscessus, M. chelonae*

11.3 Classification of NTM

The classification for NTM is based on the Runyon classification, which divides NTM into four groups: are tabulated in Table 11.2

11.4 Clinical Manifestations of NTM

The clinical signs and symptoms of NTM disease are similar to those of tuberculosis and could make a diagnosis difficult. Clinical manifestations are tabulated in Table 11.3.

NTM diseases are classified into four clinical types. These are as follows:

- Chronic pulmonary infection.
- Lymph node infection.
- Cutaneous infection.
- Disseminated infection.

11.4.1 Chronic Pulmonary Infection

Three patterns of pulmonary exposure can be seen (Table 11.4) [2].

11.4.2 Lymph Node Infection

The most commonly seen in younger children [7]. The site involved is mainly solitary lymph nodes and submandibular or cervical regions. Painless swelling is seen in the early stages, whereas in the latent phase, swelling might turn into pus which may burst later and form a sinus. There may be signs in addition to fatigue, weight loss, and fever. Because the disease is paucibacillary, cultures and smears may not be positive [8, 9].

Table 11.3 Clinical manifestations of NTM

Name of infection	Species
Chronic pulmonary infection	*M. xenopi*, *M. malmoense*, MAC, *M. kansassi*, *M. abscessus*
Lymph node infection	*M. scrofulaceum*, *M. bohemicum*, *M. kansassi*, *M. abscessus*
Cutaneous infection	*M. chelonae*, *M. xenopi*, *M. terrae*, MAC, *M. abscessus*, *M. fortuitum*
Disseminated infection	*M. avium*, *M. genavense*, M. *intracellulare*, *M. haemophilum*

Table 11.4 Pattern of pulmonary exposure and their clinical features

Patterns of pulmonary exposure	Occurrence	Commonly seen in	Clinical features
Fibro-cavitary type	Upper lobe	Seen in smokers who also have lung conditions like Chronic Obstructive Pulmonary Disease (COPD) as a result of their history of smoking	Breathing difficulty, expectorant cough, and hemoptysis
Nodular/ bronchiectatic type	With only a few pulmonary nodules in the right middle lobe and lingular bronchiectasis on the left	Majorly seen in postmenopausal and non-smoking females	Cough with prior parenchymal lung illness
Hypersensitivity pneumonitis type	Upper lobe	Due to exposure to aerosols	Fever, anorexia, progressive fatigue, weight loss, and malaise

11.4.3 Cutaneous Infection

Three types of clinical presentation can be seen with cutaneous infection.

(a) A severe cutaneous condition known as Buruli ulcer develops from nodular lesions into enormous, effortless blisters. **Mycolactone**, a poison produced most constantly by *M. ulcerans*, harms the skin [10].
(b) Fish-tank granuloma: This condition begins as a nodular, verrucous, or ulcerated granulomatous skin lesion on the hands and forearm and develops into many skin lesions. *M. marinum*, which can be caught in swimming pools, cleaning fish tanks, or any other water-related activity, is what causes the infection [11].
(c) Inflammation of the skin and delicate tissues: Both immunocompetent and immunocompromised people can get these infections from slow-growing NTM, such as *M. abscessus, M. fortuitum*, and *M. chelonae*. These organisms gain access through breaks in the skin during surgical procedures by using instruments without autoclaving, cosmetic surgeries, various implants, etc. [12, 13]

11.4.4 Disseminated Infection

It is frequently observed in HIV/AIDS patients, particularly those with CD4+ lymphocyte counts under 50 cells/L. The main cause of disseminated infection, MAC, enters patients through the colon and, sporadically, the lungs with hematogenous dissemination [2].

11.5 Laboratory Diagnosis of NTM

It is not necessary for a person with an NTM infection to be unwell or for the isolation to be of clinical importance. There are not many connotative tests for the detection of NTM infections in particular. The correct diagnosis needs to be in a correlation of the syndrome with radiological findings and then confirming it with microbiologic etiologic agents. Before interpreting NTM infections, they should first be confirmed with experienced specialists. Both diagnosing NTM and treating a patient depend heavily on laboratory techniques.

11.6 Specimen Transportation, Processing, and Collection

For the accurate identification of NTM isolates, a thorough sample collection is crucial. In the case of pulmonary samples, environmental and personal contamination should be avoided during the collection of sputum. Ideally, three early-morning sputum samples should be collected on three successive days. Extra-pulmonary samples should be directly obtained from the lesion or concerned organ. Once the sample is stored in a container, it should reach the laboratory without opening it. If delayed, should be stored at **2–8 °C** and not kept for more than a week. Decontamination should be carried out in completely sterilized environments once the sample has arrived at the lab. Since NTM are resistant to the majority of disinfectants, appropriate disinfectants should be chosen. Seventy percent alcohol and 5% phenol disinfected for cleaning and bio-safety filters are recommended.

11.7 Microbiologic Diagnosis

11.7.1 Microscopy

Carbol fuchsin stain (Ziehl-Neelson or Kinyoun method) and fluorochrome technique (auramine-O, rhodamine, auramine-rhodamine, acridine orange) are the two primary forms of Acid Fast Bacilli (AFB) stains that are used to identify mycobacteria. But it is difficult to distinguish between *Mycobacterium tuberculosis* complex (MTBC) and NTM solely on staining. Patients with extra-pulmonary TB and NTM infection have reduced smear sensitivity levels. ZN (Ziehl-Neelson) staining is a crucial microscopic technique for detecting mycobacteria, although it is still not the best way for recognizing NTM [1].

Staining method	Sample type	Findings	Examples
Ziehl-Neelson	Any clinical specimen	Beaded, bright red with blue background	All species

Staining method	Sample type	Findings	Examples
Kinyon	Solid culture media	Weak acid-fast, does not appear bright red	*M. leprae, M. abscessus,* MAC, *M. fortuitum, M. chelonae*
Auramine-O	Any clinical specimen	Bright green or yellow with black background	All species
Gomori-methenamine silver stain	Tissue specimen	Brown-black bacilli	*M. ulcerans, M. chelonae, M. kansassii*

11.7.2 Biochemical Identification

Mycobacterial group	Biochemical tests
M.tuberculosis complex	Niacin test, nitrate reduction, thiophen-2-carboxylic acid hydrazide (TCH) susceptibility testing for *M. bovis*
Photo chromogens	Niacin, urease, nitrate reduction, tween 80 hydrolysis, pyrazinamidase, 14-day arylsulfatase
Scotochromogen	Urease, nitrate reduction, tween 80 hydrolysis, 14-day arylsulfatase, semi-quantitative catalase
Non-photo chromogens	Semi-quantitative catalase, 14-day arylsulfatase, urease, nitrate reduction, heat resistance, tween 80 hydrolysis, tellurite reduction, acid phosphatase activity
Rapid growers	Nitrate reduction, tween 80 hydrolysis, 3-day arylsulfatase, iron uptake, growth on MacConkey agar

11.7.3 Culture

Media used for the culture of NTM include both media, liquid and solid, for detection and enhancement of growth. The Mycobacterial Growth Indicator Tube (MGIT) is the most typical and commonly used system. The Middlebrook 7H9 broth in MGIT has been changed, and it also includes an oxygen sensor based on fluorescence quenching that aids in the identification of mycobacterial growth. Egg-based solid media used for the growth of NTM include Lowenstein-Jensen media and agar-based media like Middlebrook 7H10 and 7H11 media. Other solid media like MYChrOme Culture Plate is a chromogenic solid culture media on which NTM colonies can be differentiated by using crystal violet dye. All non-mycobacteria colonies are turned purple/brown by the crystal violet dye, while NTM colonies remain white or non-colorized. NTM Elite agar, rapidly growing mycobacteria agar, includes four antimicrobial agents: colistin, fosfomycin, amphotericin, and C-390. Middlebrook 7H10 medium is supplemented with 10% v/v oleic acid-albumin-dextrose-catalase enrichment, 0.5% v/v glycerol, and 2 g/mL mycobactine J to create solid media like Middlebrook 7H11j. The advantage of solid media over liquid media is that they allow observation of colony morphology, growth rates, presence of mixed infection, and serve as a backup when liquid media cultures are contaminated. Cultures in liquid media have a higher yield of mycobacteria and produce results more quickly [9, 14].

11.7.4 Automated Culture Methods

In order to prevent contamination, it uses liquid media like Middlebrook 7H9 supplemented with enrichment such as Oleic acid, Albumin, Dextrose and Catalase (OADC) enrichment, which is a mixture of bovine albumin, dextrose, oleic acid, and catalase, and PANTA antibiotic (polymixin B, amhotericin B, nalidixic Acid, trimethoprim, and azlocillin), which prevents the growth of contaminants.

Principle: Different devices use different principles to detect growth.

11.7.4.1 Mycobacteria Growth Indicator Tube TB System (MGIT)

Oxygen-quenched fluorochrome is embedded in silicone at the bottom of the Mycobacteria Growth Indicator Tube. The free oxygen in the tube is used up by the bacteria as they develop and is exchanged for carbon dioxide. As the free oxygen content decreases, fluorochrome is no longer inhibited, which causes fluorescence inside the MGIT tube when seen under UV light. MGIT tubes are placed into an MGIT 960 instrument or incubated at 37 °C for an hour before being manually read under a UV lamp or checked for growing fluorescence. Most NTM (fast growers) produce growth that leaves the medium with light turbidity or a small granular/flaky appearance [15].

11.7.4.2 BacT/ALERT TB System

The BacT/ALERT System and culture bottles offer a microbial detection system as well as a culture medium containing organisms that may be present in the sample utilizing the colorimetric detection of pH change in the medium that occurs as a result of CO_2 generated by the growth of bacteria. Bottles that have been inoculated are put into the device, where they are incubated and continuously checked for the presence of bacteria that will thrive in the BacT/ALERT BPA bottles. Carbon dioxide will be created as the organism breaks down the substrates in the growth medium if microorganisms are found in the test sample. When microorganisms thrive and release carbon dioxide, the bottom of each culture bottle turns yellow. A 40 mL media reservoir and an inbuilt sensor that picks up carbon dioxide, a sign of microbial development, are both included in BacT/ALERT BPA bottles. The media contains purified water, sodium polyanethol sulfonate (SPS), pyridoxine HCl, papaic digest of soybean meal, pancreatic digest of casein, and other complicated amino acids and carbohydrate substrates [16].

11.7.4.3 Extracellular Polysaccharide (ESP) Culture TB System

It notices a change in medium pressure brought on by bacteria's generation of carbon dioxide. Before usage, 1 mL of each antibiotic dilution and 1 mL of growth supplement are added to ESP myco bottles. In place of the antibiotic solution in the control bottle, 1 cc of sterile distilled water was introduced. Each drug-containing bottle and control bottle received 0.5 mL of a 1:10 diluted 1McFarland solution as an inoculation. The ESP device is then filled with bottles. If an organism produces a signal from the drug-containing bottle within 3 days after the control, it is deemed resistant; if it does not produce a signal within 3 days, it is deemed susceptible [17].

11.7.4.4 BACTEC 460 TB System

It was predicated on using radioisotopes to measure bacterial growth, but it is no longer in use. The Middlebrook 7 H12 broth with carbon 14 labeled palmitic acid, 4 mL, is the medium used in the BACTEC 460 Tb system. The antibiotic mixture including the PANTA solution was inoculated along with the clinical sample. The BACTEC 460 equipment reads BACTEC bottles once every week for up to 6 weeks.

11.8 Radiological Diagnosis of NTM Infection

There are no radiographic pattern that shows characteristic features to a combination that differentiates NTM from mycobacteria but they are commonly encountered in pulmonary as well as extra-pulmonary NTM diseases. These are as follows:

11.8.1 Pulmonary NTM Radiological Findings

(a) ***Fibro-cavitary NTM pulmonary disease.***
    It is primarily distinguished by the top lobe, which has cavities with thin or thick walls and thick pleura around them. Middle-aged people with structural lung diseases like COPD frequently have fibro-cavitary disease [1, 18].
(b) ***Nodular bronchiectasis NTM pulmonary disease.***
    It occurs in elderly women and is linked to pectus excavatum, scoliosis, and mitral valve prolapse. Cylindrical bronchiectasis with branching centrilobular nodules can be seen on High-Resolution Computed Tomography (HRTC). The

lingual and right middle lobes are also affected, and the changes are typically bilateral [1, 19].

(c) ***Hypersensitivity pneumonitis.***

HRCT consistently shows acute hypersensitivity pneumonitis: commonly in the upper lobe, bilateral glass opacity, and centrilobular nodularity [1].

11.8.2 Extra-Pulmonary NTM Infection Radiological Findings

(a) **Inflammation of the skin and delicate tissues**—diagnosis for inflammation of the skin and delicate tissues is not required but in deeper tissues, tests like sonography, CT, and MRI are helpful in the diagnosis of such infections.
(b) **NTM Lymphadenitis**—includes ultrasound, CT, and MRIs. Common features are periostitis, osteolysis, and cystic necrosis [1, 18].

11.9 Molecular Methods for NTM

The notion of nucleic acid amplification underlies multiplex-PCR (polymerase chain reaction), whole genome sequencing (WGS), DNA microarray, line probe assay (LPA), PCR-RFLP (restriction fragment length polymorphism), and nucleic acid probes. These techniques enable more accurate identification in a short amount of time.

11.9.1 Line Probe Assay for NTM

Reverse hybridization is based on nucleic acid amplification and uses genetic probes. The GenoType® M. common mycobacteria/additional species (CM/AS) assay (Hain Lifescience, Nehren, Germany) is used in the LPA, a DNA strip-based test. The kit is utilized for NTM identification and species discrimination from culture. The 23S ribosomal RNA region is amplified using the LPA process, which is then reverse hybridized on strips of nitrocellulose membrane. There are two kits: the CM recognizes 15 mycobacterium species, including the M. TB complex, and the AS indicates 16 additional, less prevalent non-tuberculous mycobacterium species. A quick and accurate test for identifying non-tuberculous mycobacterium species is the GenoType® M. CM/AS assay [2, 4].

11.9.2 PCR-RFLP (Restriction Fragment Length Polymorphism)

The examination of restricted gene fragment band patterns, which are unique to several non-tuberculous mycobacterium species, is carried out. The method provides a quick and low-cost assay for locating non-mycobacterial species. The 16S ribosomal DNA (16SrRNA), heat shock protein 65 (*hsp*65 or *hsp*60), dnaJ, groES, 16S-23SrRNA internal transcribed spacer (ITS), and DNA-directed RNA polymerase beta chain (*rpo*B) genes have all been subjected to this method. A specific portion of the mycobacterium *hsp*65 gene is replicated and cut with restriction enzymes in the hsp65 PCR-RFLP method. The species detection algorithm then compares the obtained band patterns to the reference species [20].

11.9.3 Nucleic Acid Probes Detection

Ester-labeled genes can be processed directly on clinical samples and bind to DNA probes that are complementary to the 16S rRNA gene. MAC, *M. kansasii*, and *M. gordonae*-specific acridium ester-labeled DNA probes are presently employed in many laboratories for the identification of NTM. Within 2 h, a species can be identified using a technique based on the release of 16S rRNA from an organism isolated from solid or liquid culture media. Only a few NTM species have probes available, and it has been shown that there is a potential that the probe will react with both *M. celatum* and *M. tuberculosis* [9, 21].

11.9.4 Multiplex PCR

NTMs have emerged as a significant pathogen, and the number of diseases they cause is rising globally. MAC, *M. avium*, *M. intracellulare*, *M. kansasii*, *M. abscessus*, etc. are the most frequently isolated species from individuals with non-tubercular mycobacterial infection. Multiplex-PCR is used to find and classify NTM species, with the target genes being hsp65, rpoB, and 16S rRNA. A precise and sensitive technique for quickly identifying NTM species from clinical specimens is multiplex PCR [22].

11.10 Advance Methods for Detection of NTM

11.10.1 MALDI-TOF Method

The detection of non-tuberculous mycobacterium species has been aided by the development of an advanced technique that compares the mass spectral patterns of molecules, primarily ribosomal proteins, which are unique to non-tuberculous M. species in a library of known non-tuberculous mycobacterium strains. The basis for identification is the spectral fingerprints that extracted proteins produce. It is regarded as the fastest method and has discovered about 160 non-tuberculous mycobacterium species. Compared to the conventional microbiological identification method, the procedure is simple, rapid, and associated with much lower consumable costs [2, 9]. The technique offers rapid and accurate detection while also saving a sizable amount of money. The only challenges are optimizing the protein extraction technique and updating the database to include the most therapeutically significant species possible [10, 11].

11.10.2 Whole Genome Sequencing

The gold standard for identifying non-tuberculous mycobacterium species, which aids in comprehending their geographic and environmental distribution, is whole genome sequencing. The 16S rRNA gene is the most often investigated gene for the identification of bacteria and is the first option for sequence analysis. In order to identify NTM species, the sequences of each gene are compared with the gathered database to find commonalities. WGS can offer details on the pathogenicity and antimicrobial agent resistance traits of diverse microbes [21, 23].

11.10.3 DNA Microarray

With the use of DNA microarray technology, a large number of DNA sequences can be detected in a shorter amount of time by a single step of hybridization. The method has been applied to find resistant mutations in mycobacteria and to discover mycobacterial species. A DNA array with nucleotide probes is hybridized with fluorescence labeled amplicons produced from bacterial culture for use in a polymerase chain reaction (PCR). The scanner then picks up a fluorescent signal that is emitted by the bound amplicons. This method allows for the differentiation of 54 different mycobacterial species using 82 different 16S rRNA sequences and 51 sequences with various rpoB gene mutations [2].

11.11 Immunodiagnostic Test

11.11.1 Interferon-Gamma-Based Determination

Culture filtrate protein (CFP-10) and early secretory antigenic target (ESAT-6) are two specific antigens encoded in a region of difference 1 (RD) that are used in interferon-gamma release assays (IGRAs), which aid in the detection of NTM infection. Enzyme-linked immunospot SPOT TB assay (United Kingdom) and QuantiFERON TB, as well as its improved versions QuantiFERON TB Gold and QuantiFERON TB Gold in-tube assays (Australia), are examples of the several INF commercial kits that are currently available [1, 4].

11.11.2 Immuno-Chromatographic Test

Immuno-chromatographic testing that is quick and simple SD Bioline in South Korea created the MPT64 TB Ag kit, which makes it simple to quickly identify and distinguish between *M. tuberculosis* isolates and NTM isolates based on their MPT 64 antigen levels. *M. bovis* BCG, *M. tuberculosis*, and sub-strains of mycobacteria are all quickly identified using the extremely sensitive MPT 64 TB Ag kit [4].

11.12 Treatment of NTM

Due to a number of circumstances, such as multidrug resistance regimens, long-term therapy, and side effects brought on by frequent drug usage, the treatment of NTM infection is complicated and constrained. To handle NTM infections, a well-coordinated strategy is necessary. At the moment, ATS recommendations for assessing the clinical importance of NTM from various specimens are being followed [9].

(a) **Clinical criteria** include a chest radiograph with nodular or cavitary opacities or an HRCT scan with bronchiectasis and numerous tiny nodules are examples of clinical criteria [9].
(b) **Microbiologic criteria:** A sputum AFB smear should be repeated if the results are not diagnostic, and a culture-positive isolate from one bronchial wash, lavage, and lung biopsy with mycobacterial histological characteristics should also be examined.

The broth micro-dilution method is advised for drug susceptibility testing of NTM. The obtained MIC value aids the doctor in determining the necessary antimicrobial agent needed at the infected spot to prevent an organism's growth. At the moment, recommendations are given for MAC, *M. kansasii*, *M. marinum*, and rapidly expanding mycobacteria [9, 14].

11.13 Drug Susceptibility Testing (DST)

The manner of treatment regimens utilized for non-tubercular mycobacterial infections is guided by DST. The test culture is added to two tubes of Middlebrook 7H9 liquid medium for inoculation. One of the MGIT tubes is supplemented with a known concentration of a test substance, and the growth is compared to the un-inoculated one. If the test medicine is effective against the isolated NTM, it will suppress fluorescence and impede growth, whereas the growth control will grow unhindered and exhibit increased fluorescence. The BACTEC 960 equipment measures growth and instantly determines whether the results are resistant or sensitive.

11.13.1 M. avium *Complex*

1. **Reasons to conduct susceptibility tests** [9, 14]**.**
   (a) Patients receiving macrolide treatment who are clinically significant.
   (b) Cases with bacteremia while on macrolide prophylaxis.
   (c) Cases on macrolide treatment who experienced recurrence.
   (d) Individuals with widespread illness identified by clinically significant sputum or Bronchoalveolar Lavage (BAL) samples.

Broth-based methods like 5% OADC supplemented with cation-adjusted Mueller Hinton broth (CAMHB), are recommended to perform drug susceptibility testing. Incubation is done for 7 days at 37 °C in ambient air for micro-dilution trays. If growth is insufficient, trays should be re-incubated and read once more at 10–14 days. Drug susceptibility testing should be done after 3 months of treatment for patients with disseminated disease and after 6 months of treatment for patients with chronic pulmonary disease if the patient shows no clinical improvement and the culture is still positive.

11.13.2 M. kansasii [14]

Treatment for *M. kansasii* typically involves the use of clinically effective medications, such as isoniazid, rifampicin, and ethambutol. A different medicine combination, rifampicin, ethambutol, and clarithromycin, has been proven to be efficient. For testing with primary treatment medications, broth micro-dilution, and agar proportion have been used. It is advised to use CAMHB supplemented with 5% OADC while micro-dilution broth. The growth of micro-dilution trays is observed after 7–14 days of incubation at 37 °C and 5–10% CO_2. CO_2 must be avoided when testing macrolides. Untreated strains' MICs for these medications will fall within a specific range, so routine testing is typically unnecessary. Treatment failure happens and is frequently correlated with MIC values resistant to rifampicin, with sporadic reports of resistance to isoniazid and ethambutol as well. Susceptibility testing is crucial for patients who failed therapy or did not respond well to initial therapy, for these reasons.

11.13.3 *M. marinum* [14]

Variability in susceptibility to therapeutically effective antimicrobial drugs or a sizable risk of developing acquired mutational resistance to one or more of these agents serves as the primary justifications for susceptibility testing of NTM. *M. marinum* infections have been successfully treated with rifampicin, doxycycline, minocycline, trimethoprim-sulfamethoxazole, and clarithromycin as single agents. For individuals whose treatment has failed after several months and whose cultures are still positive, testing for *M. marinum* may be an option. It has been recommended to test for *M. marinum* using a 5% OADC supplemented with CAMHB in a broth micro-dilution. For 7 days, incubation should take place between 28 and 30 °C. A macrolide should not be tested with CO_2 and should be incubated at 5 to 10% CO_2.

11.13.4 Rapidly Growing Mycobacteria [14]

For the purpose of assessing the susceptibility of rapidly proliferating mycobacteria, the Clinical and Laboratory Standards Institute (CLSI) document M24–A2 suggests using broth micro-dilution MIC determination. Agar tests and E-tests are not advised due to inconsistent findings. *M. fortuitum*, *M. peregrinum*, *M. chelonae*, and *M. abscessus* are among the recommended strains. Amikacin, cefoxitin, ciprofloxacin, clarithromycin, co-trimoxazole, doxycycline, imipenem, linezolid, moxifloxacin, and tobramycin are medications that have been shown effective against rapidely growing bacteria (RGM). After 72 h of incubation, micro-dilution trays should be tested at a temperature of 28 to 30 °C. The MIC should be recorded if there is growth in the growth control that is at least +2, else, re-incubate the tray and read on the fourth day. Repeat testing should be done if the growth on the fifth day is insufficient. In the event that RGM does not improve after 6 months of suitable antibiotic medication, it is required to confirm species identification and repeat susceptibility testing.

11.14 Conclusion

The detection and treatment of non-tubercular mycobacteria has become a major concern worldwide. Even though the genetic diversity of NTM has implications for therapeutic effectiveness and epidemiological dominance, the clinical significance of NTM has historically been overstated. It might be able to identify and differentiate NTM with the advancement of more genetic and molecular approaches.

References

1. Pennington KM, Vu A, Challener D, Rivera CG, Shweta FNU, Zeuli JD, Temesgen Z. Approach to the diagnosis and treatment of non-tuberculous mycobacterial disease. J Clin Tuberc Other Mycobact Dis. 2021;24:100244. https://doi.org/10.1016/j.jctube.2021.100244; PMID: 34036184; PMCID: PMC8135042.
2. Sharma SK, Upadhyay V. Epidemiology, diagnosis & treatment of non-tuberculous mycobacterial diseases. Indian J Med Res. 2020;152(3):185–226. https://doi.org/10.4103/ijmr.IJMR_902_20.
3. Ryu YJ, Koh WJ, Daley CL. Diagnosis and treatment of nontuberculous mycobacterial lung disease: clinicians' perspectives. Tuberc Respir Dis (Seoul). 2016;79(2):74–84. https://doi.org/10.4046/trd.2016.79.2.74; Epub 2016 Mar 31. PMID: 27066084; PMCID: PMC4823187.
4. Maurya AK, Nag VL, Kant S, Sharma A, Gadepalli RS, Kushwaha RA. Recent methods for diagnosis of nontuberculous mycobacteria infections: relevance in clinical practice. Biomed Biotechnol Res J. 2017;1:14–8.
5. Corcoran TE, Thomas KM, Myerburg MM, Muthukrishnan A, Weber L, Frizzell R, Pilewski JM. Absorptive clearance of DTPA as an aerosol-based biomarker in the cystic fibrosis airway. Eur Respir J. 2010;35(4):781–6. https://doi.org/10.1183/09031936.00059009; Epub 2009 Aug 28. PMID: 19717485; PMCID: PMC2867481.
6. Chakrabarti A, Sharma M, Dubey ML. Isolation rates of different mycobacterial species from Chandigarh (North India). Indian J Med Res. 1990;91:111–4; PMID: 2111799.
7. Bhattacharya J, Mohandas S, Goldman DL. Nontuberculous mycobacterial infections in children. Pediatr Rev. 2019;40(4):179–90. https://doi.org/10.1542/pir.2018-0131; PMID: 30936399.
8. Deveci HS, Kule M, Kule ZA, Habesoglu TE. Diagnostic challenges in cervical tuberculous lymphadenitis: a review. North Clin Istanb. 2016;3(2):150–5. https://doi.org/10.14744/nci.2016.20982; PMID: 28058405; PMCID: PMC5206468.
9. Griffith DE, Aksamit T, Brown-Elliott BA, Catanzaro A, Daley C, Gordin F, Holland SM, Horsburgh R, Huitt G, Iademarco MF, Iseman M, Olivier K, Ruoss S, von Reyn CF, Wallace RJ Jr, Winthrop K, ATS Mycobacterial Diseases Subcommittee; American Thoracic Society; Infectious Disease Society of America. An official ATS/IDSA statement: diagnosis, treatment, and prevention of nontuberculous mycobacterial diseases. Am J Respir Crit Care Med. 2007;175(4):367–416. https://doi.org/10.1164/rccm.200604-571ST; Erratum in: Am J Respir Crit Care Med 2007 Apr 1;175(7):744–5. Dosage error in article text.
10. Deshayes C, Angala SK, Marion E, Brandli I, Babonneau J, Preisser L, Eyangoh S, Delneste Y, Legras P, De Chastellier C, Stinear TP, Jackson M, Marsollier L. Regulation of mycolactone, the M. ulcerans toxin, depends on nutrient source. PLoS Negl Trop Dis. 2013;7(11):e2502. https://doi.org/10.1371/journal.pntd.0002502; PMID: 24244764; PMCID: PMC3828164.
11. Petrini BM. Marinum: ubiquitous agent of waterborne granulomatous skin infections. Eur J Clin Microbiol Infect Dis. 2006;25(10):609–13. https://doi.org/10.1007/s10096-006-0201-4.

12. Uslan DZ, Kowalski TJ, Wengenack NL, Virk A, Wilson JW. Skin and soft tissue infections due to rapidly growing mycobacteria: comparison of clinical features, treatment, and susceptibility. Arch Dermatol. 2006;142(10):1287–92. https://doi.org/10.1001/archderm.142.10.1287.
13. Xu X, Lao X, Zhang C, Cao C, Ding H, Pang Y, Ning Q, Zou J, Zang N, Hu D, Chen M. Chronic M. avium skin and soft tissue infection complicated with scalp osteomyelitis possibly secondary to anti-interferon-γ autoantibody formation. BMC Infect Dis. 2019;19(1):203. https://doi.org/10.1186/s12879-019-3771-3; PMID: 30819109; PMCID: PMC6396482.
14. CLSI. Susceptibility testing of mycobacteria, nocardiae, and other aerobic actinomycetes; approved standard-second edition. (n.d.). www.clsi.org.
15. BD_BACTEC-MGIT_PE_TR. (n.d.) Siddiqi SH, Rüsch-Gerdes S. 2006. For BACTEC™ MGIT 960™ TB System (Also applicable for Manual MGIT) Mycobacteria Growth Indicator Tube (MGIT) Culture and Drug Susceptibility Demonstration Projects.
16. Thorpe TC, Wilson ML, Turner JE, Diguiseppi JL, Willert M, Mirrett S, Barth Reller L. BacT/Alert: an automated colorimetric microbial detection system. J Clin Microbiol. 1990;28(7):1608.
17. LaBombardi VJ. Comparison of the ESP and BACTEC systems for testing susceptibilities of M. tuberculosis complex isolates to pyrazinamide. J Clin Microbiol. 2002;40(6):2238–9. https://doi.org/10.1128/JCM.40.6.2238-2239.2002; PMID: 12037096; PMCID: PMC130663.
18. Chu HQ, Li B, Zhao L, Huang DD, Zhang ZM, Xu JF, Zhang JB, Gui T, Xu LY, Sun XW. Chest imaging comparison between non-tuberculous and tuberculosis mycobacteria in sputum acid fast bacilli smear-positive patients. Eur Rev Med Pharmacol Sci. 2015;19(13):2429–39. https://doi.org/10.1183/13993003.congress-2015.pa2674.
19. Jeong YJ, Lee KS, Koh WJ, Han J, Kim TS, Kwon OJ. Nontuberculous mycobacterial pulmonary infection in immunocompetent patients: comparison of thin-section CT and histopathologic findings. Radiology. 2004;231(3):880–6. https://doi.org/10.1148/radiol.2313030833; Epub 2004 Apr 29. PMID: 15118112.
20. Baris A, Bayraktar B. Identification of the mycobacterial strains isolated from clinical specimens using hsp65 PCR-RFLP method. Sisli Etfal Hastan Tip Bul. 2020;54(3):364–70. https://doi.org/10.14744/SEMB.2019.66587; PMID: 33312037; PMCID: PMC7729718.
21. Quan TP, Bawa Z, Foster D, Walker T, Del Ojo Elias C, Rathod P, Iqbal Z, Bradley P, Mowbray J, Walker AS, Crook DW, Wyllie DH, TEA P, Smith EG, MMM Informatics Group. Evaluation of whole-genome sequencing for mycobacterial species identification and drug susceptibility testing in a clinical setting: a large-scale prospective assessment of performance against line probe assays and phenotyping. J Clin Microbiol. 2018;56(2):e01480–17. https://doi.org/10.1128/JCM.01480-17; PMID: 29167290; PMCID: PMC5786738.
22. Chae H, Han SJ, Kim SY, Ki CS, Huh HJ, Yong D, Koh WJ, Shin SJ. Development of a one-step multiplex PCR assay for differential detection of major M. species. J Clin Microbiol. 2017;55:2736–51. https://doi.org/10.1128/JCM.00549-17; PMID: 28659320.
23. Somoskövi A, Hotaling JE, Fitzgerald M, Jonas V, Stasik D, Parsons LM, Salfinger M. False-positive results for M. celatum with the AccuProbe M. tuberculosis complex assay. J Clin Microbiol. 2000;38(7):2743–5. https://doi.org/10.1128/JCM.38.7.2743-2745.2000; PMID: 10878076; PMCID: PMC87016.

Chapter 12
Sensor and Nanotechnology-Based Diagnostics in the Field of Mycobacteriology

Mohanraj Thangarasu, Shunmuga Nathan Shunmuga Nainar, Shakkthivel Piraman, and Vasantha Vairathevar Sivasamy

Abstract For many decades, tuberculosis (TB) has remained a serious global health issue since it is causing incalculable mortality by a single infectious agent until the spread of the coronavirus (COVID-19) pandemic and ranking ahead of HIV/AIDS. According to World Health Organization (WHO) 2020, the COVID pandemic has reduced TB infection rates by around 18% worldwide from 7.1 million to 5.8 million, but the mortality rate of TB-infected people increased from 1.2 million to 1.3 million. This epidemic, often called the "Great White Plague," is projected to kill more people in the future decades. Hence a rapid, sensitive and low-cost technique is very much needed for early detection of *M. tuberculosis* for combating this severe illness. Even today, tuberculosis diagnosis relies on direct smear microscopy, solid culture, chest radiography and tuberculin-skin-testing tools that have poor performance and requires special infrastructure. These limitations led to the development of nanomaterial-based sensor platforms (NSPs) that are rapid, affordable, highly sensitive and selective and thus favorable for real-time analysis and portability. The simple operation techniques and easily tunable properties of electrochemical and optical biosensors can facilitate the fabrication of portable miniaturized sensor devices. This book chapter will cover the detailed review, opportunities based on the performance and challenges arising in the fabrication along with the future recommendations of the biosensors reported for the detection of mycobacterium.

Keywords Biosensors · Nanomaterials · Mycobacteriology · MTBC and NTM · Advantages and disadvantages · Challenges

---

M. Thangarasu · S. N. Shunmuga Nainar · V. Vairathevar Sivasamy (✉)
Department of Natural Products Chemistry, School of Chemistry, Madurai Kamaraj University, Madurai, India, Tamil Nadu

S. Piraman
Department of Nanoscience and Technology, Alagappa University, Karaikudi, Tamil Nadu, India

A. Singh, D. Sharma (eds.), *Diagnosis of Mycobacterium*,
https://doi.org/10.1007/978-981-99-5624-1_12

12.1 Introduction

Tuberculosis (TB) causing *Mycobacterium tuberculosis* complex (MTBC) is one of the top ten causes of death in humans and animals. Unfortunately, the mortality rate of TB in humans was found to be higher than HIV and AIDS in the year 2018 [1]. WHO estimated that there are about 10.4 million cases and 1.8 million deaths from TB, out of which 90% are adults [2]. India reported a sharp 19% rise in tuberculosis cases in 2021 over the previous years [3]. According to WHO modeling projections, the number of people developing TB and dying from the disease could be much higher in 2021 and 2022 and becoming a very serious threat to human health and the economy. In 2020, India, Indonesia, South Africa and the Philippines were the four countries that accounted for 44% of global TB cases [2].

Despite several drugs available on the market for TB infection, TB has re-emerged as a serious health concern and is untreatable due to the outbreak of multidrug-resistant (MDR) and extensively drug-resistant (XDR) strains [4]. WHO reported that pulmonary rifampicin-resistant tuberculosis (RR-TB) was a case in which *M. tuberculosis* became resistant to rifampicin, one of the strongest available TB medicines [5]. There are several conventional techniques available for the diagnosis of TB infection which are listed in Table 12.1 [6–8].

Even though most of the conventional techniques listed are sensitive and selective, they suffer from serious disadvantages like being expensive, time consuming, requirement of technical expertise and special infrastructures. In addition, most of these techniques suffer from poor sensitivity and less reproducibility. Therefore, a simple, accurate, reliable, cost-effective and a technique capable of discriminating between drug-sensitive and drug-resistant bacilli, MTBC and NTM (Non-tuberculosis mycobacteria) in the early stage is of an urgent need.

To overcome the limitations with conventional diagnostic methods, identification and implementation of various TB biomarkers have been investigated. Detecting these biomolecules ultimately reveals the prognosis of TB. Biomarkers of TB include host *M. tuberculosis*-specific antigens (ESAT-6, CFP-10, 28-kDa MPT64, Ag85 complexes, Hsp), LAM (Lipoarabinomannan) components of *M. tuberculosis* cell wall and conservative insertion sequence elements (IS*6110*), cytokines secreted by immune cells (IFN-gamma) and volatile organic biomarkers (VOBs) [9].

12.2 Electrochemical Biosensors for the Diagnosis of MTBC

Biosensor is an analytical device which measures the chemical/biological substances by using biomolecules as a sensing element. The biosensor consists of a biorecognition element connected to a suitable transducer, that converts the observed chemical or physical changes into a measurable signal as shown in Fig. 12.1. [10]. Among the other types of biosensors, electrochemical sensors are superior in aspects such as sensitivity, specificity, cost effectiveness, rapidness and portability [8]. This

Table 12.1 List of conventional methods available for the diagnostics of MTBC with their advantages and disadvantages

Conventional tests	Advantages	Disadvantages
Smear microscopy	• Rapid, simple and inexpensive • Higher specificity	• Low sensitive and unsuitable for watery specimen • Cannot differentiate between various mycobacteria
Solid culture microscopy	• Definitive diagnosis for all samples • Isolate available for strain typing	• Delayed diagnosis (3 to 8 weeks) and cross contamination • Requires skilled technician and special infrastructure
Liquid culture microscopy	• Rapid, higher sensitivity and specificity than solid culture • Drug susceptibility testing (DST) to all drug is possible • Growth supplement to enhance growth	• Expensive and requires BSLIII lab • Higher rate of contamination • Growth needs to be confirmed by acid fast bacilli (AFB), smear, immunochromatographic test (ICT) • Not suitable for blood and urine samples
Tuberculin skin test	• Inexpensive and widely available • Minimal infrastructure	• Low specificity and cross reactivity • False positive results and manual errors
IFN-γ release assay (IGRA)	• One-time test • Higher sensitivity and specificity • BCG vaccination does not interfere	• Costly and requires special infrastructure • Cannot distinguish active TB from LTBI • Poor reproducibility
Real-time polymerase chain reaction (PCR)-based tests	• Rapid and highly sensitive to MDR-TB/XDR-TB strains	• Technically challenging, technical expertise • High expenditure (maintenance and service protocols)
Xpert MTB/RIF assay	• Rapid and highly sensitive • Identification and DST are possible	• Costly and unsuitable for blood samples • Amplification inhibitors may give false results
Line probe assay (LPA)	• Rapid, user friendly and sensitive • Simultaneous detection of MTBC and drug-resistant strains	• Expensive and technical expertise • Recommended for smear-positive samples • Complex multistep process

(continued)

Table 12.1 (continued)

Conventional tests	Advantages	Disadvantages
Loop-based nucleic acid amplification (LAMP)	• Simple and rapid • Minimal infrastructure • Commonly used for extra-pulmonary tuberculosis (EPTB)	• Low sensitivity and specificity • Multistep process • Not suitable for drug-resistant TB
Lipoarabinomannan (LAM) test	• Rapid, cost-effective and user friendly • Highly sensitive and fast response	• Low-specificity • Digital reader is required for data interpretation
Enzyme linked immunosorbent assay (ELISA)	• Rapid, affordable and user-friendly • Higher specificity and sensitivity	• Sensitivity and specificity depends upon the antigens used • Not reliable • Cross-reactivity of the capture probe with target

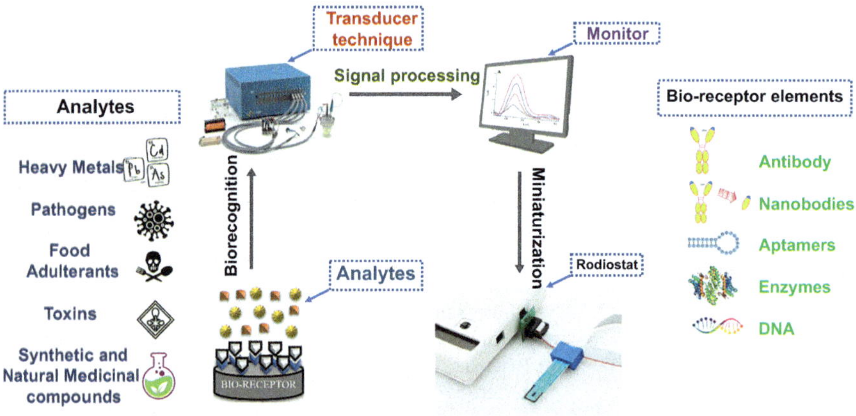

Fig. 12.1 The basic working principle and key components of a biosensor

section will focus on electrochemical biosensors reported for TB diagnosis (Fig. 12.1).

12.2.1 Nanotechnology-Based Electrochemical Biosensors

Nanomaterials are composed of nanoparticles which are less than 100 nm, at least in one dimension. In general, nanomaterials have unique characteristics like nanoscale size, electrical conductivity, increased surface to volume ratio, chemical reactivity, mechanical strength and electrocatalytic properties [11]. The development of novel and multifunctional nanomaterials require interdisciplinary efforts of the scientific community and the special properties of these materials are exploited in the development of the biosensors to improve the sensitivity and selectivity [12]. Various unique sensor platforms fabricated using nanomaterials, polymers and self-assembled monolayers (SAMs) have been reported. Among these platforms, nano-structured metal/metal oxide nanoparticles and carbonaceous materials are advantageous in sensing applications due to their nanoscale size with increased surface area, excellent electrical and optical properties, higher chemical and thermal stability [13]. Electrochemical biosensors have been fabricated by measuring some electrochemical parameters like charge accumulation, current, impedance and voltage [8].

Amperometry is one of the suitable techniques in the case of electrochemical techniques used in the biosensors due to its simple electronic circuit system and being miniaturizable. It is a highly sensitive technique that directly measures the current change through a constant potential [14]. Previous investigations suggested that different mycobacterial strains produce distinctive volatile organic biomarkers (VOBs) that can be exploited as a potential method for the identification of latent TB infection [15]. *M. tuberculosis* and *M. bovis* cultures were observed to emit four major volatile organic compounds (VOCs), including methyl phenylacetate, methyl-anisate, methyl nicotinate, and o-phenylanisole [16]. Screening these VOBs could pave way for the development of a more accurate and user-friendly indirect TB diagnosis. Dhiman Bhattacharyya et al. proposed an amperometric sensor by cobalt-functionalized TiO_2 nanotube arrays (Co-TNA) for the detection of VOBs. When VOBs were supplied at the start of each bias, the current increases and in the absence of the biomarker, the current gets reduced quickly. The limit of detection (LOD) was determined to be $\sim$18 pg/μL [17]. In a more similar work, York. R. Smith et al. utilized in situ-cobalt functionalized TiO_2 NTAs (iCo-TNA) for an enhanced sensitivity, reproducibility and response time [18]. Navin Kumar Mogha et al. developed a highly sensitive chrono-amperometric biosensor based on Au nanoparticles (AuNPs) modified reduced graphene oxide nanoribbons (rGONRs) for the detection of *M. tuberculosis* oligonucleotide sequence and achieved an LOD of 0.1 fM [19].

Several voltammetry techniques such as differential pulse voltammetry (DPV), square wave voltammetry (SWV), cyclic voltammetry (CV) and stripping voltammetry were well established. Among the various voltammetric techniques, DPV and SWV are commonly preferred for the detection of *M. tuberculosis* strains due to their high sensitivity towards TB biomarker. The DPV-based electrochemical biosensors reports are summarized as below. Iron oxide (IO) nanoparticles have unique properties like high adsorption capacity, low mass transfer resistance, exceptional water solubility, fast electron transfer capability and biocompatibility [20]. Due to their superparamagnetic nature, these magnetite nanoparticles were widely applied in electrochemical immunosensor to preconcentrate and separate biomolecules [21]. Lemma Teshome Tufa et al. fabricated an immunosensor based on a nanotriplex consisting of graphene quantum dot-coated Fe_3O_4@Ag core-shell nanostructure for sensing the culture filtrate protein-10 (CFP-10) antigen. As a label for signal amplification, AuNPs-conjugated CFP-10 antibodies are used. The LOD was found to be 0.33 ng/mL [22]. Dan Gou et al. developed an ultrasensitive electrochemical immunoassay for the determination of the *M. tuberculosis*-specific secretory protein MPT64 using graphene oxide-Fe_3O_4-Pt nanocomposite as a signal reporter. This sensor platform was applied to serum samples spiked with MPT64 protein and a detection limit of 0.34 fg/mL was obtained [23]. Ningning Li et al. developed a rapid, sensitive and highly selective aptasensor using AuNP- and horseradish peroxidase (HRP)-functionalized Prussian blue nanoparticles as signalling probes and APT-I aptamer linked Au electrode as platform. With wide linear range, this aptasensor reached the detection limit of 21 fg/mL. The results show that the aptasensor is reliable for the detection of MPT64 in the early stage.

The novel functionalized metal organic framework (P-MOF) and MXene-based nanomaterials are having effective applications in the field of electrochemical biosensors. Metal-organic salts provoked a talk among the scientific world since its discovery by Alfred Werner during his research on coordination complexes [24]. MOF-based platforms have proven to be a potential sensor for the detection of biomolecules and pathogens [25]. Yuhan Chen et al. developed a novel sandwich-type electrochemical aptasensor for the detection of MPT64 genes of *M. tuberculosis*. The sensing surface comprises a polyethyleneimine (PEI) functionalized metal organic framework (P-MOF) and the electroactive tetraoctylammonium bromide (TOAB) was used to provoke the inherent electroactivity of the tracer label. The aptasensor yielded a detection limit of about 0.33 fg/mL [26]. Ningning Li et al. used an amino modified UiO-66 MOF to develop an ultrasensitive voltammetric aptasensor for the detection of MPT64 genes of *M. tuberculosis.* Two aptamers specific for the MPT64 genes were immobilized on the modified gold electrode to achieve a detection limit of 10 fg/mL [27]. Yishi Li et al. developed an electrochemical aptasensor for the detection of MPT64 antigen using an anthraquinone derivative tagged with a mesoporous carbon material as signalling material. The aptasensor which used Ce-MOF as a sensing platform yielded an LOD value of 67.6 fg/mL [28]. Linlin Li et al. developed an aptasensor using reduced graphene oxide/Fe-MOF composite (rGO/MOF) to detect early secretory antigenic target-6 (ESAT-6) of *M. tuberculosis.* The addition of the electroactive molecule, toluidine blue facilitated the electron transfer and thus achieved a detection limit as low as 3.3×10^{-5} ng/mL in the spiked human serum [29]. Jiaojiao Xie et al. reported a label-free aptasensor utilizing bimetallic Zr- and Ce-based MOF, nitrogen-doped graphene tagged with toluidine blue nanohybrid for the detection of ESAT-6. The proposed sensor showed good reproducibility with a limit of detection of 12 fg/mL and has exceptional capability to become a diagnostic tool for tuberculosis in clinical practice [30].

MXene, a rapidly growing new family of 2D material has captivated the minds of researchers due to its unique combination of properties such as enormous surface area, high electrical conductivity, hydrophilic nature, ease of functionalization (-O, -OH, -F), compositional variability etc. Kobra Salimiyan Rizi et al. proposed a novel genosensor based on a highly conductive $Ti_3C_2T_x$ MXene polypyrrole nanocomposite for detecting a specific DNA target of the *M. tuberculosis* genomic IS*6110* sequence. With a linear range of 100 fM to 25 nM, this biosensor has accurately detected *M. tuberculosis* in human sputum samples [31]. Rizi et al. also developed a rapid, low-cost and PCR-free DNA nanobiosensor with high sensitivity and specificity for MTBC detection using a composite consist of multi-walled carbon nanotubes (MWCNTs), polypyrrole (PPy) and potassium-substituted hydroxyapatite (KHAp) nanoparticles. The obtained LOD and limit of quantification (LOQ) values are 50.3 and 167.5 pM, respectively [32]. Lijuan Bai et al. designed an ultrasensitive electrochemical aptasensor for the detection of the MPT64 antigen of *M. tuberculosis* in human serum samples. Here, AuNPs-decorated fullerene-polyaniline nanocomposite (GNPs-C_{60}-PAn) was labelled with an aptamer which was used as a tracer label. The linear range of this sensor was found from 0.02 to 1000 pg/mL [33]. Apart from that, more reports on DPV-based DNA biosensors [34–42], immunosensors [43, 44], genosensors [45] and aptasensors [46–53] were explored widely for the timely detection of MTBC in humans.

Among the SWV-based biosensor reports, Anna Miodek et al. demonstrated that the association of PPy, MWCNTs and Polyamidoamine (PAMAM) G4 nanomaterials by electrochemical patterning would be an efficient way to detect PCR samples and discriminate the rpoB genes of DR-TB. The detection limit obtained was found to be 0.3 fM [54]. Yun Wang et al. used square wave anodic stripping voltammetry (SWASV) to detect IFN-γ, a host TB specific cytokine, by labelling MNPs functionalized antibody conjugated with AuNPs, secondary antibodies and CdS NPs [55]. Marwa Haddaoui et al. reported an electrochemical biosensor using naphthoquinone redox group-labelled DNA probes. This sensor detected and distinguished the rpoB genome of *M. tuberculosis* using SNPs in real samples [56]. Some of the other SWV-based electrochemical DNA biosensors [57–60], protein nucleic acid (PNA) biosensor [61] and aptasensors [62, 63] were reported for the early diagnosis of tuberculosis disease in TB-infected patients.

The majority of antigen and antibody molecules are electrochemically inert. Hence, electrochemical impedance spectroscopy (EIS) has been used directly to identify these immunospecies by measuring the changes in R_{ct} value. EIS is a sensitive, selective, non-destructive and less expensive technique than the conventional assay methods [64]. Subash. C. B. Gopinath et al. developed an impedimetric biosensor by using a nanogapped silver-supported dielectric surface with AuNPs-coated ZnO thin film for the detection of AuNPs conjugated with 16 kDa Hsp of *M. tuberculosis* [65]. Zhang et al. fabricated an impedimetric sensor using zirconia cross-linked $Ti_3C_2T_x$ as sensing probe and AuNPs-PNA as a sensing platform as shown in Fig. 12.2., focusing on the H37Ra genome of 16 s rDNA region of *M. tuberculosis* as a biomarker. Highly electrically conductive $Ti_3C_2T_x$ was used to amplify the signal, which avoids the need of additional redox probes [66]. Several other reports on impedance-based DNA biosensors [67–70], and genosensor [71] also have been widely explored for sensing various strains of *M. tuberculosis*.

12.2.2 Non-Nanomaterial-Based Electrochemical Biosensors

Khalil Abnous et al. designed an electrochemical-based triple-helix molecular switch (THMS) with a combined amperometry sensor for the rapid (50-min) detection of IFN-γ. Under optimized conditions, this sensor can detect 3 pg/mL in spiked serum samples [72]. Sasinee Bunyarataphan et al. developed a ratiometric electrochemical biosensor with internal control integrated via duplex PCR (REC-ICdPCR) and dual signal readout. Using MB as a reporter and ferrocene as internal control, this sensor achieved an LOD of 1.26 fM. This biosensor was applied to the detection of *M. tuberculosis* in practical samples [73]. Other voltammetric-based DNA biosensors [74, 75], PNA biosensor [76], aptasensors [77–80] and immunosensor [81] have been investigated for addressing tuberculosis disease. As the earliest report on the electrochemical biosensor for *M. tuberculosis*, Joseph Wang et al. opted chronopotentiometry technique to detect *M. tuberculosis* using short oligonucleotide sequence [82] (Fig. 12.2).

Marzhan Sypabekova et al. demonstrated an aptamer-based impedimetric biosensor for the detection of MPT64 with a thiolated aptamer and achieved an LOD of 81 pM for a spiked human serum sample [83]. Kyoungin Min et al. used RNA and DNA aptamers to create a simple and direct electrochemical sensor for detecting IFN-γ. The RNA-based aptasensor showed a low detection limit of 100 fM, whereas the DNA-based aptasensor detected 1 pM of IFN-γ [84]. Several other EIS-based electrochemical DNA biosensors [85–87], aptasensor [88] and immunosensors [89, 90] were also widely explored for screening TB-infected patients with good sensitivity and LOD. Saengdee et al. constructed a label-free, ion-selective FET-based immunosensor for the detection of Ag85B. The results were demonstrated as a linear increase in the gate potential change with respect to the concentration of Ag85B and achieved LOD of 0.12 mg/mL [91]. Some of the other electrochemical sensors based on field effect transistor (FET) were reported for the timely detection of IFN-γ [92] and Ag85B of *M. tuberculosis* [93] (Table 12.2).

12.3 Optical Biosensors for the Detection of MTBC

Optical sensors gather information about a system by the measurement of photons rather than measuring electrons in the case of electrochemical sensors. These measurements depend on the absorbance/scattering of lights and fluorescence emission that occurs either in the ultraviolet (UV), visible or near-infrared region [101]. This section summarizes the optical biosensors reported for the detection of MTBC.

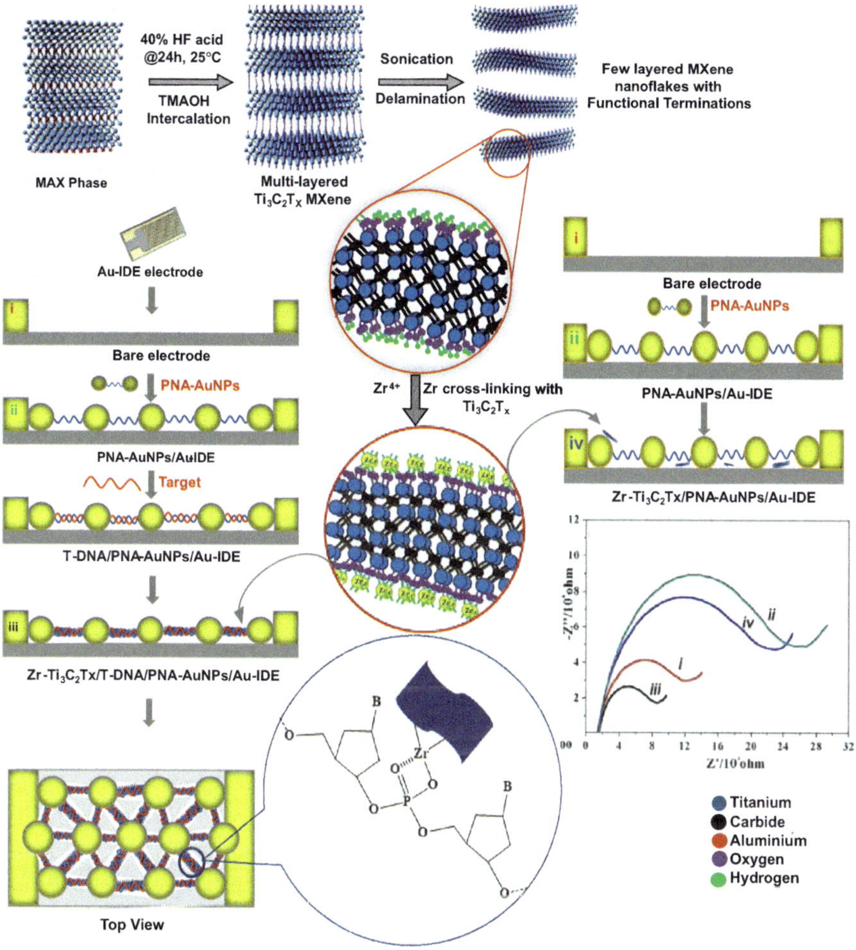

Fig. 12.2 Construction of an impedimetric biosensor for the detection of *M. tuberculosis* (H37Ra) using a $Ti_3C_2T_x$ MXene-based sensing probe [66]

12.3.1 Nanotechnology-Based Optical Biosensors

Recently, nanotechnology is being used widely in health monitoring, and treatment of deadly diseases. The nanomaterial-based biosensors with desirable properties and functions will be designed according to the interactions of the nanomaterials with the biomolecules [102]. Despite helping in the aid of an early detection of the disease, nanomaterial-based biosensors play a major role in the drug development and in biomedical research by detecting the biomolecules specific to a particular disease such as nucleic acids, proteins, pathogens etc. [103]. Nanotechnology paves a way to the development of biodiagnosis by reducing the instrument size and cost, remains user friendly and allows rapid detection and also provides high selectivity

Table 12.2 Electrochemical biosensors reported for the detection of *Mycobacterium tuberculosis*

A. Nanomaterial-based electrochemical biosensors

(i) DPV-based electrochemical biosensors

Sensing platform	Biomarker	LOD	Linear range	Reference
ssDNA-ZrO_2/Au	DNA of *M. tb*	0.065 ng/μL	641–0.065 ng/μL	[42]
Dual labelled AuNPs/gDNA/p-DNA/AuNPs/ITO	DNA of *M. tb*	1.25 ng/mL	1.25–50 ng/mL	[40]
MNPs-capture probe/t-DNA/reporter probe-AuNPs	IS*6110*	0.01 ng/μL	–	[41]
Au-SPA/anti-LAM/LAM/Ag-SPCE	Anti-LAM	5.3 ng/mL	15.6–1000 ng/mL	[44]
MHA/SAM-(GR-IFN-γ)/Au	IFN-γ	0.065 pM	0.1–0.7 pM	[53]
DNA/rGO-AuNPs/GCE	IS*6110*	–	1 Fm–1 nM	[39]
SA-BSA-AuNPs/Biotin-PEG/SPCE	DNA of *M. tb*	1 CFU/aliquot	–	[38]
BSA/Ab1/AuNPs/PDDA/ITO	IFN-γ	0.048 pg/mL	0.1–10,000 pg/mL	[94]
MCH/MBA-I/Au	MPT64	20 fg/mL	0.2–1000 pg/mL	[33]
AuNPs/PANI/rGO	IS*6110*	50 fM	0.1 pM–10 nM	[36]
p-DNA/NHS/EDC/GP/PANi/SPCE	DNA of *M. tb*	7.853 $\times 10^{-7}$ M	10^{-6}–10^{-9} M	[35]
Aptamer/SA/PEDOT-CNT/FTO	MPT64	0.5 ± 0.2 fg/mL	10^{7}–10^{3} fg/mL	[52]
Probe ssPNA/NH_2-GO/CdSQDs/SPCE	PNA of M. tb	8.95 $\times 10^{-11}$ M	10^{-11}–10^{-7} M	[95]
Biotinylated Aptamer/SA/CHIT-IO-GR/FTO	MPT64	0.9 fg/mL	10^{8}–1.0 fg/mL	[51]
BSA/anti-MPT64/His-tagged PG/Au	MPT64	0.34 fg/mL	5 fg/mL–1 ng/mL	[23]
BSA/Ab1/Fe_3O_4@Ag/GQD/GCE	CFP-10	0.33 ng/mL	0.005–500 μg/mL	[22]
MCH/aptamer/Au	MPT64	10 fg/mL	0.02–1000 pg/mL	[27]
BSA/CapAb/GR-PANi/SPCE	CFP-10	15 ng/mL	20–100 ng/mL	[43]
BSA/p-DNA/AuNPs/FC_{60}/GCE	IS*6110*	0.33 fg/mL	1 fM–10 nM	[37]
β-ME/Thiol-Ap/AuNPs/SPE	HspX	10 pg	0.01–100 ng	[50]
A-DDAH/AuNPs/SFG/GCE	IFN-γ	19 fg/mL	0.1–0.5 pg/mL	[47]
N_3-DNA/CP-DNA/CFP-10 Ap/DBCO-DNA/AU	CFP-10	10 pg/mL	10^{-2}–10^{2} pg/mL	[48]
MCH/Apt-I/Au	MPT64	21 fg/mL	1–10,000 fg/mL	[49]
BSA/MAA-II/Au@Pt/P-MOF/Au	MPT64	0.33 fg/mL	1 fg/mL–1 ng/mL	[26]

p-DNA/CTAB/NCC/MPA-Fe_2O_3/SPCE	DNA of *M. tb*	7.96×10^{-13} M	10^{-6}–10^{-12} M	[34]
ssDNA/BSA/GO-CHI/ITO	IS*6110*	3.4 pM	7.86–94.3 pM	[45]
BSA/CapApt/GR/PANI/SPGE	CFP-10-ESAT-6	1.5 ng/mL	5–500 ng/mL	[46]
p-DNA/KHAp/PPY/MWCNTs/GCE	IS*6110*	50.3 pM	100 pM–100 nM	[32]
ssDNA/PPY/MXene/GCE	IS*6110*	11.24 fM	100 fM–25 nM	[31]
Probe DNA/CS/PEI/Au NPs/GCE	DNA of *M. tb*	0.03 fM	10^{-14}–10^{-10} M	[96]
BSA/bio-CP/SA/dep-Au/Ce-MOFs/GCE	MPT64	67.6 fg/mL	10^2 fg/mL–10 ng/mL	[28]
(ii) SWV-based electrochemical biosensors				
PNA/Fe_3O_4-GOPS/ITO	DNA of *M. tb*	0.1 fM	0.1–50 fM	[61]
CdSe QDs–SA/MCH/probe-DNA/AuNPs/GCE	DNA of *M. tb*	8.7 fM	10^{-14}–10^{-9} M	[58]
Ethanolamine/ssDNA/Fc/PAMAM/PPY/MWCNT/Au	rpoB	0.3 fM	1 fM–10 pM	[54]
Probe/Au/PV-QD/GCE	DNA of *M. tb*	0.2 fM	0.5–500 fM	[57]
Probe-DNA/AuNTsA	DNA of *M. tb*	0.05 ng/μL	0.01–00 ng/μL	[59]
dAb-AuNP-CdS-NP/cAb/AuNP/MNP-PANi/SPCE	IFN-γ	0.4 pg/mL	0.4–40 pg/mL	[55]
p-DNA/spaNQ/PAMAM G4/PPy/Fe_3O_4 NPs/AuSPCE	rpoB	4 fM	1 fM–0.1 pM	[56]
MB-IFN-γ Ap/AuNPs-coated SiNWs	IFN-γ	0.14 ng/mL	0.2–1 ng/mL	[62]
AuNPs-conjugated Ap/PAMAM/MoS_2/GCE	IFN-γ	2 fg/mL	0.01–10^3 fg/mL	[63]
ssDNA/Fc/PAMAM G2/EDA/nw-PPY/Au	rpoB	0.36 aM	1 aM–100 fM	[60]
(iii) EIS-based electrochemical biosensors				
ssDNA-NanoZrO_2-CNT/ITO	DNA of *M. tb*	0.01 nM	10^{-2}–10^{-8} mM	[67]
Probe-DNA/f-Fe_3O_4NPs/MBA/Au	DNA of *M. tb*	6 ng/μL	–	[71]
Anti-16 kDa HSP antibody/Au sputtered ZnO/Si wafer	16 kDa HSP	100 fM	100 fM–1 nM	[65]
p-DNA/AuNRs/3D-GR/3D-Ni	IS*6110*	10 fM	10 fM–0.1 μM	[69]
acpaPNA/CHO-CGI/ePAD	DNA of *M. tb*	1.24 nM	2–200 nM	[68]

(continued)

Table 12.2 (continued)

PNA-AuNPs/AuSPCE	H37Ra	20 CFU/mL	10^2–10^8 CFU/mL	[66]
ssDNA/AgNPs/LSG-NF/IDE	IS*6110*	1 fM	1 fM–1 nM	[70]
(iv) CV-based electrochemical biosensors				
ssDNS/Fc/EDA/MWCNTs/Au	H37Rv	0.7 fM	0.1 fM–1 pM	[97]
BSA/EBA/Pt@AuTB/P-MOF/rGO/GCE	ESAT-6	3.3 ×10^{-5} ng/mL	10^{-4}–10^2 ng/mL	[29]
BSA/EBA/NG@Zr-MOF-on-Ce-MOF@Tb/GCE	ESAT-6	12 fg/mL	10^2 fg/mL–10 ng/mL	[30]
(v) Amperometry-based electrochemical biosensors				
iCo-TiO_2NTAs/Ti	VOCs of *M. tb*	–	275–360 ppm	[18]
Co-TiO_2NTAs/Ti	VOCs of *M. tb*	0.018 ppm	0.01–10 and 10–1000 ppm	[17]
ssDNA/AuNPs/rGONRs/ITO	DNA of *M. tb*	0.1 fM	0.1 fM–10^{-6} M	[19]
(vi) FET-based biosensors				
IFN-γ aptamer/GR/PDMS	IFN-γ	83 pM	–	[92]
Anti-Ag85B-Ab/CA/APTES/Silicon nitride-ISFET	Ag85B	0.12 μg/mL	0.12–1 μg/mL	[91]
SiNW-FET/SOI	Ag85B	0.01 fg/mL	1 fg–100 μg/mL	[93]
B. Other electrochemical biosensors				
(i) Voltammetry-based electrochemical biosensors				
21-mer PNA probe/PPY-PVS/ITO	16s–23s rRNA	0.1 fM	–	[76]
MCH/Thiolated-Aptamer/Au	IFN-γ	0.06 nM	0.06–10 nM	[80]
bioRb8-106/BSA/SA/SPCE & mAb/BSA/IgG AP/SPE	Ag360, Ag231	0.001 μg/mL	0.005–0.1 μg/mL	[81]
IFN-γ/MB/CD4 Ab/PEG-DA/glass electrode	IFN-γ	60 pM	60 pM–9 nM	[79]
(DNA-SH, PNA-SH or Sp32-SH)/SAM(JUGSH)/Au	DNA of *M. tb*	300 pM	–	[75]
Anti-CD4 or anti-CD14 A/MB-IFN-γ/TNF-γ Ap/Au	IFN-γ	0.06 nM	0.6–10 nM	[78]
MCH/MB-Ap/Au	IFN-γ	1.3 ng/mL	1–500 ng/mL	[98]
P-Fc-A-IFN-γ/Au or P-Fc-A-IFN-γ-P-MB-A-Lys/Au	IFN-γ	1.14 ×10^{-3} nM	0.01–10 nM	[77]

Biotin-DNA/avidin/PGA/PGE	H37Ra	1.3 nM	1.5–12.5 nM	[99]
IFN-γ/MB/THMS/MCH/Ap/STP/SPGE	IFN-γ	3 pg/mL	10–1500 pg/mL	[72]
ssDNA/Au/Si-substrate	16s rDNA	40 pM	100 pM–1 μM	[74]
MCH/Mtb-MB & IC-FC/SPGE	Mtb ICdPCR	1.26 fM	10 fM–10 nM	[73]
(ii) EIS-based electrochemical biosensors				
MD-2/SAM (Cys or acetyl-Cys)/Au	IFN-γ	0.02 fg/mL	0–12 pg/mL	[90]
Ab/acetylcysteine/PcAu	IFN-γ	–	10^{-18}–10^{-9} M	[89]
IFN-γ/MCE/Ap/Pentanethiol/Au	IFN-γ	100 fM, 1 pM	–	[84]
Target DNA/MCH/Thiolated P2 DNA/Au	*rpoB*	–	–	[87]
IFN-γ/HDT-MCH/ACP/Au-IDE	IFN-γ	11.56 pM	22.2 pM–0.1 nM	[88]
Ethanolamine/ssDNA/Fc-ac-OMPA/Au	rpoB	0.2 fM	1 fM–100 pM	[86]
EG-Aptamer/aptamer linker/Au	MPT64	81 pM	1–50 nM	[83]
(iii) Amperometry-based electrochemical biosensors				
SS DNA/CPE	DNA of MTB	40 ng/mL	1–4 μg/mL	[82]
SA-HRP/biotin-Ab/BSA/cAb/PS	IFN-γ	126.75 pg/mL	15–1000 pg/mL	[100]
Aptamer/NHS/EDC-ABA/SPCE	CFP-10	1.05 ng/mL	2.5–250 ng/mL	[85]

and sensitivity [104, 105]. Hence, NSPs have proven to be a promising technique in the eradication of the spread of this deadly disease by timely detection of MTBC.

The surface plasmon resonance (SPR) technique was the most exploited technique that uses nanomaterials for the detection of MTBC. The signal of this optical technique is being enhanced mostly by the metal nanoparticles [106]. Pei et al. recently developed an ultrasensitive SPR-based sensor for *M. tuberculosis*-specific DNA. The detection was based on the counting of localized AuNPs by the dark-field microscopic imaging and, rolling-circle amplification (RCA). The group obtained a good detection limit of 10 fM of MTB-DNA [107]. Priego et al. developed a cost-efficient POC detection technique for the direct and non-invasive detection of TB in human urine. This technique was designed to detect the LAM, a liposaccharide found in the cell walls of the mycobacterium. The optimization of this immunoassay provided a detection limit of 475 pg/mL of LAM in undiluted human urine [108]. Prabowo et al. fabricated a graphene-based SPR sensor for the detection of DNA strains of *M. tuberculosis*. The sensing probe comprises single strand DNA (ss-DNA) which was covalently bound with the gold nano urchins (GNu). This ss-DNA/GNu was used to achieve a detection limit of 28 fM of the complementary ss-DNA in the salt buffer [109]. Silvestri et al. developed a peptide nucleic acid-based label-free biosensor for the detection of *M. tuberculosis* DNA through an azimuthally controlled grating-coupled SPR (GC-SPR) technique. The GC-SPR sensor obtained a detection limit of 0.26 pM which was found to be one magnitude lower than that of fluorescence method (8.9 pM) [110]. Other SPR-based DNA biosensors [111–116] were explored widely throughout the time for the timely detection of MTBC in humans. In addition, other SPR-based biosensors targeting the CFP-10 antigen [117–119], the CFP-10-ESAT-6 complex [120] and the Ag85 antigen [121] were developed with high sensitivity and lower LOD values in the range of ng to μg/mL.

The strong chemisorption between the substrate and the organic moiety of interest paves the way to the formation of SAMs. It is one of the most widely used approaches to form ultra-thin organic films with controlled thickness. The moderate stability of SAMs for a reasonable period allows several reliable analysis of biomolecules [122]. Pelaez et al. developed a label-free SPR-based sensor utilizing the SAMs of thiols onto the gold surface to detect and quantify the HspX of *M. tuberculosis*. The sensor which relies highly on the specific anti-HspX monoclonal antibody had a good detection limit of 0.63 ng/mL in the pre-treated sputum samples [123]. Prabowo et al. developed a rapid, portable organic light emitting diode (OLED)-based SPR biosensor for the detection of MTBC by targeting IS*6110* genes of MTBC through nested-PCR. The binding matrix consists of a self-assembled monolayer formed by a mixture of thiols, $HS(CH_2)_3OH$ and $HS(CH_2)_{10}COOH$ over a gold substrate. The developed sensor can differentiate the MTBC strains from NTM strains at ease. The MTBC PCR products have a detection limit of 63 pg/mL in the clinical samples [124]. The formation of self-assembled monolayer in the fabrication of a nanomaterial-based biosensor also helps in the simultaneous detection of multiple MTBC antigens [125, 126].

In fluorimetry method, either the antibody or the aptamer is immobilized on the photoluminescent nanomaterials like quantum dots (QDs), upconversion nanoparticles (UCNPs), fluorophores encapsulated in metallic nanoparticles etc. The conjugated nanoparticles were used as fluorescent probes to detect the pathogens [127]. Shojaei et al. conjugated the cadmium telluride quantum dots (CdTe QDs) and gold nanoparticles with two specific oligonucleotides which forms the basis of sandwich type fluorescence resonance energy transfer (FRET)-based biosensors. The developed sensor was used to detect the 6 kDa ESAT-6 genes of MTBC. The obtained LOD value was found to be 10 fg/μL which was far lower than that of PCR and nested PCR (100 fg/μL) [128]. Other FRET-based biosensors [129, 130], and the biosensors which used quantum dots in their sensing surface [131] reported earlier for the detection of MTBC also yielded similar results. Kim et al. developed an immunofluorescence microtip sensor to detect the cells of MTBC present in the sputum samples. This novel sensing technique enables a good detection limit of 200 CFU/mL in sputum samples [132]. In another example, a fluorescence linked immunosorbent assay (FLISA) was developed by Wood et al., which was found to be an ultra-sensitive biosensor technique by targeting LAM of MTBC in clinical samples of HIV-negative patients. The detection limit was found to be 1 fg/mL [133].

The colorimetric sensors utilize the reflection and interference of the scattered light which aid in the naked-eye detection, molecular recognition and instantaneous detection. This method allows easy detection of even from smaller molecules to bulky proteins [134]. Wang et al. developed a simple and reliable method to detect the genes of MTBC by coupling LAMP method with lateral flow biosensor (LFB) assay (LAMP-LFB). To the amplified products, the test line (TL) was conjugated with anti-FITC and the control line (CL) with biotin-bovine serum albumin (BSA) on a nitrocellulose membrane. This LFB was designed to sense the IS*6110* and IS*1081* genes of MTBC which has a LOD value of 1 fg/μL [135]. Chen et al. developed a novel method consisting of multiplex LAMP combined with LFB (mLAMP-LFB). A set of primers specific to IS*6110* and gyrB genes of MTBC were designed and labelled with anti-FAM, anti-DIG, and biotin-BSA. The LOD value of the mLAMP-LFB assay was found to be 100 fg/μL [136]. Some of the other colorimetric sensors aim at simultaneous detection of more than one antigen [137–140], the CFP-10-ESAT-6 complex of *M. tuberculosis* [141] and uses two types of nanomaterials for ultra-sensitive immunoassay [142].

12.3.2 Non-Nanomaterial-Based Optical Biosensors

Trilling et al. developed two biotinylated Llama heavy chain antibodies (Biotinylated VHHs) to target 16 kDa Hsp of *M. tuberculosis*. The SPR-based biosensor yielded a detection limit of 0.4μg/mL [143]. Liu et al. designed a rapid and label-free detection of *M. tuberculosis* by using a novel sensing platform based on silicon photonic microring sensors and asymmetric isothermal amplification technique (SPMS-AIA). The detection limit was found to be 3.2 copies of DNA for IS*6110* and 12 copies of DNA for IS*1081* in the clinical sputa [144]. Schmidt et al. designed a novel, single-molecule sensitive FLISA for the detection of LAM of MTBC. The whole procedure took place just within 20 min and was also cost-effective. The designed sensor achieved an LOD of 10^{-14} M which was found to be three orders in magnitude more sensitive than the conventional ELISA method [145]. Ng et al. designed a novel colorimetric sensor to detect the ESAT-6 genes of MTBC. They extracted genomic DNA (gDNA) from the *M. tuberculosis* H37Ra strain and purified it by using surface plasmon resonance imaging (SPRi) technique. The nucleic acid was amplified with recombination polymerase assay (RPA) by the addition of 11-biotin dUTPs and a set of primers. The SA-coated magnetic beads and the HRP-labelled SA were added to these DNA amplicons. In the case of a positive reaction, addition of tetramethyl benzidine (TMB) changes the solution from colourless to blue as a result of oxidation of TMB by HRP (see Fig. 12.3). The detection limit of this naked-eye detection was found to be 1 CFU of MTBC [146]. Joon et al. combined the LAMP method with lateral flow dipstick (LAMP-LFD) to design a POC biosensor for the determination of sdaA genes of *M. tuberculosis*. This rapid and cost-effective sensor obtained a detection limit of 5 fg/μL [147]. Some of the other non-nanomaterial-based biosensors [148–151] also show good sensitivity and selectivity against MTBC (Table 12.3).

12.4 Other Biosensors Reported for MTBC

Piezoelectric quartz crystal (PQC)-based biosensors measure the variations in the oscillating frequency of a piezoelectric material as a result of the biochemical response brought on by target recognition. Biomolecules are immobilized on a quartz crystal microbalance (QCM) that is connected to an oscillating circuit and coated on both sides with a magnetic substance. Target binding causes an increase in the mass of QCM or changes in solution resistance and/or capacitance. Multichannel-series piezoelectric quartz crystal (MSPQC) allows for simultaneous analysis of multiple samples [162]. Huang et al. proposed a new clustered regularly interspaced short palindromic repeats (CRISPR/Cas9)-based MSPQC sensor for the detection of 16S rRNA of *M. tuberculosis*. When the target existed, the capture probe hybridized to form dsDNA, which could be recognized and cut by CRISPR/Cas9, resulting in a response. The detection limit was 30 CFU/mL, and the detection time was down to 2.3 h [163]. Fengjiao He et al. developed an aptamer-based MSPQC sensor using an Au-IDE for the detection of the specific fused antigen CFP-10-ESAT-6, which is secreted only by *M. tuberculosis* in its early culture period. They achieved a better detection limit of 10^3 CFU/mL [164]. Domínguez et al. proposed a label-free assay for the genomic detection of *M. tuberculosis* and rifampicin-resistant tuberculosis (RR-TB) using the IS*6110* and rpoB genes as the respective biomarkers. This method relies on the quantification of the hydration-induced stress on microcantilever functionalized with DNA probes. Without the need for long culture steps or PCR amplification, they achieved an LOD of 2 pg/mL and an analysis time of 90 min [165]. Various other MSPQC-based biosensors

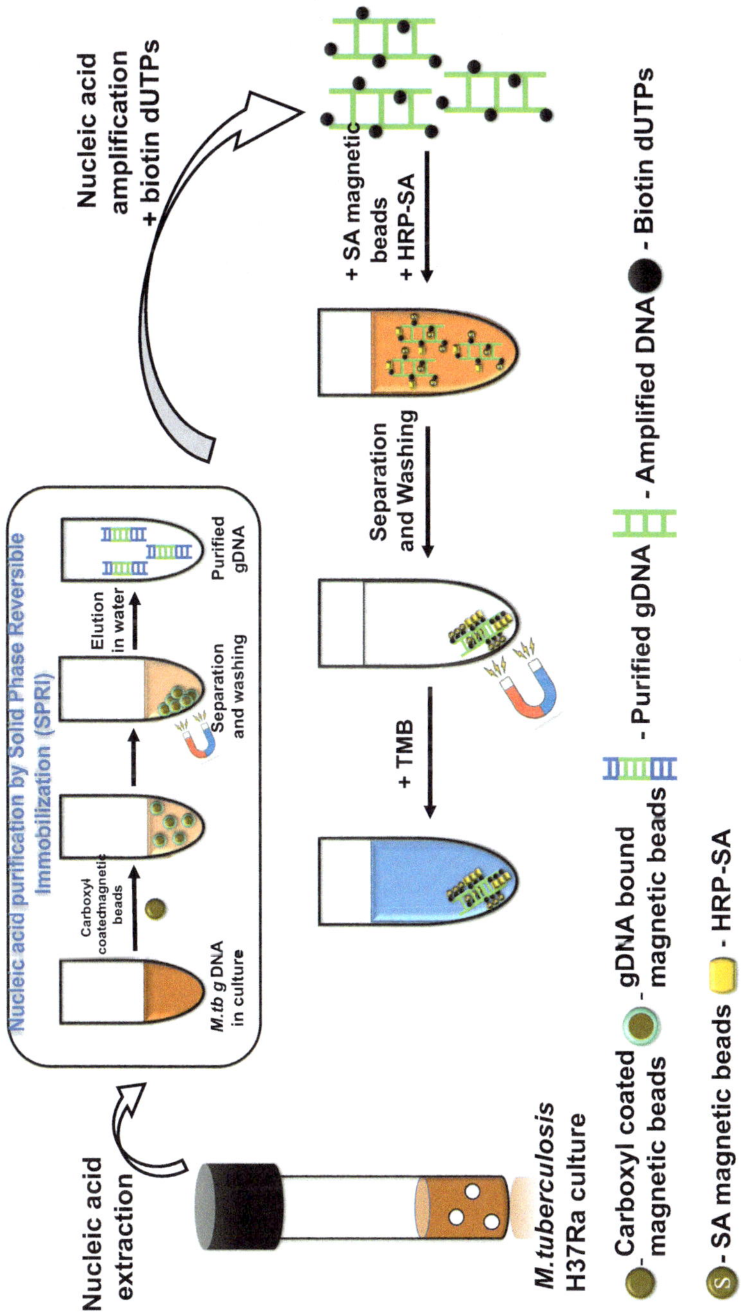

Fig. 12.3 Naked-eye colorimetric detection of *M. tuberculosis* (adapted from Ng et al. 2015)

Table 12.3 Optical biosensors available for the detection of *Mycobacterium tuberculosis*

Sensing platform	Biomarker	LOD	Linear range	Reference
A. Nanomaterial-based optical biosensor				
(i) Surface plasmon resonance-based optical biosensors				
i. Cys-PNA/Au ii. Thiolated DNA/Au	DNA of *M. tb*	i. 1.0 ng/mL ii. 3.0 ng/mL	i. 5–50 ng/mL ii. 5–50 ng/mL	[111]
Probe-ssODNs/Au	Target-ssODNs	30 ng/μL	0–1 μM	[112]
Thiolated DNA/Au	*rpoB*	10 nM	1–100 nM	[113]
AuNPs/anti-CFP-10/Cys	CFP-10	100 ng/mL	0.1–1 μg/mL	[117]
Au/ZnO/thiolated aptamers	IFN-γ	33 pM	0.3–333 nM	[152]
8-Mercaptooctanoic acid SAM/serum antibodies	TB antibodies	–	–	[125]
SiO_2 SAM/streptavidin/biotin-BSA/anti-ESAT-6	LAM, ESAT-6, Ag85	78 nM, 125 nM and 183 nM	–	[126]
Thiolated DNA/1-mercapto-6-hexanol/Au	*rpoB*	50 nM	50–500 nM	[114]
Thiolated DNA/Au	DNA of *M. tb*	115 ng/mL	1–10 μg/mL	[115]
Cr/Au/PNA	DNA of *M. tb*	0.26 pM	0–10 nM	[110]
AuNPs/SPR chips	DNA of MTBC	10^4 CFU/mL	10^3–10^9 CFU/mL	[116]
AuNRDs/complementary DNA	IFN-γ	10 pM	0.01–1 nM	[153]
Graphene/ssDNA-Gnu	IS*6110*	28 fM	0–10^8 fM	[109]
SAM-modified SPR chip/anti-DIG	IS*6110*	63 pg/mL	0.01–100 ng/mL	[124]
Amidogen/Au NRs	CFP-10-ESAT-6	–	–	[120]
Au chips/anti-Ag85	Ag85	10 ng/mL	10–1000 ng/mL	[121]
MPNs/anti-CFP-10/BSA	CFP-10	0.1 μg/mL	0.1–100 ng/mL	[119]
Nanophotonic Chip/PLL-g-PEG/anti-LAM-Ig/BSA	LAM	475 pg/mL	1–1000 ng/mL	[108]
AuNPs/Mixed SAM/mAb anti-HspXs	HspX	0.63 ng/mL	116–175 ng/mL	[123]
(ii) Fluorimetry-based optical biosensors				
Au Microtips/anti-IgY	MTB H37Ra	200 CFU/mL	10^1–10^6 CFU/mL	[132]
GO/aptamer hairpin probe	IFN-γ	1.5 fM	3.5 fM–1.0 pM	[154]
Streptavidin labelled UCNPs	IS*6110*	10^2 copies/μL	10^1–10^6 copies/μL	[129]
CdTe QDs and AuNPs conjugated with DNA	ESAT-6	10 fg/μL	0–200 fg/mL	[128]
MMSs/QDs/BSA/anti-MTB	Hsp65	10^3 CFU/mL	10^8–10^1 CFU/mL	[131]
Ap-GQDs and Ep-GQDs	IFN-γ	2 pg/mL	5–100 pg/mL	[155]

(continued)

Table 12.3 (continued)

AuNPs/Cy3	MTB16s rDNA	3 ng/μL	40–10 ng/μL	[130]
SNPs-streptavidin/SSB	IFN-γ	1 pg/mL	10 pg/mL–4 ng/mL	[156]
ReS_2 NSs/aptamer TiS_2 NSs/aptamer	IFN-γ	ReS_2–57.6 pM TiS_2–82.7 pM	0–400 pg/mL 0–300 pg/mL	[157]
PMSSQ/Ag/anti-LAM	LAM	1 fg/mL	1 fg/mL–10 ng/mL	[133]
GO/TB7.7 specific aptamers	IFN-γ	0.7089 nM	1 nM–300 nM	[158]
(iii) Colorimetry-based optical biosensors				
AuNP-DNA conjugates	DNA of *M. tb*	30 μg/mL	50–300 nM	[137]
SA-lum-AuNPs/Biotin-P1/P2	IFN-γ	0.4 nM	0.5–100 nM	[159]
Thiolated DNA/Au	IS*6110*	5 pg/μL	10^{-1}–10^{-6} copies	[138]
MMP@Au/GBP/anti-CFP-10	CFP-10	0.3 pM	10^{-7}–10^{-12} CFU/mL	[142]
AuNPs/antibody conjugates	CFP-10-ESAT-6	–	–	[141]
SA-PNPs/anti-FAM/anti-DIG/biotin-BSA	IS*6110* & gyrB	100 fg/μL	0.0001–1 fg/μL	[136]
NP-LFB/MG/anti-FITC/anti-DIG	IS*6110*, MPB64	10 fg/μL	0.0001–1 fg/μL	[139]
SA-DNPs/anti-FITC/biotin-BSA	IS*6110*, IS*1081*	1 fg/μL	1 ng/μL to 1 fg/μL	[135]
SA-DNPs/biotin-BSA/anti-FITC/anti-DIG	IS*6110*, MTP40	125 fg/μL	12.5 ng/μL to 1.25 fg/μL	[140]
B. Other optical biosensors				
(i) Surface plasmon resonance-based optical biosensors				
Biotinylated VHH A23	Hsp protein	0.4 μg/mL	0.4–50 μg/mL	[143]
ssDNA/probe	IS*6110*, IS*1081*	5 fg/μL	5 fg/μL–0.5 ng/μL	[144]
(ii) Fluorimetry-based optical biosensors				
Cy5/biotin/anti-*M. tb*	Live *M. tb* in liver	10^5 cells/mL	10^7–10^9 cells/mL	[149]
PEGMA/anti-LAM	LAM	10^{11} M	10^{-11}–10^{-15} M	[145]
TO/ssDNA	IFN-γ	2 nM	3–120 nM	[160]
TPE-aptamer	IFN-γ	2 pg/mL	0–100 pg/mL	[161]
MBs-H1	IS*6110* DNA	10 pM	0.01–100 nM	[151]
(iii) Colorimetry-based optical biosensors				
CNBr/R-NH_2	Niacin	1.25 μg/mL	0–25 μg/mL	[148]
Nitrocellulose/neutravidin/anti-DIG	DNA of *M. tb*	3 copies	3–3000 copies	[150]
SA MBs/HRP-SA/TMB	ESAT-6	1 CFU/mL	1–100 CFU/mL	[146]
SYBR Green/biotin/anti-FITC	DNA of *M. tb*	5 fg/μL	5 fg/μL–1 ng/μL	[147]

[166–170], PQC-based biosensors [171–173] and QCM-based biosensor [174] were also reported for the detection of MTBC with better performance.

12.5 Biosensors for the Detection of Non-Tuberculosis Mycobacteria (NTM)

Non-tuberculosis mycobacteria (NTM) are the species belonging to the mycobacterium family other than the *Mycobacterium Tuberculosis complex (MTBC) and Mycobacterium leprae*. There are more than 170 species of NTM identified till date [175]. NTM is found common in people more than 50 years of age. Based on the growth of this mycobacterium, NTM is classified into slow growers (growth taking more than 7 days), and rapid growers (growth taking less than 7 days) [176]. NTM was found to be habitat in the water bodies like brackish, marshy waters and even in municipal water bodies. Recently, there is an alarming increase in the contagious spread of the rapid growing *M. abscessus* group worldwide [177] (Fig. 12.4).

The major problem that arises in the detection of NTM is to note the difference between the disease caused by NTM and that caused by MTBC. NTM is environmentally ubiquitous and can easily contaminate the laboratory reagents and specimens. Thus, this contamination leads to false positive results in clinical samples [178]. Hence, the diagnosis of NTM is very less explored even though they are considered as potential pathogens equal to the MTBC.

Poch et al. immobilized the ssDNA aptamer linked with the redox-active methylene blue over a gold-coated screen-printed carbon electrode (SPCE) to serve as a transducer and successfully detect the mannosylated-Lipoarabinomannan (man-LAM), one of the specific antigens to the NTM species. The detection limit of this electrochemical DNA biosensor was found to be 10^{-11} M with a good linear range of 10^{-11} to 10^{-7} M [179]. Kumanan et al. developed an aptasensor to sense the *IS900* genes of rapidly growing *M. avium subsp. Paratuberculosis.* The aptasensor was developed using lateral flow and microtiter assays. In the microtiter assay, the surfactant induced liposome lysis followed by the release of encapsulated dye molecules emitting fluorescence which was used to detect the amplified products. The linear range and detection limit for this microtiter assay was found to be 0–1000 nM and 0.1 nM, respectively. The RT-PCR product of RNA extracted from spiked fecal samples containing 10^1 to 10^6 organisms were used for the LFB in order to assess the sensitivity of the assay. The LFB assay was found to be effective and detected ten organisms in the RNA extracted from the fecal samples [180]. Chen et al. developed an optical biosensor by combining the LAMP method with LFB (LAMP-LFB). The SA-coated gold nanoparticles were immobilized onto the conjugate pad. A different set of primers was designed and labelled with biotin at one end and FAM at the other end. To the amplified products, the test line (TL) was conjugated with anti-FAM and the control line (CL) with biotin-BSA on the nitrocellulose membrane of the LFB. The positive responses will indicate with two red lines at both TL

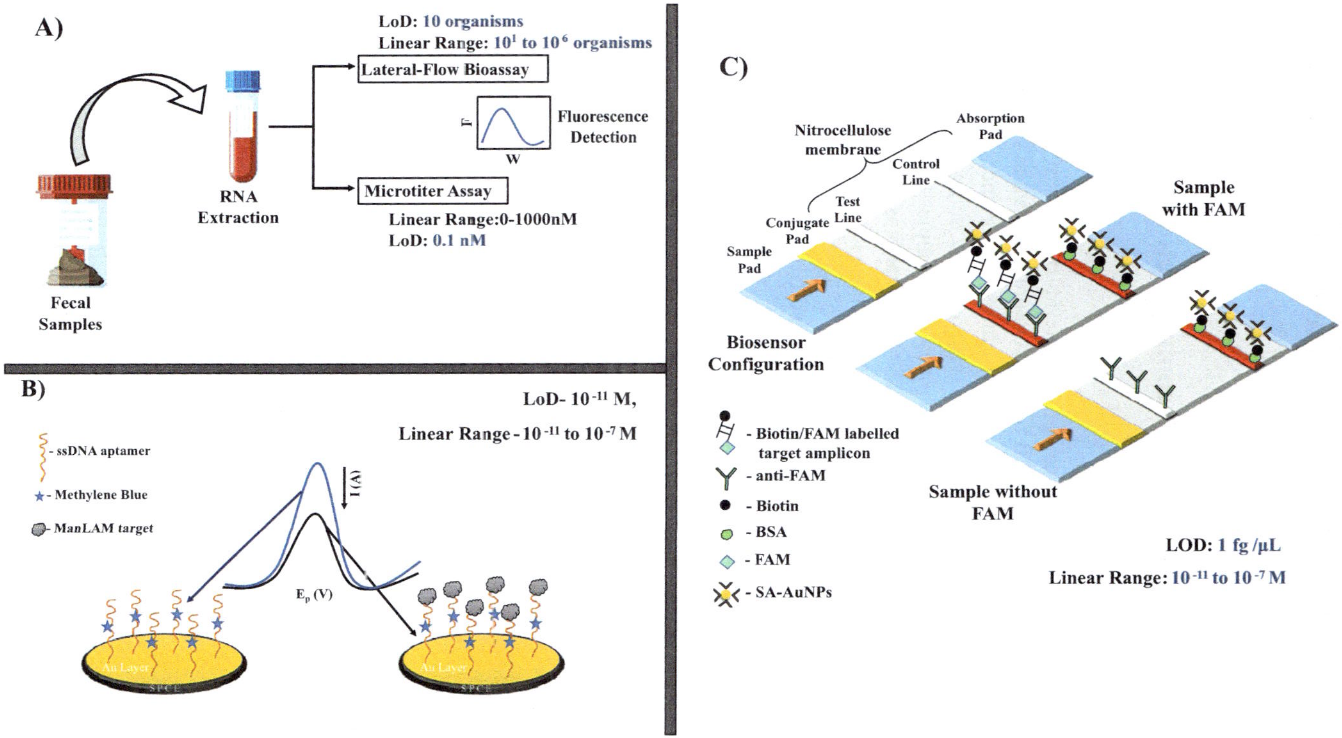

Fig. 12.4 Biosensors reported for the detection of non-tuberculous mycobacteria (NTM). (A) Electrochemical DNA biosensor for NTM, (B) A biosensor assay for the detection of *mycobacterium avium* subsp. paratuberculosis in fecal samples, (C) Detection of *mycobacterium kansasii* using combination of LAMP-LFB technique [adapted from 179, 180, 181]

and CL, whereas the negative responses will show red line only at the CL. The detection limit was found to be 1 fg/μL with a linear range of 10^{-11} to 10^{-7} M [181]. The schematic representation of the above NTM based sensors are given in Fig. 12. 4.

12.6 Opportunities, Future Recommendations and Challenges

The potentially well compacted area of nanomaterial-based biosensors aid the early diagnosis of mycobacterium. The TB biomarkers present even at femto level in the human body cause infection; hence it is of utmost importance to have detection limits as low as possible. The biosensors with an LOD below 10 fg/μL (limit of infection) have the potential to be applied on clinical samples for the early detection of mycobacterium [182]. The reported electrochemical biosensors for MTBC have detection limits in the "nano" (0.01 to 40 ng/mL, 0.06 to 1.3 nM), "pico" (0.05 to 126.75 pg/mL, 3.4 to 300 pM), "femto" (0.01 to 67.6 fg/mL, 0.03 to 100 fM), and even in "atto" level (0.36 aM). The MTBC-related optical biosensors on the other hand have a similar detection limit in the level of "nano" (1 to 115 ng/mL, 0.4 to 183 nM), "pico" (1 to 475 pg/mL, 0.26 to 82.7 pM) and in "femto" (1 to 125 fg/μL, 1.5 and 28 fM). The lower detection limit in the other MTBC biosensors were reported in the range of 10–10^6 CFU/mL, and 0.5μg/mL to 2 pg/mL. Based on the above reports, some of the reported electrochemical, optical and other biosensors have shown LOD values well below the limit of infection. Even though the detection of NTM is very less explored, the reported biosensors worked very well and reached detection limits as low as the femto level. Hence, these highly sensitive biosensors can be fabricated into miniaturized devices for the affordable and early detection of mycobacterium in clinical samples.

Over two decades of TB diagnosis, nanomaterial-based electrochemical biosensors achieved lowest detection limits (0.01 to 67.6 fg/mL, 0.03 to 100 fM) compared to the other non-nanomaterial-based electrochemical biosensors (0.02 fg/mL to 126.75 fg/mL, 0.2 fM to 81 pM). Most of the reported NSPs utilized a combination of polymers like PPy, PANi, PEDOT or Metal oxide NPs such as Fe_2O_3, Fe_3O_4, MNPs, with carbonaceous materials like GR, GO, rGO, C_{60}, CNTs, MOF, MXene and metal nanostructures like AuNPs, Au@Pt core shell and QDs. This combination synergistically increases the stability, ultra-level sensitivity and specificity, and improves the lower detection limit and response time compared to the other reported sensing platforms. This is due to their high active surface area, biocompatibility and their higher electronic conductance [31]. For example, gold nanoparticles maintain the bioactivity of immobilized biomolecules and also have higher ability to transport electrons between nanomaterials and electroactive species [183]. Out of these reports, most of the biosensors were fabricated based on voltammetric techniques (DPV and SWV) and uses aptamer as biorecognition element and MPT64 and

IS*6110* as biomarkers. Even though few of the antibody-based reports achieved lowest LOD (0.3 fg/mL), aptamer-based biosensors relatively have lower LOD (0.7 to 67.6 fg/mL) values. The reported aptamer-based biosensors have more advantages such as increase in number of loading at active sites, flexibility, cost-effectiveness, stability, higher affinity and specificity towards the target biomarker compared to the antibody-based immunoassay [23].

Among the reported optical biosensors, the DNA and nanomaterial-based optical biosensors yielded lower detection limits. Out of those reports, almost every biosensor was developed into LFB with the help of nanomaterials. These LFB-based biosensors are affordable, highly stable, allow multiplex detection and the results were obtained promptly (60 to 80 mins). However, the addition of organic probes and dyes to the LFB system can potentially interfere with the properties of the liposomal delivery system. In exchange for using these organic fluorophores as labels, the nanomaterials provide greater advantage by increasing the inertness and stability. However, the LFB-based biosensor is a qualitative assay. In order to achieve quantification, researchers have developed fluorimetry, colorimetry and other optical biosensors for the detection of mycobacterium. These optical biosensors also yielded comparable detection limits with LFB-based biosensors (LOD values of 10^4 CFU/mL, 1 ng/mL to 3000 ng/mL). The fluorimetry-based biosensors are the most commonly used optical biosensors and have higher sensitivity towards TB biomarkers among the so far reported optical biosensors. Hence, the DNA and nanomaterial-based fluorimetric biosensors will be recommended for the diagnosis of mycobacterium. The electrochemical biosensors were 20 times better than optical biosensors in terms of lower detection limit in pico level and 100 times better than the reported optical and other biosensors in terms of lower detection limit in nano and femto levels. But the qualitative multiplex detection of various TB biomarkers is very much feasible in the case of optical biosensors. However, the electrochemical biosensors, because of their high sensitivity, stand high against the optical and other biosensors for the early detection of MTBC. In addition to the other inherent advantages of electrochemical biosensors, in the case of detecting organic metabolic products (VOBs, cytokines), amperometric biosensors are more reliable and are able to integrate into continuous analysis systems with the lowest LOD values of 18 pg/μL for VOBs and 126.75 pg/mL for IFN-γ. It is suggested that rather than looking for highly pathogenic *M. tuberculosis* genomic material, the quantification of VOBs present in TB patients' breath and/or the amount of IFN-γ is preferred. This indirect method is safe, simple and affordable for determining the response of host's immune system to *M. tuberculosis*.

However, there are certain challenges turned up during the fabrication and development of nanomaterial-based biosensors. Some common demerits in the development of biosensors include the preparation of sensing platform which requires high precaution and laboratory skills for raising antibodies/sequencing aptamers and primers specific for the TB biomarkers. Generally, they are not readily available for every biomarker and the isolation/sequencing process involves challenging steps. Further, the culture media may easily get contaminated and utmost care must be taken to prevent/eliminate them. Hence, isolation of biomarkers for the development of biosensors is not very easy. But these difficult methodologies are not involved in the conventional techniques for the detection of mycobacterium. In addition, the assembly of nanomaterials for the fabrication of high-performance electrochemical/optical biosensors is another big task. Furthermore, the non-tuberculosis mycobacterium is ubiquitous and found everywhere in the atmosphere. This allows the bacterium to easily get contaminated in the NTM-negative samples and show false positive results. This is one among the reasons for which the sensors of NTM are very less explored. Despite all these challenges, still biosensor remains as the more specific, sensitive, selective and faster technique for the early detection of mycobacterium.

12.7 Conclusions

To conclude, this book chapter starts with a brief introduction to the mycobacterium and conventional diagnostic techniques of MTBC. The second part deals with a short introduction to electrochemical biosensors and a detailed review of nanomaterial and non-nanomaterial-based electrochemical DNA/aptamer/antibody-based biosensor reports for the detection of mycobacterium. Similarly, in the third part, an introduction to the optical biosensors followed by a detailed review for the detection of mycobacterium by nanomaterial and non-nanomaterial-based fluorimetric/colorimetric/SPR optical biosensors are covered. Later on, in the fourth and fifth parts, a brief review about the other types of biosensors for MTBC and biosensors for NTM are covered. Finally, opportunities, future recommendations and challenges faced during the development of biosensors for the detection of the mycobacterium are discussed.

Acknowledgement The authors would like to acknowledge TANSCHE (Project File No: RGP/2019-20/ALU/HECP-0081) and RUSA Phase 2.0 (Project File No: 007-R2/RUSA/MKU/2020-21) for the financial support.

References

1. Poorolajal J. Neglected major causes of death much deadlier than COVID-19. J Res Health Sci. 2020;20:e00478.
2. WHO. Global tuberculosis report (2020) & (2021).
3. Indian TB Report, 2022: Coming together to end TB altogether; https://tbcindia.gov.in/WriteReadData/IndiaTBReport2022/TBAnnaulReport2022.pdf.
4. Thanyani ST. An assessment of two evanescent field biosensors in the development of an immunoassay for tuberculosis. Thesis (PhD). Pretoria: University of Pretoria; 2009.
5. Nahid P, Hopewell PC. Tuberculosis treatment. In: International encyclopedia of public health (second edition). St Louis, MI: Academic Press; 2017. p. 267–76.
6. Joshi H, Kandari D, Maitra SS, Bhatnagar R. Biosensors for the detection of mycobacterium tuberculosis: a comprehensive overview. Crit Rev Microbiol. 2022;48(6):784–812.

7. Gupta S, Kakkar V. Recent technological advancements in tuberculosis diagnostics—a review. Biosens Bioelectron. 2018;115:14–29.
8. Golichenari B, Nosrati R, Farokhi-Fard A, Faal Maleki M, Gheibi Hayat SM, Ghazvini K, Behravan J. Electrochemical-based biosensors for detection of mycobacterium tuberculosis and tuberculosis biomarkers. Crit Rev Biotechnol. 2019;39(8):1056–77.
9. Uhuo OV, Waryo TT, Douman SF, Januarie KC, Nwambaekwe KC, Ndipingwi MM, Ekwere P, Iwuoha EI. Bioanalytical methods encompassing label-free and labeled tuberculosis aptasensors: a review. Anal Chim Acta. 2022;1234:340326.
10. Eggins BR. Chemical sensors and biosensors, vol. 2. Hoboken, NJ: John Wiley & Sons; 2002.
11. Baig N, Kammakakam I, Falath W. Nanomaterials: a review of synthesis methods, properties, recent progress, and challenges. Mater Adv. 2021;2(6):1821–71.
12. Malhotra BD, Ali MA. Chapter 1—nanomaterials in biosensors: fundamentals and applications, micro and nano technologies. In: Nanomaterials for biosensors; 2018. p. 1–74.
13. Kumar N, Kumbhat S. Unique properties. In: Essentials in nanoscience and nanotechnology; 2016. p. 326–60.
14. Koncki R. Recent developments in potentiometric biosensors for biomedical analysis. Anal Chim Acta. 2007;599(1):7–15.
15. Phillips M, Cataneo RN, Condos R, Ring Erickson GA, Greenberg J, La Bombardi V, Munawar MI, Tietje O. Volatile biomarkers of pulmonary tuberculosis in the breath. Tuberculosis. 2007;87(1):44–52.
16. Syhre M, Chambers ST. The scent of mycobacterium tuberculosis. Tuberculosis. 2008;88(4):317–23.
17. Bhattacharyya D, Smith YR, Mohanty SK, Misra M. Titania nanotube array sensor for electrochemical detection of four predominate tuberculosis volatile biomarkers. J Electrochem Soc. 2016;163(6):B206–14.
18. Smith YR, Bhattacharyya D, Mohanty SK, Misraa M. Anodic functionalization of titania nanotube arrays for the electrochemical detection of tuberculosis biomarker vapors. J Electrochem Soc. 2016;163(3):B83–9.
19. Mogha NK, Sahu V, Sharma RK, Masram DT. Reduced graphene oxide nanoribbon immobilized gold nanoparticles based electrochemical DNA biosensor for detection of mycobacterium tuberculosis. J Mater Chem B. 2018;6:5181–7.
20. Zhang H, Ma L, Li P, Zheng L. A novel electrochemical immunosensor based on nonenzymatic ag@au-Fe_3O_4 nanoelectrocatalyst for protein biomarker detection. Biosens Bioelectron. 2016;85:343e350.
21. Wang L, Neoh K, Kang E, Shuter B, Wang SC. Superparamagnetic hyperbranched polyglycerol-grafted Fe_3O_4 nanoparticles as a novel magnetic resonance imaging contrast agent: an in vitro assessment. Adv Funct Mater. 2009;19:2615e2622.
22. Tufa LT, Sangjin O, Tran VT, Kim J, Jeong KJ, Park TJ, Kim HJ, Lee J. Electrochemical immunosensor using nanotriplex of graphene quantum dots, Fe_3O_4, and ag nanoparticles for tuberculosis. Electrochim Acta. 2018;290:369–77.
23. Gou D, Xie G, Li Y, et al. Voltammetric immunoassay for mycobacterium tuberculosis secretory protein MPT64 based on a synergistic amplification strategy using rolling circle amplification and a gold electrode modified with graphene oxide, Fe_3O_4 and pt nanoparticles. Microchim Acta. 2018;185:436.
24. Augustus EN, Nimibofa A, Kesiye IA, Donbebe W. Metal-organic frameworks as novel adsorbents: a preview. Am J Environ Protect. 2017;5(2):61–7.
25. Kreno LE, Leong K, Farha OK, Allendorf M, Van Duyne RP, Hupp JT. Metal–organic framework materials as chemical sensors. Chem Rev. 2012;112(2):1105–25.
26. Chen Y, Liu X, Guo S, Cao J, Zhou J, Zuo J, Bai L. A sandwich-type electrochemical aptasensor for mycobacterium tuberculosis MPT64 antigen detection using C_{60}NPs decorated N-CNTs/GO nanocomposite coupled with conductive PEI-functionalized metal-organic framework. Biomaterials. 2019;216:119253.

27. Li N, Huang X, Sun D, Yu W, Tan W, Luo Z, Chen Z. Dual-aptamer-based voltammetric biosensor for the mycobacterium tuberculosis antigen MPT64 by using a gold electrode modified with a peroxidase loaded composite consisting of gold nanoparticles and a Zr(IV)/terephthalate metal-organic framework. Microchim Acta. 2018;185(12):543.
28. Yishi L, Peng D, Guo S, Yang B, Zhou J, Zhou J, Zhang Q, Bai L. Aptasensor for mycobacterium tuberculosis antigen MPT64 detection using anthraquinone derivative confined in ordered mesoporous carbon as a new redox nanoprobe. Bioelectrochemistry. 2022;147:108209.
29. Li L, Yuan Y, Chen Y, Zhang P, Bai Y, Bai L. Aptamer based voltammetric biosensor for mycobacterium tuberculosis antigen ESAT-6 using a nanohybrid material composed of reduced graphene oxide and a metal-organic framework. Microchim Acta. 2018;185(8):379.
30. Xie J, Mu Z, Yan B, Wang J, Zhou J, Bai L. An electrochemical aptasensor for mycobacterium tuberculosis ESAT-6 antigen detection using bimetallic organic framework. Microchim Acta. 2021;188(11):404.
31. Rizi KS, Hatamluyi B, Darroudi M, Meshkat Z, Aryan E, Soleimanpour S, Rezayi M. PCR-free electrochemical genosensor for mycobacterium tuberculosis complex detection based on two-dimensional Ti_3C_2 MXene-polypyrrole signal amplification. Microchem J. 2022;179:107467.
32. Rizi KS, Meshkat Z, Chankandi M, Gholami M, Sankian M, Ghazvini K, Farsiani H, Aryan E, Rezayi M. A PCR-free genome detection of mycobacterium tuberculosis complex in clinical samples using MWCNT/PPy/KHApNps modified electrochemical nano-biosensor. J Electrochem Soc. 2021;168:077501.
33. Bai L, Chen Y, Bai Y, Chen Y, Zhou J, Huang A. Fullerene-doped polyaniline as new redox nanoprobe and catalyst in electrochemical aptasensor for ultrasensitive detection of mycobacterium tuberculosis MPT64 antigen in human serum. Biomaterials. 2017;133:11–9.
34. Zaid MHM, Che-Engku-Chik CEN, Yusof NA, et al. DNA electrochemical biosensor based on iron oxide/nanocellulose crystalline composite modified screen-printed carbon electrode for detection of mycobacterium tuberculosis. Molecules. 2020;25(15):3373.
35. Mohamad FS, Mat Zaid MH, Abdullah J, Zawawi RM, Lim HN, Sulaiman Y, Abdul RN. Synthesis and characterization of polyaniline/graphene composite nanofiber and its application as an electrochemical DNA biosensor for the detection of mycobacterium tuberculosis. Sensors. 2017;17(12):2789.
36. Chen Y, Li Y, Yang Y, et al. A polyaniline-reduced graphene oxide nanocomposite as a redox nanoprobe in a voltammetric DNA biosensor for mycobacterium tuberculosis. Microchim Acta. 2017;184:1801–8.
37. Chen Y, Guo S, Zhao M, Zhang P, Xin Z, Tao J, Bai L. Amperometric DNA biosensor for mycobacterium tuberculosis detection using flower-like carbon nanotubes-polyaniline nanohybrid and enzyme-assisted signal amplification strategy. Biosens Bioelectron. 2018;119:215–20.
38. Ng BYC, Xiao W, West NP, Wee EJH, Wang Y, Trau M. Rapid, single-cell electrochemical detection of mycobacterium tuberculosis using colloidal gold nanoparticles. Anal Chem. 2015;87(20):10613–8.
39. Liu C, Jiang D, Xiang G, Liu L, Liu F, Pu X. An electrochemical DNA biosensor for the detection of mycobacterium tuberculosis, based on signal amplification of graphene and a gold nanoparticle–polyaniline nanocomposite. Analyst. 2014;139:5460–5.
40. Thiruppathiraja C, Kamatchiammal S, Adaikkappan P, Santhosh DJ, Alagar M. Specific detection of mycobacterium sp. genomic DNA using dual labelled gold nanoparticle based electrochemical biosensor. Anal Biochem. 2011;417(1):73–9.
41. Torres-Chavolla E, Alocilja EC. Nanoparticle based DNA biosensor for tuberculosis detection using thermophilic helicase-dependent isothermal amplification. Biosens Bioelectron. 2011;26(11):4614–8.
42. Das MM, Sumana G, Nagarajan R, Malhotra BD. Application of nanostructured ZnO films for electrochemical DNA biosensor. Thin Solid Films. 2010;519(3):1196–201.
43. Azmi UZM, Yusof NA, Kusnin N, Abdullah J, Suraiya S, Ong PS, Raston NHA, Rahman SFA, Fathil MFM. Sandwich electrochemical immunosensor for early detection of

tuberculosis based on graphene/polyaniline-modified screen-printed gold electrode. Sensors. 2018;18(11):3926.
44. Wang L, Leng C, Tang S, Lei J, Ju H. Enzyme-free signal amplification for electrochemical detection of mycobacterium lipoarabinomannan antibody on a disposable chip. Biosens Bioelectron. 2012;8(1):421–4.
45. Javed A, Abbas SR, Hashmi MU, Babar NUA, Hussain I. Graphene oxide based electrochemical genosensor for label free detection of mycobacterium tuberculosis from raw clinical samples. Int J Nanomedicine. 2021;16:7339–52.
46. Azmi UZM, Yusof NA, Abdullah J, Mohammad F, Ahmad SAA, Suraiya S, Raston NHA, Faudzi FNM, Khiste SK, Al-Lohedan HA. Aptasensor for the detection of mycobacterium tuberculosis in sputum utilising CFP-10-ESAT-6 protein as a selective biomarker. Nano. 2021;11(9):2446.
47. Xu L, Lei S, Liu Z, Ouyang G, Zou L, Ye B. A label-free IFN-γ aptasensor based on target-triggered allosteric switching of aptamer beacon and streptavidin-inorganic hybrid composites. Anal Chim Acta. 2019;1087:29–35.
48. Li J, Hu K, Zhang Z, Teng X, Zhang X. Click DNA cycling in combination with gold nanoparticles loaded with quadruplex DNA motifs enable sensitive electrochemical quantitation of the tuberculosis-associated biomarker CFP-10 in sputum. Microchim Acta. 2019;186(9):662.
49. Li N, Huang X, Sun D, Zhong Y, Chen Z. A sensitive and rapid electrochemical aptasensor based on au@PB for selective detection of mycobacterium tuberculosis antigen MPT64. J Electrochem Soc. 2019;166:B604.
50. Das R, Dhiman A, Mishra SK, Haldar S, Sharma N, Bansal A, Ahmad Y, Kumar A, Tyagi JS, Sharma TK. Structural switching electrochemical DNA aptasensor for the rapid diagnosis of tuberculous meningitis. Int J Nanomedicine. 2019;14:2103–13.
51. Thakur H, Kaur N, Sabherwal P, Sareen D, Prabhakar N. Aptamer based voltammetric biosensor for the detection of mycobacterium tuberculosis antigen MPT64. Microchim Acta. 2017;184:1915–22.
52. Thakur H, Kaur N, Sareen D, Prabhakar N. Electrochemical determination of M. tuberculosis antigen based on poly(3,4-ethylenedioxythiophene) and functionalized carbon nanotubes hybrid platform. Talanta. 2017;171:115–23.
53. Yan G, Wang Y, He X, Wang K, Liu J, Du Y. A highly sensitive label-free electrochemical aptasensor for interferon-gamma detection based on graphene-controlled assembly and nuclease cleavage-assisted target recycling amplification. Biosens Bioelectron. 2013;15(44):57–63.
54. Miodek A, Mejri N, Gomgnimbou M, Sola C, Korri-Youssoufi H. E-DNA sensor of mycobacterium tuberculosis based on electrochemical assembly of nanomaterials (MWCNTs/PPy/PAMAM). Anal Chem. 2015;87(18):9257–64.
55. Wang Y, Mazurek JH, Alocilja EC. Measurement of interferon gamma concentration using an electrochemical immunosensor. J Electrochem Soc. 2016;163(5):B140–5.
56. Haddaoui M, Sola C, Raouafi N, Korri-Youssoufi H. E-DNA detection of rpoB gene resistance in mycobacterium tuberculosis in real samples using Fe_3O_4/polypyrrole nanocomposite. Biosens Bioelectron. 2018;128:76–82.
57. Yan Z, Gan N, Zhang H, Wang D, Qiao L, Cao Y, Li T, Hu F. A sandwich-hybridization assay for simultaneous determination of HIV and tuberculosis DNA targets based on signal amplification by quantum dots-PowerVision™ polymer coding nanotracers. Biosens Bioelectron. 2015;71:207–13.
58. Zhang C, Lou J, Tu W, Baoa J, Dai Z. Ultrasensitive electrochemical biosensing for DNA using quantum dots combined with restriction endonuclease. Analyst. 2014;140:506–11.
59. Torati SR, Reddy V, Yoon SS, Kim C. Electrochemical biosensor for mycobacterium tuberculosis DNA detection based on gold nanotubes array electrode platform. Biosens Bioelectron. 2016;15(78):483–8.
60. Khoder R, Korri-Youssoufi H. E-DNA biosensors of M. tuberculosis based on nanostructured polypyrrole. Mater Sci Eng. 2020;C 108:110371.

61. Prabhakar N, Solanki PR, Kaushik A, Pandey MK, Malhotra BD. Peptide nucleic acid immobilized biocompatible silane nanocomposite platform for mycobacterium tuberculosis detection. Electroanalysis. 2010;22:2672–82.
62. Liu Y, Rahimian A, Krylyuk S, Vu T, Crulhas B, Stybayeva G, Imanbekova M, Shin DS, Davydov A, Revzin A. Nanowire aptasensors for electrochemical detection of cell-secreted cytokines. ACS Sens. 2017;2(11):1644–52.
63. Jin H, Gui R, Gao X, Sun Y. An amplified label-free electrochemical aptasensor of γ-interferon based on target-induced DNA strand transform of hairpin-to-linear conformation enabling simultaneous capture of redox probe and target. Biosens Bioelectron. 2019;145:111732.
64. Wang X, Han X, Ma A, Chen L, Liang H, Litifu A, Xue F. Fabrication of electrochemical immunosensor for interferon-γ determination and its application of tuberculosis diagnosis. Int J Electrochem Sci. 2017;12:7262–71.
65. Gopinath SCB, Perumal V, Kumaresan R, et al. Nanogapped impedimetric immunosensor for the detection of 16 kDa heat shock protein against mycobacterium tuberculosis. Microchim Acta. 2016;183:2697–703.
66. Zhang J, Li Y, Duan S, He F. Highly electrically conductive two-dimensional Ti_3C_2 MXenes-based 16S rDNA electrochemical sensor for detecting mycobacterium tuberculosis. Anal Chim Acta. 2020;1123:9–17.
67. Das MM, Chetna Sumana G, Srivastava A, Vijayan N, Nagarajan R, Malhotra BD. Zirconia grafted carbon nanotubes-based biosensor for M. tuberculosis detection. Appl Phys Lett. 2011;99:143702.
68. Teengam P, Siangproh W, Tuantranont A, Vilaivan T, Chailapakul O, Henry CS. Electrochemical impedance-based DNA sensor using pyrrolidinyl peptide nucleic acids for tuberculosis detection. Anal Chim Acta. 2018;1044:102–9.
69. Perumal V, Saheed MSM, Mohamed NM, Saheed MSM, Murthe SS, Gopinath SCB, Chiu JM. Gold nanorod embedded novel 3D graphene nanocomposite for selective bio-capture in rapid detection of mycobacterium tuberculosis. Biosens Bioelectron. 2018;116:116–22.
70. Tai MJY, Perumal V, Gopinath SCB, et al. Laser-scribed graphene nanofiber decorated with oil palm lignin capped silver nanoparticles: a green biosensor. Sci Rep. 2021;11:5475.
71. Costa MP, Andrade CAS, Montenegro RA, Melo FL, Oliveira MDL. Self-assembled monolayers of mercaptobenzoic acid and magnetite nanoparticles as an efficient support for development of tuberculosis genosensor. J Colloid Interface Sci. 2014;433:141–8.
72. Abnous K, Danesh NM, Ramezani M, Mona A, Koroush YH, Ahmad SE, Amirhossein B, Seyed MT. A triple-helix molecular switch-based electrochemical aptasensor for interferon-gamma using a gold electrode and methylene blue as a redox probe. Microchim Acta. 2017;184:4151–7.
73. Bunyarataphan S, Prammananan T, Japrung D. Ratiometric electrochemical biosensor based on internally controlled duplex PCR for detection of mycobacterium tuberculosis. J Electrochem Soc. 2022;169(2):027501.
74. Blair EO, Hannah S, Vincent V, Hüseyin A, Tanil K, Paul HA, Fatma GD, Damion CK. Biologically modified microelectrode sensors provide enhanced sensitivity for detection of nucleic acid sequences from mycobacterium tuberculosis. Sens Act Rep. 2020;2(1):100008.
75. Zhang QD, March G, Noel V, Piro B, Reisberg S, Tran LD, Hai LV, Abadia E, Nielsen PE, Sola C, Pham MC. Label-free and reagentless electrochemical detection of PCR fragments using self-assembled quinone derivative monolayer: application to mycobacterium tuberculosis. Biosens Bioelectron. 2012;32(1):163–8.
76. Prabhakar N, Singh H, Malhotra BD. Nucleic acid immobilized polypyrrole-polyvinylsulphonate film for mycobacterium tuberculosis detection. Electrochem Commun. 2008;10:821–6.
77. Xia J, Song D, Wang Z, Zhang F, Yang M, Gui R, Xia L, Bi S, Xia Y, Li Y, Linhua X. Single electrode biosensor for simultaneous determination of interferon gamma and lysozyme. Biosens Bioelectron. 2014;68:55–61.

78. Liu Y, Kwa T, Revzin A. Simultaneous detection of cell-secreted TNF-α and IFN-γ using micropatterned aptamer-modified electrodes. Biomaterials. 2012;33(30):7347–55.
79. Liu Y, Yan J, Howland MC, Kwa T, Revzin A. Micropatterned aptasensors for continuous monitoring of cytokine release from human leukocytes. Anal Chem. 2011;83(21):8286–92.
80. Liu Y, Tuleouva N, Ramanculov E, Revzin A. Aptamer-based electrochemical biosensor for interferon gamma detection. Anal Chem. 2010;82(19):8131–6.
81. Díaz-González M, González-García MB, Costa-García A. Immunosensor for mycobacterium tuberculosis on screen-printed carbon electrodes. Biosens Bioelectron. 2005;20(10):2035–43.
82. Wang J, Rivas G, Cai X, Dontha N, Shiraishi H, Luo D, Valera FS. Sequence-specific electrochemical biosensing of M. tuberculosis DNA. Anal Chim Acta. 1997;337(1):41–8.
83. Sypabekova M, Jolly P, Estrela P, Kanayeva D. Electrochemical aptasensor using optimized surface chemistry for the detection of mycobacterium tuberculosis secreted protein MPT64 in human serum. Biosens Bioelectron. 2018;123:141–51.
84. Min K, Cho M, Han SY, Shim YB, Ku J, Ban C. A simple and direct electrochemical detection of interferon-gamma using its RNA and DNA aptamers. Biosens Bioelectron. 2008;23(12):1819–24.
85. Yunus MH, Yusof NA, Raston NHA, Noor SSM, Sulaiman Y, Abdullah J. A novel amperometric aptamer–antibody sandwich assay for the detection of tuberculosis with diazonium electrografted enhanced modified electrode. IEEE Sensors J. 2021;21(20):22442–9.
86. Bizid S, Blili S, Mlika R, Said RH, Korri-Youssoufi H. Direct electrochemical DNA sensor based on a new redox oligomer modified with ferrocene and carboxylic acid: application to the detection of mycobacterium tuberculosis mutant strain. Anal Chim Acta. 2017;994:10–8.
87. Matsishin MJ, Rachkov O, Errachid A, Dzyadevych SV, Alexei S. Development of impedimetric DNA biosensor for selective detection and discrimination of oligonucleotide sequences of the rpoB gene of mycobacterium tuberculosis. Sensors Actuators B Chem. 2015;222:1152–8.
88. Ding S, Mosher C, Lee XY, Das SR, Cargill AA, Tang X, Chen B, McLamore ES, Gomes C, Hostetter JM, Claussen JC. Rapid and label-free detection of interferon gamma via an electrochemical aptasensor comprising a ternary surface monolayer on a gold interdigitated electrode array. ACS Sens. 2017;2(2):210–7.
89. Bart M, Stigter EC, Stapert HR, de Jong GJ, van Bennekom WP. On the response of a label-free interferon-gamma immunosensor utilizing electrochemical impedance spectroscopy. Biosens Bioelectron. 2005;1:49–59.
90. Dijksma M, Kamp B, Hoogvliet JC, van Bennekom WP. Development of an electrochemical immunosensor for direct detection of interferon-gamma at the attomolar level. Anal Chem. 2001;73(5):901–7.
91. Saengdee P, Chaisriratanakul W, Bunjongpru W, Sripumkai W, Srisuwan A, Hruanun C, Poyai A, Phunpae P, Pata S, Jeamsaksiri W, Kasinreak W, Promptmas C. A silicon nitride ISFET based immunosensor for Ag85B detection of tuberculosis. Analyst. 2021;141:5767–75.
92. Farid S, Meshik X, Choi M, Mukherjee S, Lan Y, Parikh D, Poduri S, Baterdene U, Huang C, Wang YY, Burke DM, Stroscio MA. Detection of interferon gamma using graphene and aptamer-based FET-like electrochemical biosensor. Biosens Bioelectron. 2015;71:294–9.
93. Ma J, Du M, Wang C, Xie X, Wang H, Li T, Chen S, Zhang L, Mao S, Zhou X, Wu M. Rapid and sensitive detection of mycobacterium tuberculosis by an enhanced nanobiosensor. ACS Sens. 2021;6:3367.
94. Zhang Y, Zhang B, Ye X, Yan Y, Huang L, Jiang Z, Tan S, Cai X. Electrochemical immunosensor for interferon-γ based on disposable ITO detector and HRP-antibody-conjugated nano gold as signal tag. Mater Sci Eng C. 2016;59:577–84.
95. Zaid MHM, Abdullah J, Yusof NA, Sulaiman Y, Wasoh H, Noh MFM, Issa R. PNA biosensor based on reduced graphene oxide/water soluble quantum dots for the detection of mycobacterium tuberculosis. Sensors Actuators B Chem. 2017;241:1024–34.
96. Wu C, Ma Y, Chen L. A low-fouling and reusable biosensor for the sensitive electrochemical assay of tuberculosis gene. J Electroanal Chem. 2022;907:116013.

97. Zribi B, Roy E, Pallandre A, Chebil S, Koubaa M, Mejri N, Haghiri-Gosnet AM. A microfluidic electrochemical biosensor based on multiwall carbon nanotube/ferrocene for genomic DNA detection of mycobacterium tuberculosis in clinical isolates. Biomicrofluidics. 2016;10(1):014115.la.
98. Chen Y, Pui TS, Kongsuphol P, Tang KC, Arya SK. Aptamer-based array electrodes for quantitative interferon-γ detection. Biosens Bioelectron. 2013;53:257–62.
99. Yesil M, Donmez S, Arslan F. Development of an electrochemical DNA biosensor for detection of specific mycobacterium tuberculosis sequence based on poly (L-glutamic acid) modified electrodes. J Chem Sci. 2016;128:1823–9.
100. Moschou D, Greathead L, Pantelidis P, Kelleher P, Morgan H, Prodromakis T. Amperometric IFN-γ immunosensors with commercially fabricated PCB sensing electrodes. Biosens Bioelectron. 2016;86:805–10.
101. Borisov SM, Wolfbeis OS. Optical biosensors. Chem Rev. 2008;108(2):423–61.
102. Huang X, Zhu Y, Kianfar E. Nano biosensors: properties, applications and electrochemical techniques. J Mater Res Technol. 2021;12:1649–72.
103. Touhami A. Biosensors and nanobiosensors: design and applications. Nanomedicine. 2014;15:374–403.
104. Parolo C, Merkoçi A. Paper-based nanobiosensors for diagnostics. Chem Soc Rev. 2013;42:450–7.
105. Ukhurebor KE, Onyancha RB, Aigbe UO, Uk-Eghonghon G, Kerry RG, Kusuma HS, Balogun VA. A methodical review on the applications and potentialities of using nanobiosensors for disease diagnosis. Biomed Res Int. 2022;
106. Hutter E, Fendler JH. Exploitation of localized surface plasmon resonance. Adv Mater. 2004;16(19):1685–706.
107. Pei X, Hong H, Liu S, Li N. Nucleic acids detection for mycobacterium tuberculosis based on gold nanoparticles counting and rolling-circle amplification. Biosensors. 2022;12(7):448.
108. Ramirez-Priego P, Martens D, Elamin AA, Soetaert P, Van Roy W, Vos R, Anton B, Bockstaele R, Becker H, Singh M, Beinstman P, Lechuga LM. Label-free and real-time detection of tuberculosis in human urine samples using a nanophotonic point-of-care platform. ACS Sens. 2018;3(10):2079–86.
109. Prabowo BA, Alom A, Secario MK, Masim FCP, Lai HC, Hatanaka K, Liu KC. Graphene-based portable SPR sensor for the detection of mycobacterium tuberculosis DNA strain. Proc Engin. 2016;168:541–5.
110. Silvestri D, Sonato A, Ruffato G, Meneghello A, Antognoli A, Cretaio E, Dettin M, Zamuner A, Casarin E, Zacco G, Romanato F, Morpurgo M. A peptide nucleic acid label-free biosensor for mycobacterium tuberculosis DNA detection via azimuthally controlled grating-coupled SPR. Anal Methods. 2015;7(10):4173–80.
111. Prabhakar N, Arora K, Arya SK, Solanki PR, Iwamoto M, Singh H, Malhotra BD. Nucleic acid sensor for M. tuberculosis detection based on surface plasmon resonance. Analyst. 2008;133(11):1587–92.
112. Duman M, Piskin E. Detection of mycobacterium tuberculosis complex and Mycobacterium gordonae on the same portable surface plasmon resonance sensor. Biosens Bioelectron. 2010;26(2):908–12.
113. Rachkov A, Patskovsky S, Soldatkin A, Meunier M. Surface plasmon resonance detection of oligonucleotide sequences of the rpoB genes of mycobacterium tuberculosis. Talanta. 2011;85(4):2094–9.
114. Rachkov A, Patskovsky S, Soldatkin A, Meunier M. Discrimination of single base mismatched oligonucleotides related to the rpoB gene of mycobacterium tuberculosis using a surface plasmon resonance biosensor. Biotechnol Appl Biochem. 2013;60(4):453–8.
115. Hsu SH, Lin YY, Lu SH, Tsai IF, Lu YT, Ho HT. Mycobacterium tuberculosis DNA detection using surface plasmon resonance modulated by telecommunication wavelength. Sensors. 2013;14(1):458–67.

116. Xiang Y, Zhu X, Huang Q, Zheng J, Fu W. Real-time monitoring of mycobacterium genomic DNA with target-primed rolling circle amplification by an au nanoparticle-embedded SPR biosensor. Biosens Bioelectron. 2015;66:512–9.
117. Hong SC, Chen H, Lee J, Park HK, Kim YS, Shin HC, Kim CM, Park TJ, Lee SJ, Koh K, Kim HJ, Chang CL, Lee J. Ultrasensitive immunosensing of tuberculosis CFP-10 based on SPR spectroscopy. Sensors Actuators B Chem. 2011;156(1):271–5.
118. Chen H, Liu F, Koh K, Lee J, Ye Z, Yin T, Sun L. Sensitive detection of tuberculosis using nanoparticle-enhanced surface plasmon resonance. Microchim Acta. 2013;180(5):431–6.
119. Zou F, Wang X, Qi F, Koh K, Lee J, Zhou H, Chen H. Magneto-plasmonic nanoparticles enhanced surface plasmon resonance TB sensor based on recombinant gold binding antibody. Sensors Actuators B Chem. 2017;250:356–63.
120. Sun W, Yuan S, Huang H, Liu N, Tan Y. A label-free biosensor based on localized surface plasmon resonance for diagnosis of tuberculosis. J Microbiol Methods. 2017;142:41–5.
121. Trzaskowski M, Napiórkowska A, Augustynowicz-Kopeć E, Ciach T. Detection of tuberculosis in patients with the use of portable SPR device. Sensors Actuators B Chem. 2018;260:786–92.
122. Chaki NK, Vijayamohanan K. Self-assembled monolayers as a tunable platform for biosensor applications. Biosens Bioelectron. 2002;17(1–2):1–12.
123. Peláez Gutiérrez EC, Estévez MC, Mongui Á, Menéndez MC, Toro C, Herrera-Sandoval OL, Robledo J, Garcia MJ, Portillo PD, Lechuga LM. Detection and quantification of HspX antigen in sputum samples using plasmonic biosensing: toward a real point-of-Care (POC) for tuberculosis diagnosis. ACS Infect Dis. 2020;6(5):1110.
124. Prabowo BA, Chang YF, Lai HC, Alom A, Pal P, Lee YY, Chiu NF, Hatanaka K, Su LC, Liu KC. Rapid screening of mycobacterium tuberculosis complex (MTBC) in clinical samples by a modular portable biosensor. Sensors Actuators B Chem. 2018;254:742–8.
125. Hsieh SC, Chang CC, Lu CC, Wei CF, Lin CS, Lai HC, Lin CW. Rapid identification of mycobacterium tuberculosis infection by a new array format-based surface plasmon resonance method. Nanoscale Res Lett. 2012;7(1):1–6.
126. Mukundan H, Kumar S, Price DN, Ray SM, Lee YJ, Min S, Eum S, Kubicek-Sutherland J, Resnick JM, Grace WK, Anderson AS, Hwang SH, Cho SN, Via LE, Barry C, Sakamuri R, Swanson BI. Rapid detection of mycobacterium tuberculosis biomarkers in a sandwich immunoassay format using a waveguide-based optical biosensor. Tuberculosis. 2012;92(5):407–16.
127. Sutarlie L, Ow SY, Su X. Nanomaterials-based biosensors for detection of microorganisms and microbial toxins. Biotechnol J. 2017;12(4)
128. Shojaei TR, Salleh MAM, Tabatabaei M, Ekrami A, Motallebi R, Rahmani-Cherati T, Hajalilou A, Jorfi R. Development of sandwich-form biosensor to detect mycobacterium tuberculosis complex in clinical sputum specimens. Braz J Infect Dis. 2014;18:600–8.
129. Hwang SH, Im SG, Sung H, Hah SS, Cong VT, Lee DH, Son SJ, Oh HB. Upconversion nanoparticle-based Förster resonance energy transfer for detecting the IS*6110* sequence of mycobacterium tuberculosis complex in sputum. Biosens Bioelectron. 2014;53:112–6.
130. Mansour A, Tammam S, Althani A, Azzazy HM. A single tube system for the detection of mycobacterium tuberculosis DNA using gold nanoparticles-based FRET assay. J Microbiol Methods. 2017;139:165–7.
131. Yang X, Ebrahimi A, Li J, Cui Q. Fullerene–biomolecule conjugates and their biomedicinal applications. Int J Nanomedicine. 2014;9:77.
132. Kim JH, Yeo WH, Shu Z, Soelberg SD, Inoue S, Kalyanasundaram D, Ludwig J, Furlong CE, Riley JJ, Weigel KM, Cangelosi GA, Oh K, Lee KH, Gao D, Chung JH. Immunosensor towards low-cost, rapid diagnosis of tuberculosis. Lab Chip. 2012;12(8):1437–40.
133. Wood A, Barizuddin S, Darr CM, Mathai CJ, Ball A, Minch K, Somoskovi A, Hamasur B, Connelly JT, Weigl B, Andama A, Cattamanchi A, Gangopadhyay K, Bok S, Gangopadhyay S. Ultrasensitive detection of lipoarabinomannan with plasmonic grating biosensors in clinical samples of HIV negative patients with tuberculosis. PLoS One. 2019;14(3):e0214161.

134. Piriya A, Joseph P, Daniel K, Lakshmanan S, Kinoshita T, Muthusamy S. Colorimetric sensors for rapid detection of various analytes. Mater Sci Eng C. 2017;78:1231–45.
135. Wang X, Wan G, Wang Y, Quan S, Qi H, Sun L, Shen C, Huang H, Jiao W, Shen A. Development and preliminary application of multiplex loop-mediated isothermal amplification coupled with lateral flow biosensor for detection of mycobacterium tuberculosis complex. Front Cell Infect Microbiol. 2021;11:666492.
136. Chen X, Huang J, Xiao Z, Yang X, Chen Y, Zheng W, Chen W, Chen H, Li S. Highly specific and sensitive detection of the mycobacterium tuberculosis complex using multiplex loop-mediated isothermal amplification combined with a nanoparticle-based lateral flow biosensor. Braz J Microbiol. 2021;52(3):1315–25.
137. Silva LB, Veigas B, Doria G, Costa P, Inácio J, Martins R, Fortunato E, Baptista PV. Portable optoelectronic biosensing platform for identification of mycobacteria from the mycobacterium tuberculosis complex. Biosens Bioelectron. 2011;26(5):2012–7.
138. Kaewphinit T, Santiwatanakul S, Chansiri K. Colorimetric DNA based biosensor combined with loop-mediated isothermal amplification for detection of mycobacterium tuberculosis by using gold nanoprobe aggregation. Sens Transducers. 2013;149(2):123.
139. Huang J, Xiao Z, Yang X, Chen X, Wang X, Chen Y, Zheng W, Chen W, Chen H, Li S. Two target genes based multiple cross displacement amplification combined with a lateral flow biosensor for the detection of mycobacterium tuberculosis complex. BMC Microbiol. 2021;21(1):1–11.
140. Yang X, Huang J, Chen X, Xiao Z, Wang X, Chen Y, Zheng W, Chen W, Chen H, Lei S, Hu Y, Li S. Rapid and visual differentiation of mycobacterium tuberculosis from the mycobacterium tuberculosis complex using multiplex loop-mediated isothermal amplification coupled with a nanoparticle-based lateral flow biosensor. Front Microbiol. 2021;12:708658.
141. Ariffin N, Yusof NA, Abdullah J, Abd Rahman SF, Ahmad Raston NH, Kusnin N, Suraiya S. Lateral flow immunoassay for naked eye detection of mycobacterium tuberculosis. J Sens. 2020;2020:1.
142. Kim J, Lee KS, Kim EB, Paik S, Chang CL, Park TJ, Kim HJ, Lee J. Early detection of the growth of mycobacterium tuberculosis using magnetophoretic immunoassay in liquid culture. Biosens Bioelectron. 2017;96:68–76.
143. Trilling AK, Harmsen MM, Ruigrok VJ, Zuilhof H, Beekwilder J. The effect of uniform capture molecule orientation on biosensor sensitivity: dependence on analyte properties. Biosens Bioelectron. 2013;40(1):219–26.
144. Liu Q, Lim BKL, Lim SY, Tang WY, Gu Z, Chung J, Barkham T. Label-free, real-time and multiplex detection of mycobacterium tuberculosis based on silicon photonic microring sensors and asymmetric isothermal amplification technique (SPMS-AIA). Sensors Actuators B Chem. 2017;255:1595–603.
145. Schmidt R, Jacak J, Schirwitz C, Stadler V, Michel G, Marmé N, Schutz GJ, Hoheisel JD, Knemeyer JP. Single-molecule detection on a protein-array assay platform for the exposure of a tuberculosis antigen. J Proteome Res. 2011;10(3):1316–22.
146. Ng BY, Wee EJ, West NP, Trau M. Naked-eye colorimetric and electrochemical detection of mycobacterium tuberculosis-toward rapid screening for active case finding. ACS Sens. 2016;1(2):173–8.
147. Joon D, Nimesh M, Gupta S, Kumar C, Varma-Basil M, Saluja D. Development and evaluation of rapid and specific sdaA LAMP-LFD assay with Xpert MTB/RIF assay for diagnosis of tuberculosis. J Microbiol Methods. 2019;159:161–6.
148. Pariwono AM, Lo T, Lim CS, Wang SX, Chan YW. Rapid tuberculosis detection technique for on-site patient screening. J Biomed Pharm Eng. 2007;1(1):27–33.
149. Denton KA, Kramer MF, Lim DV. Rapid detection of mycobacterium tuberculosis in lung tissue using a fiber optic biosensor. J Rapid Meth Aut Mic. 2009;17(1):17–31.
150. Roskos K, Hickerson AI, Lu HW, Ferguson TM, Shinde DN, Klaue Y, Niemz A. Simple system for isothermal DNA amplification coupled to lateral flow detection. PLoS One. 2013;8(7):e69355.

151. Chu ZJ, Xiao SJ, Liu YH, Xiong GL, Huang DJ, Wang SP, Zhao XJ, Zhang ZB. Rapid and sensitive detection of the IS6110 gene sequences of mycobacterium tuberculosis based on hybridization chain reaction and reusable magnetic particles. Sensors Actuators B Chem. 2019;282:904–9.
152. Chang C, Lin S, Lee C, Chuang T, Hsueh P, Lai H, Lin C. Amplified surface plasmon resonance immunosensor for interferon-gamma based on a streptavidin- incorporated aptamer. Biosens Bioelectron. 2012;37(1):68–74.
153. Lin DZ, Chuang PC, Liao PC, Chen JP, Chen YF. Increasing the spectral shifts in LSPR biosensing using DNA-functionalized gold nanorods in a competitive assay format for the detection of interferon-γ. Biosens Bioelectron. 2016;81:221–8.
154. Hu K, Liu J, Chen J, Huang Y, Zhao S, Tian J, Zhang G. An amplified graphene oxide-based fluorescence aptasensor based on target-triggered aptamer hairpin switch and strand-displacement polymerization recycling for bioassays. Biosens Bioelectron. 2013;42(1):598–602.
155. Liu G, Zhang K, Ma K, Care A, Hutchinson MR, Goldys EM. Graphene quantum dot based "switch-on" nanosensors for intracellular cytokine monitoring. Nanoscale. 2017;9(15):4934–43.
156. Taghdisi SM, Danesh NM, Ramezani M, Yazdian-Robati R, Abnous K. An amplified fluorescent aptasensor based on single-stranded DNA binding protein, copper and silica nanoparticles for sensitive detection of interferon-gamma. Anal Chim Acta. 2017;984:162–7.
157. Dhenadhayalan N, Sriram MI, Lin KC. Aptamer-based fluorogenic sensing of interferon-gamma probed with ReS2 and TiS2 nanosheets. Sens Actuators B Chem. 2018;258:929–36.
158. Jeon J, Lee J, So J, Lee JB, Lee H, Chang Y, Shin S, Jo J, Ban C. Homogeneous fluorescent aptasensor for active tuberculosis diagnosis by direct quantification of circulating TB7.7 based on aptamer beacon with graphene oxide. Sens Actuators B Chem. 2020;317:128126.
159. Jiang J, He Y, Yu X, Zhao J, Cui H. A homogeneous hemin/G-quadruplex DNAzyme based turn-on chemiluminescence aptasensor for interferon-gamma detection via in-situ assembly of luminol functionalized gold nanoparticles, deoxyribonucleic acid, interferon-gamma and hemin. Anal Chim Acta. 2013;791:60–4.
160. Pan L, Huang Y, Wen C, Zhao S. Label-free fluorescence probe based on structure-switching aptamer for the detection of interferon gamma. Analyst. 2013;138(22):6811–6.
161. Ma K, Zhang F, Sayyadi N, Chen W, Anwer AG, Care A, Xu B, Tian W, Goldys EM, Liu G. "Turn-on" fluorescent aptasensor based on AIEgen labeling for the localization of IFN-γ in live cells. ACS Sens. 2018;3(2):320–6.
162. Uhou OV, Waryo TT, Douman SF, Januarie KC, Nwambaekwe KC, Ndipingwi MM, Iwuoha EI. Bioanalytical methods encompassing label-free and labeled tuberculosis aptasensors: a review. Anal Chim Acta. 2022;1234:340326.
163. Huang J, Liang Z, Liu Y, Zhou J, He F. Development of an MSPQC nucleic acid sensor based on CRISPR/Cas9 for the detection of mycobacterium tuberculosis. Anal Chem. 2022;94(32):11409–15.
164. He F, Xiong Y, Liu J, Tong F, Yan D. Construction of au-IDE/CFP-10-ESAT-6 aptamer/DNA-AuNPs MSPQC for rapid detection of mycobacterium tuberculosis. Biosens Bioelectron. 2016;77:799–804.
165. Domínguez CM, Kosaka PM, Sotillo A, Mingorance J, Tamayo J, Calleja M. Label-free DNA-based detection of mycobacterium tuberculosis and rifampicin resistance through hydration induced stress in microcantilevers. Anal Chem. 2015;87(3):1494–8.
166. Zhang X, Feng Y, Yao Q, He F. Selection of a new mycobacterium tuberculosis H37Rv aptamer and its application in the construction of a SWCNT/aptamer/au-IDE MSPQC H37Rv sensor. Biosens Bioelectron. 2017;98:261–6.
167. Zhang J, Huang J, He F. The construction of mycobacterium tuberculosis 16S rDNA MSPQC sensor based on exonuclease III-assisted cyclic signal amplification. Biosens Bioelectron. 2019;138:111322.

168. Ren J, Ma L, Li Z, Huang H, Yi S. Simultaneous and early detection of mycobacterium tuberculosis resistance to antituberculosis drugs using an indirect series piezoelectric system. Biosens Bioelectron. 2013;43:115–9.
169. Mi X, He F, Xiang M, Lian Y, Yi S. Novel phage amplified multichannel series piezoelectric quartz crystal sensor for rapid and sensitive detection of mycobacterium tuberculosis. Anal Chem. 2012;84(2):939–46.
170. Ren J, He F, Yi S, Cui X. A new MSPQC for rapid growth and detection of mycobacterium tuberculosis. Biosens Bioelectron. 2008;24(3):403–9.
171. Kaewphinit T, Santiwatanakul S, Promptmas C, Chansiri K. Detection of non-amplified mycobacterium tuberculosis genomic DNA using piezoelectric DNA-based biosensors. Sensors. 2010;10(3):1846–58.
172. Kaewphinit T, Santiwatanakul S, PromptmasC CK. Development of piezoelectric DNA-BASED biosensor for direct detection of Mycobaterium tuberculosis in clinical specimens. Sens Transducers. 2010;113:115–26.
173. Hiatt LA, Cliffel DE. Real-time recognition of mycobacterium tuberculosis and lipoarabinomannan using the quartz crystal microbalance. Sensors Actuators B Chem. 2012;174:245–52.
174. Kaewphinit T, Santiwatanakul S, Chansiri K. Gold nanoparticle amplification combined with quartz crystal microbalance DNA based biosensor for detection of mycobacterium tuberculosis. Sens Transducers. 2012;146:156–63.
175. Ratnatunga CN, Lutzky VP, Kupz A, Doolan DL, Reid DW, Field M, Bell SC, Thomson RM, Miles JJ. The rise of non-tuberculosis mycobacterial lung disease. Front Immunol. 2020;11:303.
176. Alffenaar JW, Märtson AG, Heysell SK, Cho JG, Patanwala A, Burch G, Kim HY, Sturkenboom MG, Byrne A, Marriott D, Sandaradura I. Therapeutic drug monitoring in non-tuberculosis mycobacteria infections. Clin Pharmacokinet. 2021;60:711–25.
177. Bryant JM, Grogono DM, Rodriguez-Rincon D, Everall I, Brown KP, Moreno P, Verma D, Hill E, Drijkoningen J, Gilligan P, Esther CR. Population-level genomics identifies the emergence and global spread of a human transmissible multidrug-resistant nontuberculous mycobacterium. Science (New York, NY). 2017;6313:751.
178. Brown-Elliott BA, Griffith DE, Wallace RJ. Diagnosis of nontuberculous mycobacterial infections. Clin Lab Med. 2002;22(4):911–25.
179. Poch D, Sodia T, Honda JR, Bonham AJ. Electrochemical DNA biosensors for detecting nontuberculous mycobacteria (NTM). NTM. 2007;2007:6.
180. Kumanan V, Nugen SR, Baeumner AJ, Chang YF. A biosensor assay for the detection of Mycobacterium avium subsp. paratuberculosis in fecal samples. J Vet Sci. 2009;10(1):35–42.
181. Chen C, Lu J, Long B, Rao Z, Gao Y, Wang W, Gao W, Yang J, Zhang S. Detection of mycobacterium kansasii using a combination of loop-mediated isothermal amplification (LAMP) and lateral flow biosensors. Int Microbiol. 2021;24:75–82.
182. Sankar S, Ramamurthy M, Nandagopal B, Sridharan G. An appraisal of PCR-based technology in the detection of mycobacterium tuberculosis. Mol Diagn Ther. 2011;15(1):1–11.
183. Ramulu TS, Venu R, Sinha B, Lim B, Jeon SJ, Yoon SS, Kim CG. Nanowires array modified electrode for enhanced electrochemical detection of nucleic acid. Biosens Bioelectron. 2013;40(1):258–64.

Chapter 13
Nontuberculous Mycobacterium Infections in Lung Disease and Medical Interventions

Anjana Suresh, Gayathri S. Kamath, Fidha Latheef, Greeshma Sasikumar, and Sreejith Parameswara Panicker

Abstract Infection with nontuberculous mycobacteria (NTM) is on the rise all across the world. Lung disease due to NTM is more prevalent and frequent and is increasingly recognized worldwide. Primarily because approximately 200 different species of NTM have been identified, *Mycobacterium avium* complex (MAC), *Mycobacterium abscessus* complex (MABC) and *Mycobacterium kanasii* are mostly responsible for lung infections. Nontuberculous mycobacteria are opportunistic pathogens that harm humans with underlying lung disease or weak immune systems, although anyone can get an NTM infection. There are three main patient groups that are more susceptible to getting NTM lung disease: those with immunologic diseases or suspected genetic conditions that increase the risk of developing bronchiectasis and/or lung infections; those with anatomic lung abnormalities; and people with no significant lung or immunological issues. Previously, it was supposed that NTM transmissions between individuals were physically impossible. Recent research has shown that this is actually feasible. Two different types of NTM lung disease appear rapidly progressive cavitary disease and less severe nodular bronchiectasis. Diagnosis method is different for rapid growing and slow growing NTM. Molecular techniques have replaced traditional biochemical testing. The treatment of NTM disorders requires long-term, targeted antibiotic medication. However, determining the most effective course of action and length of therapy is difficult given the present limitations of prospective, controlled, and randomized treatment studies. Due to the lack of prospective, controlled, and randomized treatment studies, determining the optimal treatment plans and durations for NTM disorders is difficult. This article summarizes the current understanding of NTM lung illness and treatment options.

Keywords Lung disease · NTM · Diagnosis · Treatment

A. Suresh · G. S. Kamath · F. Latheef · G. Sasikumar · S. P. Panicker (✉)
Department of Zoology, Advanced Centre for Regenerative Medicine and Stem Cell in Cutaneous Research (AcREM-Stem), University of Kerala, Thiruvananthapuram, Kerala, India
e-mail: psreejith@keralauniversity.ac.in

A. Singh, D. Sharma (eds.), *Diagnosis of Mycobacterium*,
https://doi.org/10.1007/978-981-99-5624-1_13

13.1 Introduction

Nontuberculous mycobacterium (NTM), commonly known as environmental mycobacteria, environmental opportunistic pathogens, atypical mycobacteria, and MOTT (mycobacteria other than tuberculosis) are types of mycobacteria that do not develop in leprosy or tuberculosis found in humans and animals [1, 2]. Nontuberculous mycobacteria, often called environmental mycobacteria, are found in large quantities in moist soil, marshland, and streams, including natural and municipal water sources [3]. Different NTM species favor various types of environments. Mycobacterium organisms other than tuberculosis were discovered rapidly after Koch discovered tuberculosis in 1882, but it was not until the 1950s it was recognized that these organisms might cause disease in humans [4]. Unlike tuberculosis and leprosy, NTM rarely spreads from animal to human or human to human. Everyone inhales NTM, however only some people are affected by it [2]. If we already have health issues or a compromised immune system, we may contract NTM, which can infect several body parts [5]. These infections can arise in other organs but typically start in the lungs [6]. There have been identified over 200 different forms of NTM mycobacteria [7]. Nontuberculous mycobacterial lung disease (NTM-LD) is caused by three major human pathogen groups: *Mycobacterium avium* complex (MAC), *Mycobacterium abscessus* complex (MABC), and *Mycobacterium kansasii* [8–10]. MAC mainly includes *Mycobacterium intracellulare* and *Mycobacterium avium* [11], and MABC primarily includes *Mycobacterium massiliense* and *Mycobacterium abscessus* [12]. The other most frequent causes of NTM disease are *M. fortuitum*, and *M. kansasii*, which are very challenging to treat [8, 13]. NTM infection affects individuals with gastroesophageal reflux illness, lung disorder-related emphysema, bronchiectasis, cystic fibrosis, chronic obstructive pulmonary disease (COPD), and asthma [14, 15]. It also affects those with weakened immune systems, such as the elderly, and those on immunosuppressant medications for medical purposes [5]. NTM disease affects older women who are tall and slender and have inherited problems such as scoliosis, mitral valve prolapse, or pectus excavatum (for those who develop pulmonary MAC) [16]. Lungs are the location of most NTM infections [6, 17]. However, NTM lung infections are mainly brought on by MAC, *M. kasassii and M. abscessus* [18]. Possibly other NTM potentially lead to pulmonary disease in immune-competent as well as immune-compromised individuals [6, 19].

Blood, bones, lymph nodes, skin, and soft and smooth tissue (skin nodules or infections in surgical incisions after an operation) can all get infected with NTM [5, 20]. Many people obtain NTM infections by inhaling mycobacteria from the environment, drinking, bathing, or showering in water contaminated by NTM, or being exposed to them through a surgical instrument [6, 21]. NTM are present in all areas of the environment, with soil and water sources containing most of them. They are connected to the development of biofilms [22], which leads to antibiotic resistance and disinfectants [23, 24]. Due to NTM's hydrophobicity, several of these organisms typically aerosolize from water and are resistant to high temperatures and relatively low pH [25]. NTM are most frequently categorized according to their low or fast growth pace. MAC is a slow-growing NTM. Even though it is a slow-growing organism, *Mycobacterium kasassii* is the second most prevalent reason for lung infections in the United States. It is also the cause of some infections in England's pockets [18, 26]. The third most common cause of pulmonary disease is *M. abscessus* and the most commonly isolated rapidly developing NTM [27]. Other NTM species can cause lung illness in immune-competent as well as immuno-impaired hosts [27, 28].

The host, pathogen, and environment interact in a complicated way during NTM infection. These organisms' hydrophobic cell walls with high lipid content are particularly suited to creating biofilms, which allow bacterial colonies to survive over time while remaining resistant to antiseptics and generating aerosols [22, 29]. Aerosols from showers have a substantially greater organism density, which is regarded as the main cause of NTM lung infection [30, 31]. It is entirely unclear whether an infection begins to transmit and how it accomplishes so. Identifying NTM at the subspecies level and characterizing novel species are parts of the NTM lung disease diagnosis [32]. Treating NTM lung disease is difficult, and each patient's risks and benefits must be carefully considered. An NTM-LD diagnosis requires particular symptoms, a clinical assessment, and laboratory tests. Individuals with NTM lung disease receive different treatments [32, 33].

13.2 Epidemiology of NTM Lung Disease

NTM infection is becoming more frequent and more prevalent [34]. Surveillance and claim data collected in laboratories have been used to identify the majority of the epidemiology of NTM-PD. The probability of definitive was NTM disease confirmed to be 1.20 cases/100,000 population in an evaluation of NTM patients identified at the Danish reference mycobacteriology lab, with no notable change over 25 years [35]. The most prevalent NTM diseases worldwide are MAC organisms, though prevalence varies widely by age, gender, and geographic location. MABS (*M. abscessus* complex) is a significant issue because of the extremely high levels of antibiotic resistance and the increasing prevalence of the disease in East Asian nations, including Japan, Korea, and Taiwan [36]. Patients with NTM face a particularly difficult condition if they suffer from cystic fibrosis, the most frequent genetic condition among Caucasians, who are more susceptible to MABS infection [37]. A sufficient understanding of epidemiology is difficult since NTM infections, unlike TB, are unrelated to public health reporting. The majority of epidemiologic data also originates from the United States, Japan, and Europe, therefore it's possible it doesn't accurately reflect changes in prevalence around the globe. According to numerous studies conducted in five US states, from 8.2 cases per 100,000 people in 1994 to 16 cases per 100,000 people in 2014, the rate of NTM-positive cultures has increased [38]. Similar findings were discovered in 2017, according to a Canadian researcher. The prevalence of the disease increased from 4.65 incidents per 100,000 people in 1998 to 9.08 cases per 100,000 people in 2010 [39]. The laboratory isolation rate increased to 22 isolates per 100,000 persons in 2010. In Queensland (QLD), Australia, NTM disease has been particularly severe since the TB control program began in the 1960s. From 1999 to 2005, it increased from 9.1 to 13.6 incidences per 100,000. From the same time in 2012 to the same time in 2016, the number of reported isolates went from 672 to 1171 [40, 41]. Recent clinical data suggest that the increased incidence and prevalence of NTM-PD in women aged 50 and older may be exaggerated among patients without adequate testing, according to epidemiological research of nontuberculous mycobacterial lung disease in South Korea in 2022 [42]. Another study by Victor et al. (2022) reveals that NTM infection and absolute disease numbers generally changed annually by 2.0 (95% CI: 1.6–2.3) and 0.5 (95% CI: 0.3–0.7). Along with infection by *Mycobacterium abscessus* complex ($n = 15/23$, 65.2%) and disease ($n = 2/8$, 25.0%), infection by *Mycobacterium avium* complex ($n = 15/19$, 78.9%) and disease ($n = 10/12$, 83.9%) also demonstrated an increase [43].

13.3 Pathogenesis and Virulence Mechanism of NTM

Patients with immunologic diseases or suspected genetic conditions that increase their risk of developing bronchiectasis and/or lung infections [44–48], patients with anatomic lung abnormalities [31, 49, 50], and patients without known lung or immunologic abnormalities [51–53] are the three main patient groups most likely to develop NTM lung disease. Previously, it was thought that NTM transmissions from person to person were impossible [6]. The research results of a study performed by Bryant et al. in 2013 on 1080 clinical isolates of the *M. abscessus* complex from 517 patients in cystic fibrosis hospitals in seven different countries revealed that the majority of these infections were spread from person to person, potentially by fomites and aerosols [54, 55]. The most obvious method to get pulmonary illness is by aerosolizing droplets penetrating the alveoli. Airway-specific NTM infection affects bronchial epithelial cells in the lungs. Fibronectin attachment protein of NTM facilitates more targeted mycobacterial engagement to integrin receptors by attaching to fibronectin on mucosal surfaces [56]. The NTM creation of biofilms and the suppression of inflammatory cytokine production by NTM facilitate the bronchial epithelium's invasion. After establishing a link with the complement receptors on the phagocytes, NTM was opsonized with C3b and C4b and could enter alveolar macrophages. Immune-evasion strategies are used by NTM to persist inside macrophages. NTM-related gene activation enhances replication and the activation of apoptosis in macrophages [56, 57]. Phospholipids, apolar lipids, amphipathic lipids, and glycolipids found in the NTM envelope inhibit the synthesis of interferon gamma (IFN-γ) and tumor necrosis factor-alpha (TNF-α), two cytokines necessary for suppressing NTM infections [58, 59]. Several studies show that NTM lipids can stimulate TH2 immune responses, and decrease TH1 immunological responses. Even though lipids produced from NTM can modify immunological responses, it improves the bacilli's survival [59, 60]. The glycopeptidolipids (GPL), which are only produced by NTM organisms and not by other mycobacterium species [61, 62], are crucial for NTM's sliding phenotype and biofilm production [63]. Lipoarabinomannans (LAM) is another NTM virulence factor that has been identified. Differentiated THP-1 human macrophages produce IL-12, TNF-α, and IL-89 as a result of PILAM, one of the types of LAM [64]. Purified AraLAM, a different type of LAM, causes an acute infection in the lungs of genetically altered mice [65]. The pattern-recognition receptor dendritic cell-specific intercellular adhesion molecule-3-grabbing non-integrin (DC-SIGN) rapidly recognizes ManLAM, other PLAM [66]. According to certain research, rough NTM variations are associated with more severe and chronic lung disease [64]. The NTM cell envelope's lipids and lipoproteins may contribute to the enhancement of pathogenesis and disease development. NTM lipids' functional relevance is still mostly unclear. NTM-LD pathogenesis is largely determined by infection of airway epithelial cells. However, other cell types must also be thought about because cytokines and other mediators, such as prostaglandins secreted by NTM-infected epithelial cells, may affect the ability of dendritic cells, lymphocytes, and macrophages to prevent infection [67].

13.4 NTM Infection and Lung Disease

NTM lung disease can manifest in two distinct ways. Both the less progressive and more progressive forms nodular bronchiectasis is a term used to describe the less severe form of the disease. Most severe variant of NTM lung disease is cavitary disease. Airway inflammation caused by NTM infection is a less severe but still serious condition because it damages and destroys the airways over time [68]. Continued mucus buildup and subsequent recurrence of respiratory infections like bronchitis and pneumonia are hallmarks of chronic obstructive pulmonary disease. Many elderly women who have never smoked develop nodular bronchiectatic NTM disease. As a result of the NTM infection, lung tissue can become scarred, fibrotic, or pitted and eventually fail to function properly. This damage could potentially cause respiratory failure. The most of these cases are identified in smokers with a lung condition like bronchiectasis or chronic obstructive pulmonary disease. Figure 13.1 shows the disease leads to the development of NTM lung disease.

Many different diseases can have the same symptoms as NTM lung disease. These diseases include consolidation tuberculosis, atelectasis, groundglass opacities, bronchiectasis, aspergillosis, bronchiectasis, recurrent aspiration pneumonitis, nodules, tree-in-bud opacities (branching centrilobular nodules) with and without cavities, blastomycosis, histoplasmosis, and coccidiomycos [69, 70]. A complicated interaction between exposure host-related variables (particle size, number of organisms, and time) and host-related variables is expected to have an effect on the effects of inhaling NTM (immune status, genetic background, presence of localized or generalized lung damage) [71].

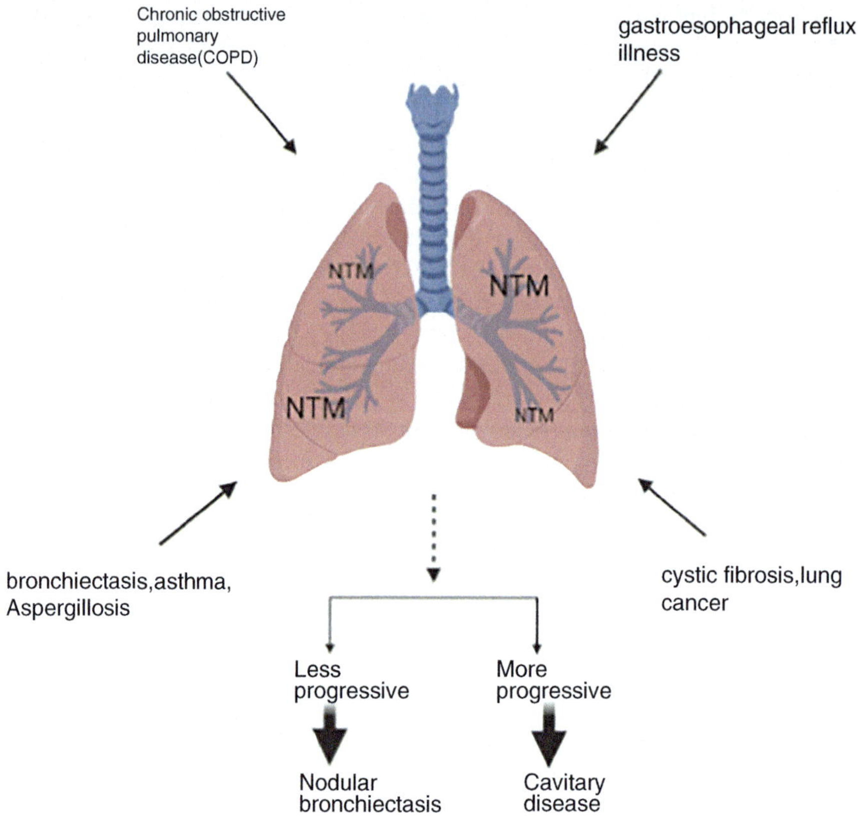

Fig. 13.1 Disease leads to the development of nontuberculosis mycobacterium infection and two forms of NTM lung disease

Mycobacteria-caused pulmonary lung disease includes the Rapidly growing Mycobacterium (RGM), particularly *Mycobacterium abscessus*, *Mycobacterium szulgai*, *Mycobacterium kansasii*, and *Mycobacterium xenopi* (*M. abscessus* and *M. bolletii*) [72, 73]. The biggest factor contributing to NTM lung disease globally is MAC. The MAC includes three mycobacterial species responsible for over half of all mycobacterial infections: *M. avium*, *M. intracellulare*, and *M. chimaera* [6, 74]. NTM isolation from respiratory specimens may have therapeutic significance due to changes in the distribution of ambient microbes and the frequency of host risk factors. Quantitative mycobacterial exposure is linked to clinical respiratory disease development [75, 76].

More than half of those with *M. abscessus* infection or the nodular bronchiectatic form of MAC do not smoke or suffer from other respiratory issues [77, 78]. Scoliosis, mitral valve prolapses, a sunken chest pectus excavatum, and recessive alterations in the cystic fibrosis regulator gene are among other possible symptoms. NTM infections, especially MAC or *M. abscessus*, *M. kansasii* in patients with underlying

COPD, drug-induced immune system suppression, HIV infection, or cancer are more prevalent in people with cystic fibrosis (CF) bronchiectasis and non-CF bronchiectasis [79–81]. The chance of getting NTM is also higher in people who lack the interferon gamma receptor, STAT-1 (signal transducer and activator of transcription), GATA2 (GATA-binding factor 2), and auto-antibodies to interferon gamma [82, 83]. The drug used to treat rheumatoid arthritis and other connective tissue diseases, known as a tumour necrosis factor-alpha antagonist, also increases patients risk for NTM infection [6, 84]. Congenital contractural arachnodactyly, hyper-IgE syndrome, and Marfan syndrome have all been associated with NTM lung disease. NORD (National Organization for Rare Disorders) has more details on some of these conditions [69, 85].

The disease is more likely linked to NTM infection in people with diffuse bronchiectasis [86]. The prevalence of NTM pulmonary infection related to bronchiectasis is rising globally. Most individuals with NTM lung infections also have bronchiectasis, or enlarged airways [87]. Additionally, to make breathing even more difficult, bronchiectasis also brings on other painful symptoms, including coughing, fatigue, wheezing, fevers, and chills. High-resolution computed tomography (HRCT) scanning shows abnormalities that differ in pattern and distribution based on the underlying aetiology of bronchiectasis. There are reports of many tiny nodules along with diffuse bronchiectasis being present in the usual HRCT results of NTM lung infection [88–90], which was also suggested by Wickremasinghe et al [91]. About 34–50% of individuals with these typical HRCT characteristics also had active NTM lung infection, particularly infection with *Mycobacterium avium* complex [89, 91]. The lingular region of the left upper lobe and the right middle lobe exhibits the most severe abnormalities when NTM lung infection is present (nodular bronchiectatic disease). On the HRCT scan, numerous tiny nodules were reported to be peribronchial granulomas and caseous material. According to research by Kwak et al., developing NTM-PD in patients with previously NTM-negative bronchiectasis is associated with increasing radiographic findings [92]. NTM pulmonary disease was seen in 23.3% of bronchiectasis patients, according to the current study by Hongjun Yin et al. According to HRCT pictures, the majority of patients (72.3%) with NTM pulmonary disease received standard anti-NTM treatment [93]. Once bronchiectasis-related structural lung disease and airflow restriction occur, bacterial colonization and chronic infection coupled with inadequate mucus clearance results in persistent inflammatory disease and further tissue deterioration [94]. The fibrocavity subtype of NTM lung disease affects older males with underlying COPD and a habit of smoking. It has holes and patches of enhanced opacity (upper lobes) [95]. It is uncommon to see traction bronchiectasis pleural thickening and volume loss due to fibrosis. Unlike pulmonary tuberculosis, NTM lung disease frequently affects the pleura and results in thin-walled cavities without atelectasis or rigidity of the lymph nodes [96]. The middle and lower lung fields are affected by bilateral multilobar bronchiectasis, characterized by tiny nodules [32]. This subtype, also referred to as Lady Windermere Syndrome, primarily affects aged nonsmoking women without chronic pulmonary problems. Low body mass index, white race, pectus excavatum, scoliosis, and mitral valve prolapse are other features that are usually present [21, 97]. Chest radiographs may show NTM and nodular bronchiectasis, however, HRCT shows these conditions better.

Clusters of small nodules, or the "tree-in-bud sign," typically measuring less than 0.5 mm, are identifiable findings. Larger nodules that may have a cavitation or not and are suspicious of cancer may develop. Two chest radiologists examined patients with NTM lung disease. They discovered that the lungs contain branching centrilobular nodules [27]. In nodules caused by NTM, fluorodeoxyglucose (FDG) uptake during a positron emission tomography (PET) has been observed [98]. Atelectasis or cystic or saccular bronchiectasis may be seen in infected lung parenchyma [99].

In 2.0–8.5% of people with lung cancer, nontuberculous mycobacterium (NTM) pulmonary illness has been identified [39, 100–102]. Although lung cancer and NTM pulmonary illness coexist in the same nodule or mass, it is rare [101–103]. NTM lung disease, which can present as a discrete pulmonary nodule or a mass that mimics lung cancer, can be detected by computed tomography (CT) [103]. Case report research by Naohiro Taira et al. recommends that the risk of lung cancer and NTM infection is thought to be higher in patients with a solitary lung mass and positive sputum or bronchial lavage fluid culture result [104]. The research by Kusumoto et al. revealed the clinical and radiological aspects of eight lung cancer patients with MAC-PD as well as the prevalence of lung cancer development in NTM-LD patients [105]. He suggests that NTM-LD may increase the risk of developing lung cancer. Furthermore, their research emphasized the value of routine chest CT follow-up in detecting lung cancer early and resulting in better outcomes. In the Atsuhisa Tamura et al. research, 25 (2.0%) of the 1258 lung cancer patients had bronchial washings that tested positive for the MAC but negative for other NTM markers [106]. Lung cancer is positively correlated with MAC culture positive bronchial washing. The most recent study by J. Conic et al. (2022) showed that common symptoms such as coughing, shortness of breath, hemoptysis, and weight loss could be signs of NTM and lung cancer in their six-patient case series [107]. The most frequent NTM symptoms were ground glass opacities and bronchiectasis, while masses and cavitation could be signs of NTM and lung cancer.

Aspergilloma, ABPA, and invasive aspergillosis are the three classic subgroups of lung disease caused by *A. fumigates* [108]. Two more types of invasive aspergillosis are angio-invasive aspergillosis and chronic, necrotizing pulmonary aspergillosis (CNPA) [109, 110]. The colonization and infection of *Aspergillus* are also more likely to occur in patients with chronic lung illness [111]. Insufficient research has been done to determine the exact cause of the link between NTM and Aspergillus lung infection. Two main hypotheses are as follows: patients with NTM infection frequently use long-term broad-spectrum antibiotics, which, because of selective pressure, raise the danger of fungus colonization. Eventually, this might result in an Aspergillus lung infection, particularly in individuals with underlying structural lung disorders [111]. NTM lung disease is also commonly identified in immunocompromised people, particularly those with chronic lung problems and those who concurrently used immunosuppressive medications that made them more susceptible to Aspergillus infections [112, 113]. NTM lung disease causes severe lung lesions, such as lung cavitation, which enhances the incidence of chronic pulmonary aspergillosis (CPA) [112, 113].

Aspergillomas may arise from large cavity lesions of an inactive *M. kansasii* infection [114], which can complicate lung disease carried on by *M. xenopi* [115]. CNPA may coexist with (MAC) infection in patients. Invasive aspergillosis' indolent form has been linked to NTM infection [116]. Allergy-related bronchopulmonary aspergillosis (ABPA) and NTM infection may coexist [117].

In their case-control investigation, Kunst et al. found that radiological evidence of *Aspergillus*-related lung illness and *Aspergillus* serology positive were more common in bronchiectasis with NTM than in controls [118]. In comparison to patients without the nontuberculous mycobacterial disease, those with the frequency of coexisting Aspergillus-related lung illness are greater than those with bronchiectasis. A rapidly evolving infectious lung disease with a greater mortality rate is chronic pulmonary aspergillosis (CPA) following NTM lung disease. The highest risk group includes people with fibrocavitary disease. Numerous research supports NTM's significance as one of the new CPA risk factors [119]. More people are becoming aware of chronic pulmonary aspergillosis (CPA) after nontuberculous mycobacterial (NTM) lung illness, especially in nations where TB is not widespread. The prevalence of it varies from 3.9 to 16.7%.

In the host immunological response to NTM, interferon gamma (IFN-γ) and interleukin-12 (IL-12) both play significant roles. When these pathways are compromised, NTM infections are more likely to be acquired [120]. In both isolated cases and epidemic families, NTM infections are brought on by faulty IFN-receptors. However, no successful clinical trials of aerosolized IFN-therapy have been performed [121].

Macrophages and monocytes generate tumour necrosis factor-alpha (TNF-α), which is regulated more by INF-γ and IL-12, which plays a significant part in managing mycobacteria (NTM). The use of TNF-blocking medications demonstrates the crucial role TNF plays in preventing and treating intracellular infections. TNF-α blocking medications should only be administered to patients with active NTM disease if they are also obtaining appropriate treatment [122, 123]. TNF-blocking drugs increase the risk of invasive fungal infections such as aspergillosis, histoplasmosis, and coccidioidomycosis in individuals [124]. According to the most recent study by Dong Won Park et al. (2022), TNF inhibitors enhance the risk of nontuberculous mycobacteria in individuals with seropositive rheumatoid arthritis who reside in areas where *Mycobacterium tuberculosis* is prevalent [4].

Pulmonary alveolar proteinosis, the use of anti-TNF medications, and aspiration from swallowing difficulties or laryngopharyngeal reflux are all serious risks of NTM lung disease [125, 126]. According to increasing data, NTM lung infection may be produced by aspiration induced by spillage of oropharyngeal secretions due to swallowing difficulties brought on by GER (Gastroesophageal reflux) of materials from the esophagus or stomach. GER was detected in 12–28% of non-NTM-infected controls and 26–44% of NTM lung disease patients among three trials [127–130]. Every individual experiences NTM-LD symptom differently [131]. While others may experience significant issues, some might have minor symptoms.

One of the severe symptoms that might appear along with a cough that occasionally contains blood is excessive mucus production [131, 132]. Other severe symptoms can include shortness of breath, a lack of appetite, loss of weight, and sleeping sweats. Ninety four percent of patients have a chronic nontuberculous mycobacterium pulmonary infection, which can lead to long-term lung damage and impaired lung function [132].

13.5 Diagnosis of NTM-LD

Specific symptoms, a clinical examination, and laboratory testing are also necessary to diagnose NTM-LD. The most current update of the joint guidelines from the American Thoracic Society (ATS) and the Infectious Disease Society of America (IDSA) in 2020 describes the diagnostic criteria for NTM-PD [133]. To be diagnosed with NTM, the infected people must meet clinical, radiological, and microbiologic criteria. Chest X-rays can evaluate lung health, although high-resolution CT imaging is recommended since it can provide better and more detailed pictures of the lungs. The precise identification of NTM species and drug susceptibility testing in vitro is required for the microbiologic validation of the diagnosis of NTM-PD. Mycobacteria with rapid and sluggish growth have been identified in NTM infections [133]. Fast-growing mycobacteria in culture, like *M. abscessus*, *M. chelonae*, and *M. fortuitum*, typically grow in a weak environment. The three subspecies of *M. abscessus* are *M. abscessus*, *M. bolletii*, and *M. massiliense*. The subspecies of *M. abscessus* must be distinguished because some of them carry genes for mutational resistance, impacting available treatments. The two most prevalent species of mycobacteria with a slow growth rate are MAC and *M. kansasii*. On liquid media, they develop in 10–14 days, while on solid media, they may require up to 2–6 weeks to grow. Pathogenic *M. kansasii* frequently manifests as upper lobe-dominant cavitary disease [133]. They can also contaminate lab specimens because of their prevalence in the environment.

13.5.1 Radiological and Clinical Indications

NTM lung disease should be diagnosed over a lengthy period of time because of its slow growth and the possibility that it would be confused for TB or another AFB-positive bacillus [134]. These factors frequently result in a delayed diagnosis, particularly when combined with a low clinical suspicion index. A persistent cough, increased sputum production, dyspnea, low-grade fever, tiredness, and weight loss are among the symptoms and indications that are typically similar to those of pulmonary tuberculosis [134, 135]. Bronchitis, nodular lesions, cavitary lesions, and parenchymal consolidation are the radiographic indications of NTM lung disease [136]. The two main symptoms of NTM lung illness are nodular bronchiectatic and fibrocavitary types [18]. The fibrocavitary type frequently affects older men with underlying lung diseases and mimics pulmonary TB. Cavities with areas of greater opacity often present in the upper lobes help to recognize this type. The most frequent radiologic sign of pulmonary tuberculosis is cavitation. However, NTM lung disease usually involves the pleura, produces thin-walled cavities, lacks lymph node calcification, lacks atelectasis, and generally progresses more slowly than pulmonary TB [113, 114]. The nodular bronchiectatic type appears as bilateral, multilobar bronchiectasis (middle and lower lung fields) with tiny nodules on chest radiography and HRCT [90, 137]. The majority of elderly nonsmoking women without underlying lung disease demonstrate this type of NTM lung illness [138, 139]. Clinicians must be aware of the connection between NTM lung disease and bronchiectasis. Based on radiologic patterns, it is difficult to differentiate between various species of NTM lung disease because of the significant overlap in common HRCT results [140, 141].

13.5.2 Laboratory Outcomes

There are NTM in the surroundings; therefore, high-quality respiratory specimens should be methodically obtained to prevent contamination. Additionally, temporary airway NTM species present in the environment may result in positive samples [142, 143]. For the diagnosis of NTM lung disease, three early morning specimens should be collected on different days [18]. Patients unable to make sputum alone might utilize hypertonic saline to enhance sputum production. The fluorochrome technique and carbol fuchsin stain (Ziehl-Neelsen or Kinyoun method) are the two AFB stains that are most frequently employed (auramine O alone or in combination with rhodamine B). Compared to the Ziehl-Neelsen and fluorochrome procedures, Kinyoun's method seems inferior [144, 145]. Because AFB staining cannot distinguish between NTM and MTB, for the diagnosis of *M. tuberculosis*, nucleic acid amplification (NAA) assays are necessary. Compared to AFB smear microscopy, NAA testing has a better positive predictive value (>95%). Many commercial tests, such as the Cobas TaqMan MTB test (Switzerland) and the amplified *M. tuberculosis* direct test (USA), are routinely utilized [146]. The culture medium comprises solid media, such as Middlebrook 7H10 and 7H11 agar or egg-based Löwenstein-Jensen agar, the same as those utilized for *M. tuberculosis*. For the detection and stimulation of growth, all mycobacteria cultures should include both solid and liquid media, which have been found to increase the sensitivity of NTM detection by an average of 15% [68]. In a laboratory context, a liquid culture incubator system using Middlebrook 7H9 broth that has been supplemented may automatically identify the development of mycobacteria, including NTM [147]. Molecular techniques like DNA sequencing, real-time PCR (Polymerase Chain Reaction), RFLP (Restriction Fragment Length Polymorphism), and line probe hybridization have replaced traditional biochemical testing. Commercial kits are available for NTM-LD identification [148]. NTM's 16S rRNA gene method to examine for species-level identification [149]. Since this gene is highly conserved across animals, relatively minor differences distinguish them [150]. Accurate species differentiation is not possible with single-target sequencing. Therefore, it is necessary to sequence the genes of multiple targets using important genes like hsp65, rpoB, and the 16S–23S internal transcribed spacer [151–155]. The MALDI-TOF MS method is also used to detect NTM. The MALDI-TOF MS (matrix assisted laser desorption ionization-time of flight mass spectrometry) technology compares the mass spectrum patterns of chemicals, primarily ribosomal proteins, distinctive to NTM species, in a library of known NTM strains to identify target bacterial species. This technique is undoubtedly accurate and cost-effective [156–159]. Sequencing requires a lot of growth, but MALDI-TOF MS requires many organisms [157]. Mycobacteria can be divided into two categories to study their drug susceptibility. They are rapidly growing mycobacteria and slow growing mycobacteria. In rapidly growing mycobacteria the drug susceptibility can be studied using various methods. Antimicrobial sensitivity patterns enable the distinction between MAC isolates and other slowly growing NTM, such as *M. simiae*, which often exhibit higher levels of drug resistance than MAC strains. The susceptibility of the *M. chelonae-M. abscessus* group to cefoxitin and tobramycin is an effective screening technique. The zones of inhibition for amikacin and kanamycin by agar diffusion are equal to those of *M. immunogenum*, in contrast to isolates of the *M. chelonae-M. abscessus* group, which show wider zones of inhibition with kanamycin than with amikacin [160, 161]. For RGM susceptibility testing, the standard is the broth microdilution method. Agar disk elution method uses round-well tissue culture plates with commercial antimicrobial disks eluted into molten agar. Like the proportional method used for *M. tuberculosis* complex isolates, susceptibility is interpreted (MTBC).

The disk diffusion method in Agar is a modified Kirby-Bauer technique that can detect mixed cultures that might be undetectable in broth. It takes very little time to set up. The E-test generates an MIC via the straightforward agar disk diffusion technique and an exponential gradient of antimicrobial concentrations [87, 161]. *Mycobacterium avium* complex (MAC) is a type of mycobacterium that grows slowly in broth, so the microdilution or macro dilution method should be used to perform AST on MAC isolates [7]. *Mycobacterium kansasii* can be treated in seven out of ten instances with rifampin, ethambutol, isoniazid (INH), or another regimen. Isolates of *M. kansasii* are tested using either the proportional or macro- or microdilution broth methods [23].

13.6 Treatment

13.7 *Mycobacterium avium* Complex Lung Disease

Because of the uncertainty around the optimal time to initiate treatment and which regimen has the highest probability of eliciting a favorable response, the NTM lung condition is challenging to treat [8]. The international respiratory medicine and infectious diseases societies such as American Thoracic Society (ATS)/ European Respiratory Society (ERS)/European Society of Clinical Microbiology and Infectious Diseases (ESCMID)/Infectious Diseases Society of America (IDSA) / British Thoracic Society (BTS) guidelines on NTM diseases (ATS (2007) and BTS (2017)) should be followed. Treatment outcomes differ among NTM species and subspecies [161]. After considering the patient's age, accompanying conditions, and disease kinds, individualized NTM treatment should be initiated. Patients with fibrocavitary disease typically need proper intervention since the condition is associated with a higher mortality rate [162, 163]. Nodular bronchiectatic disease typically develops without severe symptoms and slowly progresses [164]. A treatment regimen consisting of three to four antibiotics is commonly used to treat NTM–PD, depending on the illness' severity, the patient's drug tolerance, and side effects. After sputum conversion, the therapy is continued for at least 12 months [160, 161]. After the transition from positive to negative sputum cultures, antibiotic medication should be continued for at least 12 months. Rather than daily therapy, intermittent, three-times weekly therapy is recommended for non-cavitary nodular bronchiectatic MAC lung disease in order to enhance drug tolerance [18]. In order to increase treatment tolerance, intermittent, three-times weekly treatment is advised for non-cavitary nodular bronchiectatic MAC lung disease [165, 166]. The two major antibiotics used to treat MAC are azithromycin and clarithromycin. Currently, MAC lung disease is managed using a three-drug macrolide-based treatment consisting of macrolides, rifampin, and ethambutol [18]. Patients with severe diseases are advised to take streptomycin, especially if they have a fibrocavitary form [18]. Drug interactions between macrolides and rifampin and clarithromycin and rifampin may reduce the plasma concentrations of macrolides in MAC patients, decreasing the efficiency of antibiotic therapy for MAC lung disease [167, 168]. The reason for this is yet unknown. Treatment for MAC lung disease has a miserable success rate because of pharmaceutical side effects, prolonged therapy, and reinfection rather than remission [165, 169, 170]. Only 13% of the antibiotic regimens are recommended to patients with MAC lung infections in this study survey of Americans with MAC lung disease, despite 30% of those drugs being linked to a higher risk of acquiring macrolide resistance [171]. According to a study, the currently recommended combination of macrolide, rifampin, and ethambutol is not more effective than the combination of clofazimine, clarithromycin, and ethambutol [172, 173]. Ethambutol is

Table 13.1 Treatment for NTD

NTM species	Indications	Drugs	Therapy duration
Mycobacterium avium complex	Nodular bronchiectatic form (non-cavitary)	Clarithromycin 1000 mg TIW, rifampin 600 mg TIW, azithromycin 5 mg TIW, and ethambutol 25 mg/kg TIW	12-month conversion of negative sputum
	Cavitary nodular bronchiectatic or fibrocavitary form	Amikacin 25 mg/kg TIW or imipenem 500 mg, preferably once or twice per week; amikacin 25 mg/kg TIW or imipenem 500 mg, preferably once or twice per week	12-month conversion of negative sputum
	Macrolide-resistant	Moxifloxacin 400 mg daily, ethambutol 15 mg/kg daily, clofazimine 100 mg daily, amikacin inhaled, along with bedaquiline 450–600 mg rifampin each day	12 months after achieving sputum culture negativity
Mycobacterium abscessus complex		25 mg/kg TIW of amikacin Amikacin 25 mg/kg TIW or imipenem 500 mg either once or twice a week Cefoxitin (up to 12 g/day in split doses) 1000 mg of clarithromycin or 250 mg of azithromycin each day 600 mg/day of rifampin Ethanol (15 mg/kg/day) 300 mg/ day of isoniazid or Rifampin 10 mg/kg daily up to 600 mg, 1000 mg of clarithromycin, 250 mg of azithromycin, and 15 mg/kg of ethambutol every day	12 months after sputum culture negative is achieved

used with rifampicin because it weakens the mycobacterial cell wall, which makes it simpler for the former to penetrate the mycobacteria and reach its target site, the RNA polymerase, and because it also prevents the emergence of macrolide resistance [7, 88]. Moxifloxacin gets effective for those with MAC lung disease. But it is still unclear as to what these agents' functions are [174]. Ethambutol 25 mg/kg, rifampin 600 mg thrice weekly, clarithromycin 1000 mg, or azithromycin 500 mg are among the treatments for MAC non-cavitary nodular bronchiectatic type [8]. Streptomycin or amikacin, 1000 mg of either azithromycin or clarithromycin, 15 mg/kg of ethambutol, and 450–600 mg of rifampin are the main medications used to treat fibrocavitary form or cavitary nodular bronchiectatic form (daily plus) [8]. After the conversion of negative sputum, therapy lasts for 12 months. Because of their weak sensitivity and cross-reactivity, immunological methods employed for immunodiagnostic testing have limited use [9, 14] (Table 13.1).

13.8 *Mycobacterium abscessus* Complex Lung Disorders

Treatment of MABC lung disease is more difficult than that of MAC lung disease because there aren't many efficient antibiotics available. Because no effective antibiotics are available, treating MABC lung disease is more difficult than treating MAC lung illness [18]. The parenteral drugs amikacin, cefoxitin, and imipenem, as well as the oral macrolides clarithromycin and azithromycin are the only ones that are effective against MABC in vitro. Due to the dearth of information on this regimen's safety and effectiveness, as well as its expensive cost and numerous adverse effects, its usage may be restricted. Recent studies have reported on the efficacy of therapy for MABC lung disease, with success rates ranging from 25 to 88% [175–179]. Many patients receive therapy in conjunction with surgery, and different medications have varying treatment lengths. This cannot be applied to all MABC-infected individuals. According to Kang et al. (2015), individuals with minor diseases and those receiving chemotherapy and surgical resection may be the only ones who can receive curative treatment for MABC lung disease (2015) [180, 181]. For MABC lung illness, amikacin, cefoxitin or imipenem, and clarithromycin or azithromycin were prescribed. Weng et al. 2020's latest, recent review indicates that different advanced drugs are applied to treat MABC lung disease. With a 97–99% resistance rate, clofazimine is a novel drug with potential in vitro activity against MABC [179, 182]. After 12 months of treatment, clofazimine and amikacin have cured all cases of MABC, with 82–100% of isolates showing susceptibility [183–185]. Clofazimine-containing regimens have been linked to a number of undesirable effects, such as stomach pain and a ruddy complexion in some patients [185]. British Thoracic Society guidelines advocate nebulized amikacin therapy for patients with MABC-lung illness rather than intravenous amikacin therapy because of its higher documented sputum conversion rates (13–100%) [186–188]. Recently, liposomal amikacin was proposed as a new inhalation therapy treatment strategy [189]. When tigecycline was coupled with clarithromycin, 92.9% of patients had *M. abscess* gone, 68.8% had *M. massiliense* gone, and 100% had *M. bolletii* gone [190]. The antagonistic effects of tigecycline against quickly developing NTM when used in conjunction with amikacin are more pronounced. Linezolid is an effective alternative to traditional multidrug therapy for the treatment of NTM disease. Clinical improvements were efficiently accomplished with a daily dose of 600 mg of linezolid and a low toxicity rate during treatment [191, 192] (Table 13.1).

13.9 *Mycobacterium kansasii* Lung Disease

Once a negative sputum culture has been established, isoniazid (INH) (300 mg/day), rifampin (600 mg/day), and ethambutol (15 mg/kg/day) are the recommended therapy for pulmonary *Mycobacterium kansasii* infections for 12 months [193]. The main medication used at this dosage is rifampin. Since rifampin has a low rate of long-term relapse following drug introduction and a high rate of culture conversion, it serves as the principal medication in this regimen [18]. Patients taking rifampin-based regimens had a very high conversion rate for sputum cultures, reaching 95%, according to a Korean study. Ethambutol likely prevents the development of rifampin-resistant organisms. Infections with *M. kansasii* have not yet been adequately studied, even though new medication classes, including quinolones, macrolides, nitroimidazoles, diarylquinolines, and clofazimine, have been demonstrated to have antimycobacterial actions against *M. tuberculosis* infections [194]. In addition to Isoniazid (INH), the fourth-generation fluoroquinolone moxifloxacin and macrolides like clarithromycin exhibit strong in vitro action against *M. kansasii* [195] (Table 13.1).

13.10 Alternative Therapy for NTM Lung Infections

Even though NTM-PD is a global problem, relapses are common even when complications do not prevent the full course of treatment. NTM-PD prevalence is a problem on a worldwide scale. Guideline-based therapy for NTM-PD commonly leads to relapsing, even while problems need not impede the completion of treatment. Antibiotics are necessary for managing NTM-PD, but host risk factors, including aging, being physically lean, and immunosuppressive diseases must also be treated [5]. Multiple treatments for MAC-PD are difficult due to the cessation of multi-drug regimens and the development of macrolide resistance. Additionally, since it leads to macrolide resistance, the limited compliance with guidelines-based medication for MAC-PD treatment globally is concerning. When used alone or in addition to the standard multidrug therapy for NTM-PD, Kampo medicine, a traditional Japanese herbal mixture, has been helpful for such situations in Japan [196]. Several investigations using macrolide-resistant MAC-PD, aminoglycoside-containing therapy for over 6 months, and surgery demonstrated a significant frequency of culture conversion [197]. According to the most recent ATS/ERS/ESCMID/IDSA clinical practice recommendations, amikacin liposome inhalation solution (ALIS) combined with standard guideline-based treatment (GBT) is advised for patients with refractory MAC-PD, but as the first line of therapy [198, 199]. A therapeutic treatment for persistent mycobacterial infections is available in countries like China that employ traditional medications, including medicinal plants. They employ curcumin from *curcuma longa*, which potentiates macrophage anti-MTB activity and blocks M1 macrophage polarization due to decreased macrophage TLR2 activation [174, 198]. Astragalus polysaccharide from *Astragalus membranaceus* increases the synthesis of inflammatory cytokines by macrophages [199]. The macrophages' production of pro-inflammatory cytokines is increased by the *Astragalus membranaceus* polyphenol astragaloside [200]. Triptolide (diterpene triepoxide) from *Tripterygium regelii* reduces the generation of IL-12 and IL-23 by macrophages and prevents T-cells from producing IL-2 [201]. Osthole (coumarin), a compound found in *Cnidium monnieri*, hinders dendritic cell maturation and suppresses Th2-type cytokines. This enhances the expression of IL-10 while decreasing the expression of inflammatory cytokines [202, 203]. Several studies are being done on using herbal treatments to treat both NTM and TB. Numerous research is now being conducted on herbal remedies for treatment of both NTM and TB.

13.11 Conclusion

Infections with NTM are spreading more widely than ever. Patients with anatomic lung abnormalities, immunologic diseases, or suspected genetic conditions that increase the risk of bronchiectasis and/or lung infections are at increased risk of developing NTM lung disease. Patients without overt lung or immunologic abnormalities are also at increased risk. Individual-to-individual NTM transmissions were previously believed to be physically impractical. Recent research has demonstrated that this is not only conceivable but also highly likely. Due to the dearth of prospective, controlled, and randomized therapy studies, it is challenging to determine the best treatment strategies and timelines for NTM diseases. This article offers a succinct summary of what is now understood about NTM lung disease and the treatments that are offered. It is envisaged that cutting-edge therapies for NTM lung infections will be created in the future.

References

1. Centers for Disease Control and Prevention. Centers for Disease Control and Prevention, National Center for Emerging and Zoonotic Infectious Diseases (NCEZID), Division of Healthcare Quality Promotion (DHQP). *Nontuberculous Mycobacteria* (NTM) Infections.2019
2. Zulu M, Monde N, Nkhoma P, Malama S, Munyeme M. *Nontuberculous mycobacteria* in humans, animals, and water in Zambia: a systematic review. Front Trop Dis. 2021;2:679501.
3. Blanc SM, Robinson D, Fahrenfeld NL. Potential for *nontuberculous mycobacteria* proliferation in natural and engineered water systems due to climate change: a literature review. City Environ Interact. 2021;11:100070.
4. Park DW, Kim YJ, Sung YK, Chung SJ, Yeo Y, Park TS, Lee H, Moon JY, Kim SH, Kim TH, Yoon HJ. TNF inhibitors increase the risk of *nontuberculous mycobacteria* in patients with seropositive rheumatoid arthritis in a mycobacterium tuberculosis endemic area. Sci Rep. 2022;12(1):4003.
5. Sharma SK, Upadhyay V. Epidemiology, diagnosis & treatment of non-tuberculous mycobacterial diseases. Indian J Med Res. 2020;152(3):185.
6. Johnson MM, Odell JA. Nontuberculous mycobacterial pulmonary infections. J Thorac Dis. 2014;6(3):210.
7. Donohue MJ, Wymer L. Increasing prevalence rate of *nontuberculous mycobacteria* infections in five states, 2008–2013. Ann Am Thorac Soc. 2016;13(12):2143–50.
8. Kwon YS, Koh WJ. Diagnosis and treatment of nontuberculous mycobacterial lung disease. J Korean Med Sci. 2016;31(5):649–59.

9. To K, Cao R, Yegiazaryan A, Owens J, Venketaraman V. General overview of nontuberculous mycobacteria opportunistic pathogens: *Mycobacterium avium* and *mycobacterium abscessus*. J Clin Med. 2020;9(8):2541.
10. Stout JE, Koh WJ, Yew WW. Update on pulmonary disease due to non-tuberculous mycobacteria. Int J Infect Dis. 2016;45:123–34.
11. Philley JV, Griffith DE. Treatment of slowly growing mycobacteria. Clin Chest Med. 2015;36(1):79–90.
12. Koh WJ, Stout JE, Yew WW. Advances in the management of pulmonary disease due to *mycobacterium abscessus* complex. Int J Tuberc Lung Dis. 2014;18(10):1141–8.
13. Cowman S, van Ingen J, Griffith DE, Loebinger MR. Non-tuberculous mycobacterial pulmonary disease. Eur Respir J. 2019;54(1):1900250.
14. Faverio P, De Giacomi F, Bodini BD, Stainer A, Fumagalli A, Bini F, Luppi F, Aliberti S. Nontuberculous mycobacterial pulmonary disease: an integrated approach beyond antibiotics. ERJ Open Res. 2021;7(2):00574.
15. Swenson C, Zerbe CS, Fennelly K. Host variability in NTM disease: implications for research needs. Front Microbiol. 2018;9:2901.
16. Jeong BH, Kim SY, Jeon K, Lee SY, Shin SJ, Koh WJ. Serodiagnosis of *Mycobacterium avium* complex and *mycobacterium abscessus* complex pulmonary disease by use of IgA antibodies to glycopeptidolipid core antigen. J Clin Microbiol. 2013;51(8):2747–9.
17. Mirabal AB, Ferrer G. Lung nontuberculous mycobacterial infections. In: StatPearls. Treasure Island, FL: StatPearls; 2022. p. 28.
18. Griffith DE, Aksamit T, Brown-Elliott BA, Catanzaro A, Daley C, Gordin F, Holland SM, Horsburgh R, Huitt G, Iademarco MF, Iseman M. An official ATS/IDSA statement: diagnosis, treatment, and prevention of nontuberculous mycobacterial diseases. Am J Respir Crit Care Med. 2007;175(4):367–416.
19. Khoor A, Leslie KO, Tazelaar HD, Helmers RA, Colby TV. Diffuse pulmonary disease caused by nontuberculous mycobacteria in immunocompetent people (hot tub lung). Am J Clin Pathol. 2001;115(5):755–62.
20. Piersimoni C, Scarparo C. Extrapulmonary infections associated with nontuberculous mycobacteria in immunocompetent persons. Emerg Infect Dis. 2009;15(9):1351.
21. Dowdell K, Haig SJ, Caverly LJ, Shen Y, LiPuma JJ, Raskin L. Nontuberculous mycobacteria in drinking water systems - the challenges of characterization and risk mitigation. Curr Opin Biotechnol. 2019;57:127–36.
22. Nishiuchi Y, Iwamoto T, Maruyama F. Infection sources of a common non-tuberculous mycobacterial pathogen, *Mycobacterium avium* complex. Front Med. 2017;4:27.
23. Jarlier V, Nikaido H. Mycobacterial cell wall: structure and role in natural resistance to antibiotics. FEMS Microbiol Lett. 1994;123(1–2):11–8.
24. Falkinham JO III. Growth in catheter biofilms and antibiotic resistance of *Mycobacterium avium*. J Med Microbiol. 2007;56(2):250–4.
25. Parker BC, Ford MA, Gruft H, Falkinham JO III. Epidemiology of infection by nontuberculous mycobacteria: IV. Preferential aerosolization of mycobacterium intracellulare from natural waters. Am Rev Respir Dis. 1983;128(4):652–6.
26. Yates MD, Grange JM, Collins CH. The nature of mycobacterial disease in south East England, 1977-84. J Epidemiol Community Health. 1986;40(4):295–300.
27. Research Committee of the British Thoracic Society. Pulmonary disease caused by M. malmoense in HIV negative patients: 5-yr follow-up of patients receiving standardised treatment. Eur Respir J. 2003;21(3):478–82.
28. Jenkins PA, Campbell IA. Pulmonary disease caused by mycobacterium xenopi in HIV-negative patients: five year follow-up of patients receiving standardised treatment. Respir Med. 2003;97(4):439–44.
29. Morimoto K, Aono A, Murase Y, Sekizuka T, Kurashima A, Takaki A, Sasaki Y, Igarashi Y, Chikamatsu K, Goto H, Yamada H. Prevention of aerosol isolation of *nontuberculous mycobacterium* from the patient's bathroom. ERJ Open Res. 2018;4(3):00150.

30. Falkinham JO III. Ecology of nontuberculous mycobacteria—where do human infections come from? In: Seminars in respiratory and critical care medicine, vol. 34. New York: Thieme Medical; 2013. p. 095.
31. Kim YM, Kim M, Kim SK, Park K, Jin SH, Lee US, Kim Y, Chae GT, Lee SB. Mycobacterial infections in coal workers' pneumoconiosis patients in South Korea. Scand J Infect Dis. 2009;41(9):656–62.
32. Ryu YJ, Koh WJ, Daley CL. Diagnosis and treatment of nontuberculous mycobacterial lung disease: clinicians' perspectives. Tuberc Respir Dis. 2016;79(2):74–84.
33. Pathak K, Hart S, Lande L. *Nontuberculous mycobacteria* lung disease (NTM-LD): current recommendations on diagnosis, treatment, and patient management. Int J Gen Med. 2022;1:7619–29.
34. Prevots DR, Marras TK. Epidemiology of human pulmonary infection with nontuberculous mycobacteria: a review. Clin Chest Med. 2015;36(1):13–34.
35. Hermansen TS, Ravn P, Svensson E, Lillebaek T. Nontuberculous mycobacteria in Denmark, incidence and clinical importance during the last quarter-century. Sci Rep. 2017;7(1):6696.
36. Simons S, Van Ingen J, Hsueh PR, Van Hung N, Dekhuijzen PR, Boeree MJ, van Soolingen D. *Nontuberculous mycobacteria* in respiratory tract infections, eastern Asia. Emerg Infect Dis. 2011;17(3):343.
37. Moore JE, Kruijshaar ME, Ormerod LP, Drobniewski F, Abubakar I. Increasing reports of non-tuberculous mycobacteria in England, Wales and Northern Ireland, 1995-2006. BMC Public Health. 2010;10(1):1–6.
38. Donohue MJ. Increasing nontuberculous mycobacteria reporting rates and species diversity identified in clinical laboratory reports. BMC Infect Dis. 2018;18:1–9.
39. Tamura A, Hebisawa A, Sagara Y, Suzuki J, Masuda K, Baba M, et al. Pulmonary nontuberculous mycobacteriosis in patients with lung cancer. [Kekkaku]. Tuberculosis. 2004;79(6):367–73.
40. Thomson RM. Changing epidemiology of pulmonary nontuberculous mycobacteria infections. Emerg Infect Dis. 2010;16(10):1576.
41. Thomson R, Donnan E, Konstantinos A. Notification of nontuberculous mycobacteria: an Australian perspective. Ann Am Thorac Soc. 2017;14(3):318–23.
42. Lee H, Myung W, Koh WJ, Moon SM, Jhun BW. Epidemiology of nontuberculous mycobacterial infection, South Korea, 2007–2016. Emerg Infect Dis. 2019;25(3):569.
43. Kim YJ, Park EJ, Lee SH, Silwal P, Kim JK, Yang JS, Whang J, Jang J, Kim JM, Jo EK. Dimethyl itaconate is effective in host-directed antimicrobial responses against mycobacterial infections through multifaceted innate immune pathways. Cell Biosci. 2023;13(1):49.
44. Chan ED, Kaminska AM, Gill W, Chmura K, Feldman NE, Bai X, Floyd CM, Fulton KE, Huitt GA, Strand MJ, Iseman MD. Alpha-1-antitrypsin (AAT) anomalies are associated with lung disease due to rapidly growing mycobacteria and AAT inhibits *mycobacterium abscessus* infection of macrophages. Scand J Infect Dis. 2007;39(8):690–6.
45. Noone PG, Leigh MW, Sannuti A, Minnix SL, Carson JL, Hazucha M, Zariwala MA, Knowles MR. Primary ciliary dyskinesia: diagnostic and phenotypic features. Am J Respir Crit Care Med. 2004;169(4):459–67.
46. Tomii K, Iwata T, Oida K, Kohri Y, Taguchi Y, Nanbu Y, Kubo Y, Yaba Y, Mino M, Kuroda Y. A probable case of adult Williams-Campbell syndrome incidentally detected by an episode of atypical mycobacterial infection. Jpn J Thorac Dis. 1989;27(4):518–22.
47. Uji M, Matsushita H, Watanabe T, Suzumura T, Yamada M. A case of primary Sjögren's syndrome presenting with middle lobe syndrome complicated by nontuberculous mycobacteriosis. Nihon Kokyuki Gakkai Zasshi. 2008;46(1):55–9.
48. Witty LA, Tapson VF, Piantadosi CA. Isolation of mycobacteria in patients with pulmonary alveolar proteinosis. Medicine. 1994;73(2):103–9.
49. Rosenzweig DY. Pulmonary mycobacterial infections due to mycobacterium intracellulare-avium complex: clinical features and course in 100 consecutive cases. Chest. 1979;75(2):115–9.

50. Sonnenberg P, Murray J, Thomas RG, Godfrey-Faussett P, Shearer S. Risk factors for pulmonary disease due to culture-positive M. tuberculosis or nontuberculous mycobacteria in south African gold miners. Eur Respir J. 2000;15(2):291–6.
51. Griffith DE, Girard WM, Wallace RJ Jr. Clinical features of pulmonary disease caused by rapidly growing mycobacteria. Am Rev Respir Dis. 1993;147:1271–8.
52. Okumura M, Iwai K, Ogata H, Ueyama M, Kubota M, Aoki M, Kokuto H, Tadokoro E, Uchiyama T, Saotome M, Yoshiyama T. Clinical factors on cavitary and nodular bronchiectatic types in pulmonary *Mycobacterium avium* complex disease. Intern Med. 2008;47(16):1465–72.
53. Prince DS, Peterson DD, Steiner RM, Gottlieb JE, Scott R, Israel HL, Figueroa WG, Fish JE. Infection with *Mycobacterium avium* complex in patients without predisposing conditions. N Engl J Med. 1989;321(13):863–8.
54. Bryant JM, Grogono DM, Greaves D, Foweraker J, Roddick I, Inns T, Reacher M, Haworth CS, Curran MD, Harris SR, Peacock SJ. Whole-genome sequencing to identify transmission of *mycobacterium abscessus* between patients with cystic fibrosis: a retrospective cohort study. Lancet. 2013;381(9877):1551–60.
55. Bryant JM, Grogono DM, Rodriguez-Rincon D, Everall I, Brown KP, Moreno P, Verma D, Hill E, Drijkoningen J, Gilligan P, Esther CR. Emergence and spread of a human-transmissible multidrug-resistant nontuberculous mycobacterium. Science. 2016;354(6313):751–7.
56. McGarvey J, Bermudez LE. Pathogenesis of nontuberculous mycobacteria infections. Clin Chest Med. 2002;23(3):569–83.
57. Rose SJ, Bermudez LE. *Mycobacterium avium* biofilm attenuates mononuclear phagocyte function by triggering hyperstimulation and apoptosis during early infection. Infect Immun. 2014;82(1):405–12.
58. Puzo G. The carbohydrate-and lipid-containing cell wall of mycobacteria, phenolic glycolipids: structure and immunological properties. Crit Rev Microbiol. 1990;17(4):305–27.
59. Pourshafie MR, Sonnenfeld G, Barrow WW. Immunological and ultrastructural disruptions of T lymphocytes following exposure to the glycopeptidolipid isolated from the *Mycobacterium avium* complex. Scand J Immunol. 1999;49(4):405–10.
60. Barrow WW, De Sousa JP, Davis TL, Wright EL, Bachelet M, Rastogi N. Immunomodulation of human peripheral blood mononuclear cell functions by defined lipid fractions of *Mycobacterium avium*. Infect Immun. 1993;61(12):5286–93.
61. Enomoto K, Oka S, Fujiwara N, Okamoto T, Okuda Y, Maekura R, Kuroki T, Yano I. Rapid serodiagnosis of *Mycobacterium avium*-intracellulare complex infection by ELISA with cord factor (trehalose 6, 6′-dimycolate), and serotyping using the glycopeptidolipid antigen. Microbiol Immunol. 1998;42(10):689–96.
62. Freeman R, Geier H, Weigel KM, Do J, Ford TE, Cangelosi GA. Roles for cell wall glycopeptidolipid in surface adherence and planktonic dispersal of *Mycobacterium avium*. Appl Environ Microbiol. 2006;72(12):7554–8.
63. Howard ST, Rhoades E, Recht J, Pang X, Alsup A, Kolter R, Lyons CR, Byrd TF. Spontaneous reversion of *mycobacterium abscessus* from a smooth to a rough morphotype is associated with reduced expression of glycopeptidolipid and reacquisition of an invasive phenotype. Microbiology. 2006;152(6):1581–90.
64. Vignal C, Guérardel Y, Kremer L, Masson M, Legrand D, Mazurier J, Elass E. Lipomannans, but not lipoarabinomannans, purified from *mycobacterium chelonae* and mycobacterium kansasii induce TNF-α and IL-8 secretion by a CD14-toll-like receptor 2-dependent mechanism. J Immunol. 2003;171(4):2014–23.
65. Wieland CW, Knapp S, Florquin S, de Vos AF, Takeda K, Akira S, Golenbock DT, Verbon A, van der Poll T. Non–mannose-capped Lipoarabinomannan induces lung inflammation via toll-like receptor 2. Am J Respir Crit Care Med. 2004;170(12):1367–74.
66. Maeda N, Nigou J, Herrmann JL, Jackson M, Amara A, Lagrange PH, Puzo G, Gicquel B, Neyrolles O. The cell surface receptor DC-SIGN discriminates between mycobacterium

species through selective recognition of the mannose caps on Lipoarabinomannan. J Biol Chem. 2003;278(8):5513–6.
67. Matsuyama M, Martins AJ, Shallom S, Kamenyeva O, Kashyap A, Sampaio EP, Kabat J, Olivier KN, Zelazny AM, Tsang JS, Holland SM. Transcriptional response of respiratory epithelium to nontuberculous mycobacteria. Am J Respir Cell Mol Biol. 2018;58(2):241–52.
68. Henkle E. Spotlight on NTM lung disease.
69. Sexton P, Harrison AC. Susceptibility to nontuberculous mycobacterial lung disease. Eur Respir J. 2008;31(6):1322–33.
70. Honda JR, Knight V, Chan ED. Pathogenesis and risk factors for nontuberculous mycobacterial lung disease. Clin Chest Med. 2015;36(1):1.
71. Cowman S, Burns K, Benson S, Wilson R, Loebinger MR. The antimicrobial susceptibility of non-tuberculous mycobacteria. J Infect. 2016;72(3):324–31.
72. Lee MR, Sheng WH, Hung CC, Yu CJ, Lee LN, Hsueh PR. *Mycobacterium abscessus* complex infections in humans. Emerg Infect Dis. 2015;21(9):1638.
73. Chin KL, Sarmiento ME, Alvarez-Cabrera N, Norazmi MN, Acosta A. Pulmonary nontuberculous mycobacterial infections: current state and future management. Eur J Clin Microbiol Infect Dis. 2020;39:799–826.
74. Thornton CS, Mellett M, Jarand J, Barss L, Field SK, Fisher DA. The respiratory microbiome and *nontuberculous mycobacteria*: an emerging concern in human health. Eur Respir Rev. 2021;30(160):200299.
75. Zweijpfenning SM, van Ingen J, Hoefsloot W. Geographic distribution of nontuberculous mycobacteria isolated from clinical specimens: a systematic review. In: Seminars in respiratory and critical care medicine, vol. 39. New York: Thieme Medical; 2018. p. 336–42.
76. Modrá H, Ulmann V, Caha J, Hübelová D, Konečný O, Svobodová J, Weston RT, Pavlík I. Socio-economic and environmental factors related to spatial differences in human nontuberculous mycobacterial diseases in The Czech Republic. Int J Environ Res Public Health. 2019;16(20):3969.
77. Kotilainen H, Valtonen V, Tukiainen P, Poussa T, Eskola J, Järvinen A. Clinical findings in relation to mortality in non-tuberculous mycobacterial infections: patients with *Mycobacterium avium* complex have better survival than patients with other mycobacteria. Eur J Clin Microbiol Infect Dis. 2015;34:1909–18.
78. Victoria L, Gupta A, Gómez JL, Robledo J. *Mycobacterium abscessus* complex: a review of recent developments in an emerging pathogen. Front Cell Infect Microbiol. 2021;11:659997.
79. Park IK, Olivier KN. Nontuberculous mycobacteria in cystic fibrosis and non–cystic fibrosis bronchiectasis. In: Seminars in respiratory and critical care medicine, vol. 36. New York: Thieme Medical; 2015. p. 217–24.
80. Ratnatunga CN, Lutzky VP, Kupz A, Doolan DL, Reid DW, Field M, Bell SC, Thomson RM, Miles JJ. The rise of non-tuberculosis mycobacterial lung disease. Front Immunol. 2020;11:303.
81. Bonaiti G, Pesci A, Marruchella A, Lapadula G, Gori A, Aliberti S. Nontuberculous mycobacteria in noncystic fibrosis bronchiectasis. Biomed Res Int. 2015;2015:1.
82. Valour F, Perpoint T, Sénéchal A, Kong XF, Bustamante J, Ferry T, Chidiac C, Ader F. Interferon-γ autoantibodies as predisposing factor for nontuberculous mycobacterial infection. Emerg Infect Dis. 2016;22:1124.
83. Wu UI, Holland SM. A genetic perspective on granulomatous diseases with an emphasis on mycobacterial infections. In: Seminars in immunopathology, vol. 38. Berlin, Heidelberg: Springer; 2016. p. 199–212.
84. Gerriets V, Goyal A, Khaddour K. Tumor necrosis factor inhibitors. In: StatPearls. Treasure Island, FL: StatPearls; 2021.
85. Olivier KN. Lady Windermere dissected: more form than fastidious. Ann Am Thorac Soc. 2016;13(10):1674–6.
86. O'Donnell AE, Barker AF, Ilowite JS, Fick RB, rhDNase Study Group. Treatment of idiopathic bronchiectasis with aerosolized recombinant human DNase I. Chest. 1998;113(5):1329–34.

87. Olivier KN, Weber DJ, Wallace RJ Jr, Faiz AR, Lee JH, Zhang Y, Brown-Elliot BA, Handler A, Wilson RW, Schechter MS, Edwards LJ. Nontuberculous mycobacteria: I: multicenter prevalence study in cystic fibrosis. Am J Respir Crit Care Med. 2003;167(6):828–34.
88. Tanaka E, Amitani R, Niimi A, Suzuki K, Murayama T, Kuze F. Yield of computed tomography and bronchoscopy for the diagnosis of *Mycobacterium avium* complex pulmonary disease. Am J Respir Crit Care Med. 1997;155(6):2041–6.
89. Chung MJ, Lee KS, Koh WJ, Lee JH, Kim TS, Kwon OJ, Kim S. Thin-section CT findings of nontuberculous mycobacterial pulmonary diseases: comparison between *Mycobacterium avium*-intracellulare complex and mycobacterium abscessus infection. J Korean Med Sci. 2005;20(5):777–83.
90. Koh WJ, Lee KS, Kwon OJ, Jeong YJ, Kwak SH, Kim TS. Bilateral bronchiectasis and bronchiolitis at thin-section CT: diagnostic implications in nontuberculous mycobacterial pulmonary infection. Radiology. 2005;235(1):282–8.
91. Wickremasinghe M, Ozerovitch LJ, Davies G, Wodehouse T, Chadwick MV, Abdallah S, Shah P, Wilson R. Non-tuberculous mycobacteria in patients with bronchiectasis. Thorax. 2005;60(12):1045–51.
92. Kwak N, Lee JH, Kim HJ, Kim SA, Yim JJ. New-onset nontuberculous mycobacterial pulmonary disease in bronchiectasis: tracking the clinical and radiographic changes. BMC Pulm Med. 2020;20(1):1–7.
93. Yin H, Gu X, Wang Y, Fan G, Lu B, Liu M, Wang C, Cao B, Wang C. Clinical characteristics of patients with bronchiectasis with nontuberculous mycobacterial disease in mainland China: a single center cross-sectional study. BMC Infect Dis. 2021;21(1):1–2.
94. Eisenberg I, Yasin A, Fuks L, Stein N, Saliba W, Kramer MR, Adir Y, Shteinberg M. Radiologic characteristics of non-tuberculous mycobacteria infection in patients with bronchiectasis. Lung. 2020;198:715–22.
95. Mirsaeidi M, Machado RF, Garcia JG, Schraufnagel DE. Nontuberculous mycobacterial disease mortality in the United States, 1999–2010: a population-based comparative study. PloS One. 2014;9(3):e91879.
96. Ballarino GJ, Olivier KN, Claypool RJ, Holland SM, Prevots DR. Pulmonary nontuberculous mycobacterial infections: antibiotic treatment and associated costs. Respir Med. 2009;103(10):1448–55.
97. Szymanski EP, Leung JM, Fowler CJ, Haney C, Hsu AP, Chen F, Duggal P, Oler AJ, McCormack R, Podack E, Drummond RA. Pulmonary nontuberculous mycobacterial infection. A multisystem, multigenic disease. Am J Respir Crit Care Med. 2015;192(5):618–28.
98. Bandoh S, Fujita J, Ueda Y, Tojo Y, Ishii T, Kubo A, Yamamoto Y, NISHIYAMA Y, Ishida T. Uptake of fluorine-18-fluorodeoxyglucose in pulmonary *Mycobacterium avium* complex infection. Intern Med. 2003;42(8):726–9.
99. Lacasse Y, Girard M, Cormier Y. Recent advances in hypersensitivity pneumonitis. Chest. 2012;142(1):208–17.
100. Winthrop KL, McNelley E, Kendall B, Marshall-Olson A, Morris C, Cassidy M, et al. Pulmonary nontuberculous mycobacterial disease prevalence and clinical features: an emerging public health disease. Am J Respir Crit Care Med. 2010;182(7):977–82.
101. Tamura A, Hebisawa A, Kusaka K, Hirose T, Suzuki J, Yamane A, et al. Relationship between lung cancer and *Mycobacterium avium* complex isolated using bronchoscopy. Open Respir Med J. 2016;10:20–8.
102. Lande L, Peterson DD, Gogoi R, Daum G, Stampler K, Kwait R, et al. Association between pulmonary *Mycobacterium avium* complex infection and lung cancer. J Thorac Oncol. 2012;7(9):1345–51.
103. Hong SJ, Kim TJ, Lee JH, Park JS. Nontuberculous mycobacterial pulmonary disease mimicking lung cancer: clinicoradiologic features and diagnostic implications. Medicine. 2016;95(26):e3978.

104. Taira N, Kawasaki H, Takahara S, Chibana K, Atsumi E, Kawabata T. The presence of coexisting lung cancer and non-tuberculous mycobacterium in a solitary mass. Am J Case Rep. 2018;19:748–51.
105. Kusumoto T, Asakura T, Suzuki S, Okamori S, Namkoong H, Fujiwara H, et al. Development of lung cancer in patients with nontuberculous mycobacterial lung disease. Respir Investig. 2019;57(2):157–64.
106. Inoue R, Watanabe K, Saigusa Y, Hirama N, Hara Y, Kobayashi N, et al. Effect of coexisting advanced extrapulmonary solid cancer on progression of *Mycobacterium avium* complex lung disease. J Bras Pneumol. 2021;47(2):e20200520.
107. Conic J, Lapinel N, Ali J, Boulmay B. Association between non-tuberculous mycobacterial infection and aerodigestive cancers: a case series highlighting different features, sequence and association. Respir Med Case Rep. 2022;40:101751.
108. Buckingham SJ, Hansell DM. Aspergillus in the lung: diverse and coincident forms. Eur Radiol. 2003;13(8):1786–800.
109. Gefter WB, Weingrad TR, Epstein DM, Ochs RH, Miller WT. 'Semi-invasive' pulmonary aspergillosis: a new look at the spectrum of aspergillus infections of the lung. Radiology. 1981;140(2):313–21.
110. Böllert FG, Sime PJ, Macnee W, Crompton GK. Pulmonary *mycobacterium malmoense* and aspergillus infection: a fatal combination? Thorax. 1994;49(5):521–2.
111. Henkle E, Winthrop KL. Nontuberculous mycobacteria infections in immunosuppressed hosts. Clin Chest Med. 2015;36(1):91–9.
112. Fowler SJ, French J, Screaton NJ, Foweraker J, Condliffe A, Haworth CS, et al. Nontuberculous mycobacteria in bronchiectasis: prevalence and patient characteristics. Eur Respir J. 2006;28(6):1204–10.
113. Damaraju D, Jamieson F, Chedore P, Marras TK. Isolation of non-tuberculous mycobacteria among patients with pulmonary tuberculosis in Ontario, Canada. Int J Tuberc Lung Dis. 2013;17(5):676–81.
114. Maliwan N, Zvetina JR. Pulmonary mycetoma following *mycobacterium kansasii* infection: report of seven cases. Arch Intern Med. 1985;145(12):2180–3.
115. Johnston ID. Mycobacterium xenopi infection and aspergilloma. Tubercle. 1988;69(2):139–43.
116. Hafeez I, Muers MF, Murphy SA, Evans EG, Barton RC, McWhinney P. Non-tuberculous mycobacterial lung infection complicated by chronic necrotising pulmonary aspergillosis. Thorax. 2000;55(8):717–9.
117. Mussaffi H, Rivlin J, Shalit I, Ephros M, Blau H. Nontuberculous mycobacteria in cystic fibrosis associated with allergic bronchopulmonary aspergillosis and steroid therapy. Eur Respir J. 2005;25(2):324–8.
118. Kunst H, Wickremasinghe M, Wells A, Wilson R. Nontuberculous mycobacterial disease and aspergillus-related lung disease in bronchiectasis. Eur Respir J. 2006;28(2):352–7.
119. Phoompoung P, Chayakulkeeree M. Chronic pulmonary aspergillosis following nontuberculous mycobacterial infections: an emerging disease. J Fungi (Basel). 2020;6(4):346.
120. Dorman SE, Holland SM. Interferon-γ and interleukin-12 pathway defects and human disease. Cytokine Growth Factor Rev. 2000;11(4):321–33.
121. Field SK, Fisher D, Cowie RL. *Mycobacterium avium* complex pulmonary disease in patients without HIV infection. Chest. 2004;126(2):566–81.
122. Winthrop KL. Risk and prevention of tuberculosis and other serious opportunistic infections associated with the inhibition of tumor necrosis factor. Nat Clin Pract Rheumatol. 2006;2(11):602–10.
123. Keane J. Tumor necrosis factor blockers and reactivation of latent tuberculosis. Clin Infect Dis. 2004;39(3):300–2.
124. Wallis RS, Broder MS, Wong JY, Hanson ME, Beenhouwer DO. Granulomatous infectious diseases associated with tumor necrosis factor antagonists. Clin Infect Dis. 2004;38(9):1261–5.

125. Chan ED. Vulnerability to nontuberculous mycobacterial lung disease or systemic infection due to genetic/heritable disorders. In: Nontuberculous mycobacterial disease; 2019. p. 89–110.
126. Griffith DE, editor. Nontuberculous mycobacterial disease: a comprehensive approach to diagnosis and management. Berlin: Springer; 2018.
127. Choi S, Potts KJ, Althoff MD, Jimenez G, Bai X, Calhoun KM, et al. Histopathologic analysis of surgically resected lungs of patients with non-tuberculous mycobacterial lung disease: a retrospective and hypothesis-generating study. Yale J Biol Med. 2021;94(4):527–35.
128. Piccione JC, McPhail GL, Fenchel MC, Brody AS, Boesch RP. Bronchiectasis in chronic pulmonary aspiration: risk factors and clinical implications. Pediatr Pulmonol. 2012;47(5):447–52.
129. Koh WJ, Lee JH, Kwon YS, Lee KS, Suh GY, Chung MP, et al. Prevalence of gastroesophageal reflux disease in patients with nontuberculous mycobacterial lung disease. Chest. 2007;131(6):1825–30.
130. Thomson RM, Armstrong JG, Looke DF. Gastroesophageal reflux disease, acid suppression, and *Mycobacterium avium* complex pulmonary disease. Chest. 2007;131(4):1166–72.
131. Kang JB, Lee DH, Kwon SH, Kim N, Park YS, Yoon H, et al. The prevalence of nontuberculous mycobacterial lung disease with orwithout reflux esophagitis. Korean J Gastroenterol. 2018;71(1):18–23.
132. Carbajal M, Teneback CC. Environmental and infectious causes of bronchiectasis. In: Bronchiectasis. Cham: Springer International; 2022. p. 85–115.
133. Lipman M, Kunst H, Loebinger MR, Milburn HJ, King M. Non tuberculous mycobacteria pulmonary disease: patients and clinicians working together to improve the evidence base for care. Int J Infect Dis. 2021;113(Suppl 1):S73–7.
134. Daley CL, Iaccarino JM, Lange C, Cambau E, Wallace RJ Jr, Andrejak C, et al. Treatment of nontuberculous mycobacterial pulmonary disease: an official ATS/ERS/ESCMID/IDSA clinical practice guideline. Clin Infect Dis. 2020;71(4):e1–e36.
135. Kwon YS, Koh WJ. Diagnosis of pulmonary tuberculosis and nontuberculous mycobacterial lung disease in Korea. Tuberc Respir Dis. 2014;77(1):1–5.
136. Koh WJ, Yu CM, Suh GY, Chung MP, Kim H, Kwon OJ, et al. Pulmonary TB and NTM lung disease: comparison of characteristics in patients with AFB smear-positive sputum. Int J Tuberc Lung Dis. 2006;10(9):1001–7.
137. Jeong YJ, Lee KS, Koh WJ, Han J, Kim TS, Kwon OJ. Nontuberculous mycobacterial pulmonary infection in immunocompetent patients: comparison of thin-section CT and histopathologic findings. Radiology. 2004;231(3):880–6.
138. Lee G, Lee KS, Moon JW, Koh WJ, Jeong BH, Jeong YJ, et al. Nodular bronchiectatic *Mycobacterium avium* complex pulmonary disease. Natural course on serial computed tomographic scans. Ann Am Thorac Soc. 2013;10(4):299–306.
139. Kim RD, Greenberg DE, Ehrmantraut ME, Guide SV, Ding L, Shea Y, et al. Pulmonary nontuberculous mycobacterial disease: prospective study of a distinct preexisting syndrome. Am J Respir Crit Care Med. 2008;178(10):1066–74.
140. Kartalija M, Ovrutsky AR, Bryan CL, Pott GB, Fantuzzi G, Thomas J, et al. Patients with nontuberculous mycobacterial lung disease exhibit unique body and immune phenotypes. Am J Respir Crit Care Med. 2013;187(2):197–205.
141. Hollings NP, Wells AU, Wilson R, Hansell DM. Comparative appearances of non-tuberculous mycobacteria species: a CT study. Eur Radiol. 2002;12(9):2211–7.
142. Song JW, Koh WJ, Lee KS, Lee JY, Chung MJ, Kim TS, et al. High-resolution CT findings of *Mycobacterium avium*-intracellulare complex pulmonary disease: correlation with pulmonary function test results. AJR Am J Roentgenol. 2008;191(4):W160.
143. van Ingen J. Microbiological diagnosis of nontuberculous mycobacterial pulmonary disease. Clin Chest Med. 2015;36(1):43–54.
144. van Ingen J. Diagnosis of nontuberculous mycobacterial infections. Semin Respir Crit Care Med. 2013;34(1):103–9.

145. Pfyffer GE. Mycobacterium: general characteristics, laboratory detection, and staining procedures. In: Manual of clinical microbiology; 2015. p. 536–69.
146. Somoskövi Á, Hotaling JE, Fitzgerald M, O'Donnell D, Parsons LM, Salfinger M. Lessons from a proficiency testing event for acid-fast microscopy. Chest. 2001;120(1):250–7.
147. Norrby M. Clinical and prophylactic studies of human tuberculosis in a low-endemic setting. Doctoral dissertation. Sweden: Karolinska Institutet.
148. Guglielmetti L, Mougari F, Lopes A, Raskine L, Cambau E. Human infections due to nontuberculous mycobacteria: the infectious diseases and clinical microbiology specialists' point of view. Future Microbiol. 2015;10(9):1467–83.
149. Somoskovi A, Salfinger M. Nontuberculous mycobacteria in respiratory infections: advances in diagnosis and identification. Clin Lab Med. 2014;34(2):271–95.
150. Macheras E, Roux AL, Ripoll F, Sivadon-Tardy V, Gutierrez C, Gaillard JL, et al. Inaccuracy of single-target sequencing for discriminating species of the *mycobacterium abscessus* group. J Clin Microbiol. 2009;47(8):2596–600.
151. Tortoli E. Impact of genotypic studies on mycobacterial taxonomy: the new mycobacteria of the 1990s. Clin Microbiol Rev. 2003;16(2):319–54.
152. Frothingham R, Wilson KH. Sequence-based differentiation of strains in the *Mycobacterium avium* complex. J Bacteriol. 1993;175(10):2818–25.
153. Ben Salah IB, Adékambi T, Raoult D, Drancourt M. rpoB sequence-based identification of *Mycobacterium avium* complex species. Microbiology. 2008;154(12):3715–23.
154. Zelazny AM, Root JM, Shea YR, Colombo RE, Shamputa IC, Stock F, et al. Cohort study of molecular identification and typing of *mycobacterium abscessus*, *Mycobacterium massiliense*, and *mycobacterium bolletii*. J Clin Microbiol. 2009;47(7):1985–95.
155. Macheras E, Roux AL, Bastian S, Leão SC, Palaci M, Sivadon-Tardy V, et al. Multilocus sequence analysis and rpoB sequencing of *mycobacterium abscessus* (sensu lato) strains. J Clin Microbiol. 2011;49(2):491–9.
156. Jang MA, Koh WJ, Huh HJ, Kim SY, Jeon K, Ki CS, et al. Distribution of nontuberculous mycobacteria by multigene sequence-based typing and clinical significance of isolated strains. J Clin Microbiol. 2014;52(4):1207–12.
157. Seng P, Rolain JM, Fournier PE, La Scola B, Drancourt M, Raoult D. MALDI-TOF-mass spectrometry applications in clinical microbiology. Future Microbiol. 2010;5(11):1733–54.
158. Traxler RM, Bell ME, Lasker B, Headd B, Shieh WJ, McQuiston JR. Updated review on Nocardia species: 2006-2021. Clin Microbiol Rev. 2022;35(4):e0002721.
159. Kodana M, Tarumoto N, Kawamura T, Saito T, Ohno H, Maesaki S, et al. Utility of the MALDI-TOF MS method to identify nontuberculous mycobacteria. J Infect Chemother. 2016;22(1):32–5.
160. Mediavilla-Gradolph MC, De Toro-Peinado I, Bermúdez-Ruiz MP, García-Martínez MD, Ortega-Torres M, Montiel Quezel-Guerraz N, et al. Use of MALDI-TOF MS for identification of nontuberculous mycobacterium species isolated from clinical specimens. Biomed Res Int. 2015;2015:854078.
161. File TM Jr, Marrie TJ. Burden of community-acquired pneumonia in north American adults. Postgrad Med. 2010;122(2):130–41.
162. Jethva K, Bhatt D, Zaveri M. Epidemiology of non-tuberculous mycobacteria in India: a review. J Pharmacogn Phytochem. 2019;8(3):954–9.
163. Brown-Elliott BA, Vasireddy S, Vasireddy R, Iakhiaeva E, Howard ST, Nash K, et al. Utility of sequencing the erm (41) gene in isolates of *Mycobacterium abscessus* subsp. abscessus with low and intermediate clarithromycin MICs. J Clin Microbiol. 2015;53(4):1211–5.
164. Ito Y, Hirai T, Maekawa K, Fujita K, Imai S, Tatsumi S, et al. Predictors of 5-year mortality in pulmonary *Mycobacterium avium*-intracellulare complex disease. Int J Tuberc Lung Dis. 2012;16(3):408–14.
165. Park TY, Chong S, Jung JW, Park IW, Choi BW, Lim C, et al. Natural course of the nodular bronchiectatic form of *Mycobacterium avium* complex lung disease: long-term radiologic change without treatment. PloS One. 2017;12(10):e0185774.

166. Chae G, Park YE, Chong YP, Lee HJ, Shim TS, Jo KW. Treatment outcomes of cavitary nodular bronchiectatic-type *Mycobacterium avium* complex pulmonary disease. Antimicrob Agents Chemother. 2022;66(9):e0226121.
167. Jeong BH, Jeon K, Park HY, Kim SY, Lee KS, Huh HJ, et al. Intermittent antibiotic therapy for nodular bronchiectatic *Mycobacterium avium* complex lung disease. Am J Respir Crit Care Med. 2015;191(1):96–103.
168. Daley CL. *Mycobacterium avium* complex disease. In: Tuberculosis and non-tuberculous mycobacteria infections; 2017. p. 663–701.
169. Koh WJ, Jeong BH, Jeon K, Lee SY, Shin SJ. Therapeutic drug monitoring in the treatment of *Mycobacterium avium* complex lung disease. Am J Respir Crit Care Med. 2012;186(8):797–802.
170. Lee BY, Kim S, Hong Y, Lee SD, Kim WS, Kim DS, et al. Risk factors for recurrence after successful treatment of *Mycobacterium avium* complex lung disease. Antimicrob Agents Chemother. 2015;59(6):2972–7.
171. Park HE, Lee W, Choi S, Jung M, Shin MK, Shin SJ. Modulating macrophage function to reinforce host innate resistance against *Mycobacterium avium* complex infection. Front Immunol. 2022;13:931876.
172. Adjemian J, Prevots DR, Gallagher J, Heap K, Gupta R, Griffith D. Lack of adherence to evidence-based treatment guidelines for nontuberculous mycobacterial lung disease. Ann Am Thorac Soc. 2014;11(1):9–16.
173. Jo KW, Kim S, Lee JY, Lee SD, Kim WS, Kim DS, et al. Treatment outcomes of refractory MAC pulmonary disease treated with drugs with unclear efficacy. J Infect Chemother. 2014;20(10):602–6.
174. Jarand J, Davis JP, Cowie RL, Field SK, Fisher DA. Long-term follow-up of *Mycobacterium avium* complex lung disease in patients treated with regimens including clofazimine and/or rifampin. Chest. 2016;149(5):1285–93.
175. Koh WJ, Hong G, Kim SY, Jeong BH, Park HY, Jeon K, et al. Treatment of refractory *Mycobacterium avium* complex lung disease with a moxifloxacin-containing regimen. Antimicrob Agents Chemother. 2013;57(5):2281–5.
176. Koh WJ, Jeon K, Lee NY, Kim BJ, Kook YH, Lee SH, et al. Clinical significance of differentiation of *Mycobacterium massiliense* from *mycobacterium abscessus*. Am J Respir Crit Care Med. 2011;183(3):405–10.
177. Jeon K, Kwon OJ, Lee NY, Kim BJ, Kook YH, Lee SH, et al. Antibiotic treatment of *mycobacterium abscessus* lung disease: a retrospective analysis of 65 patients. Am J Respir Crit Care Med. 2009;180(9):896–902.
178. Lyu J, Jang HJ, Song JW, Choi CM, Oh YM, Do Lee SD, et al. Outcomes in patients with *mycobacterium abscessus* pulmonary disease treated with long-term injectable drugs. Respir Med. 2011;105(5):781–7.
179. Jarand J, Levin A, Zhang L, Huitt G, Mitchell JD, Daley CL. Clinical and microbiologic outcomes in patients receiving treatment for *mycobacterium abscessus* pulmonary disease. Clin Infect Dis. 2011;52(5):565–71.
180. Lyu J, Kim BJ, Kim BJ, Song JW, Choi CM, Oh YM, et al. A shorter treatment duration may be sufficient for patients with *Mycobacterium massiliense* lung disease than with *mycobacterium abscessus* lung disease. Respir Med. 2014;108(11):1706–12.
181. Kang HK, Park HY, Kim D, Jeong BH, Jeon K, Cho JH, et al. Treatment outcomes of adjuvant resectional surgery for nontuberculous mycobacterial lung disease. BMC Infect Dis. 2015;15(1):76.
182. van Ingen J, Totten SE, Helstrom NK, Heifets LB, Boeree MJ, Daley CL. In vitro synergy between clofazimine and amikacin in treatment of nontuberculous mycobacterial disease. Antimicrob Agents Chemother. 2012;56(12):6324–7.
183. Shen GH, Wu BD, Hu ST, Lin CF, Wu KM, Chen JH. High efficacy of clofazimine and its synergistic effect with amikacin against rapidly growing mycobacteria. Int J Antimicrob Agents. 2010;35(4):400–4.

184. Martiniano SL, Wagner BD, Levin A, Nick JA, Sagel SD, Daley CL. Safety and effectiveness of clofazimine for primary and refractory nontuberculous mycobacterial infection. Chest. 2017;152(4):800–9.
185. Yang B, Jhun BW, Moon SM, Lee H, Park HY, Jeon K, et al. Clofazimine-containing regimen for the treatment of *mycobacterium abscessus* lung disease. Antimicrob Agents Chemother. 2017;61(6):e02052–16.
186. Carey GB, Tebas P, Vinnard C, Kim D, Hadjiliadis D, Hansen-Flaschen J, Dorgan D, Glaser L, Barton G, Hamilton KW. Clinical outcomes of clofazimine use for rapidly growing mycobacterial infections. In: InOpen forum infectious diseases, vol. 6. Oxford: Oxford University Press; 2019. p. ofz456.
187. Yagi K, Ishii M, Namkoong H, Asami T, Iketani O, Asakura T, et al. The efficacy, safety, and feasibility of inhaled amikacin for the treatment of difficult-to-treat non-tuberculous mycobacterial lung diseases. BMC Infect Dis. 2017;17:1–9.
188. Olivier KN, Shaw PA, Glaser TS, Bhattacharyya D, Fleshner M, Brewer CC, et al. Inhaled amikacin for treatment of refractory pulmonary nontuberculous mycobacterial disease. Ann Am Thorac Soc. 2014;11(1):30–5.
189. Jhun BW, Yang B, Moon SM, Lee H, Park HY, Jeon K, et al. Amikacin inhalation as salvage therapy for refractory nontuberculous mycobacterial lung disease. Antimicrob Agents Chemother. 2018;62(7):e00011–8.
190. Zhang J, Leifer F, Rose S, Chun DY, Thaisz J, Herr T, et al. Amikacin liposome inhalation suspension (ALIS) penetrates non-tuberculous mycobacterial biofilms and enhances amikacin uptake into macrophages. Front Microbiol. 2018;9:915.
191. Huang CW, Chen JH, Hu ST, Huang WC, Lee YC, Huang CC, et al. Synergistic activities of tigecycline with clarithromycin or amikacin against rapidly growing mycobacteria in Taiwan. Int J Antimicrob Agents. 2013;41(3):218–23.
192. Cavusoglu C, Soyler I, Akinci P. Activities of linezolid against nontuberculous mycobacteria. New Microbiol. 2007;30(4):411–4.
193. Winthrop KL, Ku JH, Marras TK, Griffith DE, Daley CL, Olivier KN, et al. The tolerability of linezolid in the treatment of nontuberculous mycobacterial disease. Eur Respir J. 2015;45(4):1177–9.
194. MS DS, Shoen CM, Cynamon MH. Therapy for *mycobacterium kansasii* infection: beyond 2018. Front Microbiol. 2018;9:2271.
195. Alcaide F, Calatayud L, Santín M, Martín R. Comparative in vitro activities of linezolid, telithromycin, clarithromycin, levofloxacin, moxifloxacin, and four conventional antimycobacterial drugs against *Mycobacterium kansasii*. Antimicrob Agents Chemother. 2004;48(12):4562–5.
196. Suzuki T, Uneda K, Aoyagi R, Kobayashi T, Mitsuma T, Nakamoto H. Case report: Kampo medicine for non-tuberculous mycobacterium pulmonary disease. Front Nutr. 2021;8:761934.
197. Ito M, Koga Y, Hachisu Y, Murata K, Sunaga N, Maeno T, et al. Treatment strategies with alternative treatment options for patients with *Mycobacterium avium* complex pulmonary disease. Respir Investig. 2022;60(5):613–24.
198. Nasiri MJ, Calcagno T, Hosseini SS, Hematian A, Nojookambari NY, Karimi-Yazdi M, et al. Role of clofazimine in treatment of *Mycobacterium avium* complex. Front Med. 2021;8:638306.
199. Bai X, Oberley-Deegan RE, Bai A, Ovrutsky AR, Kinney WH, Weaver M, et al. Curcumin enhances human macrophage control of mycobacterium tuberculosis infection. Respirology. 2016;21(5):951–7.
200. Karuppagounder V, Arumugam S, Thandavarayan RA, Sreedhar R, Giridharan VV, Afrin R, et al. Curcumin alleviates renal dysfunction and suppresses inflammation by shifting from M1 to M2 macrophage polarization in daunorubicin induced nephrotoxicity in rats. Cytokine. 2016;84:1–9.

201. Xu HD, You CG, Zhang RL, Gao P, Wang ZR. Effects of Astragalus polysaccharides and astragalosides on the phagocytosis of mycobacterium tuberculosis by macrophages. J Int Med Res. 2007;35(1):84–90.
202. Qiu D, Zhao G, Aoki Y, Shi L, Uyei A, Nazarian S, et al. Immunosuppressant PG490 (triptolide) inhibits T-cell interleukin-2 expression at the level of purine-box/nuclear factor of activated T-cells and NF-κB transcriptional activation. J Biol Chem. 1999;274(19):13443–50.
203. Chiang CY, Lee CC, Fan CK, Huang HM, Chiang BL, Lee YL. Osthole treatment ameliorates Th2-mediated allergic asthma and exerts immunomodulatory effects on dendritic cell maturation and function. Cell Mol Immunol. 2017;14(11):935–47.

Chapter 14
Laboratory Diagnosis of Zoonotic Tuberculosis: An Update

Anil Kumar Gupta, Amit Singh, Saumya Srivastava, Anvita Gupta Malhotra, and Bipul Kumar

Abstract Zoonotic tuberculosis (ZTB) is a worldwide disease caused by *Mycobacterium tuberculosis* complex (MTC) members that can infect humans and a wide range of domestic and wild mammals. The majority of zoonoses occur when humans come into contact with relatively abundant animal species, and they continue to pose a genuine threat to public health. Bovine tuberculosis (BTB), caused by *Mycobacterium bovis (M. bovis)*, is one of the world's most economically significant zoonotic diseases. The common transmission routes from animals to humans are airborne transmission, unpasteurized milk consumption, and direct contact with untreated animal products or infected animals. The diagnosis of *M. bovis*

The original version of this chapter has been revised. The affiliation of author Anil Kumar Gupta has been updated. The correction to this chapter can be found at
https://doi.org/10.1007/978-981-99-5624-1_15

A. K. Gupta (✉)
IRCH, All India Institute of Medical Sciences, New Delhi, India

A. Singh (✉)
Department of Gastroenterology & HNU, All India Institute of Medical Sciences, New Delhi, India

S. Srivastava
Department of Ocular Pharmacology, All India Institute of Medical Sciences, New Delhi, India

Department of Microbiology, Central University of Punjab, Bathinda, India
e-mail: amit.singh@cup.edu.in

A. G. Malhotra
All India Institute of Medical Sciences, Bhopal, Madhya Pradesh, India

B. Kumar
Department of Gastroenterology & HNU, All India Institute of Medical Sciences, New Delhi, India

CSIR-Institute of Genomics and Integrative Biology, New Delhi, India

A. Singh, D. Sharma (eds.), *Diagnosis of Mycobacterium*,
https://doi.org/10.1007/978-981-99-5624-1_14

infection in animal species is critical for limiting disease propagation and management. The detection of *M. bovis*-infected individuals is difficult because only severely diseased animals show clinical manifestations, limiting its early-stage controls. The utility of conventional and immunological diagnostic tools in detecting infection at an early stage is limited. However, developing novel reagents and technologies for detecting *M. bovis* infection in domestic species is helping advance bovine TB diagnosis.

Keywords Zoonotic tuberculosis (ZTB) · Bovine tuberculosis (BTB) · Diagnostics

14.1 Introduction

Zoonosis are contagious diseases that spread naturally from animals to people and vice versa. They are responsible for more than 61% of the new diseases that are currently spreading around the world [1]. The main cause of the rapid spread of these diseases is an increase in human activity. Herders, veterinarians, foresters, and farmers are among those who are vulnerable to zoonoses. The most common zoonoses are rabies, avian influenza, leishmaniasis, brucellosis, and bovine tuberculosis, which have the greatest impact on global health [2].

Members of the *Mycobacterium tuberculosis* complex (MTC) causes zoonotic tuberculosis (ZTB) in a broad range of domestic and wild mammals including *M. bovis* (cattle), *M. caprae* (sheep and goats), *M. microti* (rodents), *M. mungi* (banded mongooses), *M. orygis* (members of the Bovidae family) and *M. pinnipedii* (seals and sea lions), of which bovine tuberculosis (BTB) is a major zoonosis that has high risk of interspecies contamination and can induce respiratory disorders in both cattle and humans [3]. It endangers public health and generates financial losses owing to decreased productivity, the mandatory killing of test-positive animals, and the expense of preventative measures. Furthermore, it offers protection measures in and around protected natural areas in jeopardy. Likewise, it jeopardizes conservation measures in and near protected natural areas. In affluent nations, the direct relationship between *M. bovis* infection in cattle and sickness in people has been thoroughly demonstrated, but information from underdeveloped nations is limited [4].

In the past two decades, bovine tuberculosis (BTB) has gained attention as a growing public health threat, particularly in low- and middle-income countries where the incidence of BTB is unknown. BTB is becoming progressively important due to the susceptibility of humans, especially in immunocompromised persons in developing countries [5]. However, infection is currently a major concern in the developing world's human population because humans and animals share the same microenvironment and living quarters, particularly in rural areas [6].

The epidemiology of *M. bovis*-caused bovine TB has been impacted by both natural and anthropogenic animal movement [7]. The natural mobility of wildlife reservoir animals accelerates the disease's transmission among domestic animals and, as a result, the disease's public health effect. *M. bovis* can be spread from animal to animal, and sometimes from human to human [8]. *M. bovis* illness can also be spread directly from cattle to human beings via an aerogenous route, as well as by coming into contact with goods contaminated with the infected herd's nose and mouth discharges [7].

The majority of *M. bovis*-infected cattle seem normal. They can carry and shed the organisms for years without presenting any symptoms of sickness. Only in the advanced stages of disease the clinical signs were seen; therefore, most infected cattle are either slaughtered or culled for other reasons before they reach the advanced stages [9]. The clinical symptoms of disease include emaciation, fever, coughing, laboured breathing, reduced milk production infrequently, and diarrhoea. Swollen superficial lymph nodes or rupture or drain to outside [10].

The diagnosis of ZTB is a prime step in not only control of the disease and its management but also essential towards the evaluation of surveillance strategies, However, it's difficult because of the broad taxonomic plurality, the capture and restraint challenges associated with sample collection, lack of gold standard diagnostic techniques, lack of evidence about the true infection status, difficulties in understanding and conducting experimental studies, and the lack of resources and finances [4]. In recent years, many technologies for diagnosis have been contributing to more successful disease avoidance, control, and eradication programmes [10]. Several method have been employed for determining the presence of the etiologic agent in biological specimens, as well as indirect detection by determining the immune response of the host to the etiologic agent [11]. This chapter discusses the presently available tests for ZTB diagnostics and also highlights the prospects of advances in zoonotic TB diagnostic methods.

14.2 Clinical Presentation

14.2.1 In Animals

Bovine tuberculosis is a chronic devastating illness marked by the growth of nodular granulomas known as tubercles. In many animals, the infection is persistent, and symptoms may be absent, even in late instances when many organs are affected. Subclinical symptoms include vulnerability, dyspnoea, anorexia, emaciation, lymph node enlargement, and cough, particularly in progressive TB [12]. Lymph nodes, primarily in the head and thorax, lungs, intestines, liver, spleen, pleura, and peritoneum, are frequently affected. Head and neck lymph nodes may become obviously impacted, burst, drain, and in severe instances become substantially swollen and clog air passages, the alimentary canal, or blood arteries. Clinical indicators may vary depending on the involvement of the lung, displayed as cough, dyspnoea, and other signs of low-grade pneumonia that can be triggered by changes in temperature or physical pressure on the trachea. The involvement of digestive tract is revealed by intermittent diarrhoea or constipation and extreme emaciation, and acute respiratory distress may arise during the terminal phases of tuberculosis [10].

14.2.2 In Humans

M. bovis infection in humans shows clinical manifestations comparable to *M. tuberculosis* infection. The majority of research investigations have indicated the most common clinical sign of *M. bovis* infection in humans is the extrapulmonary form of the disease; however, approximately half of the post-primary cases include the lung, which is linked to the human-to-human spread of tuberculosis caused by *M. bovis* infection [13]. The primary infection of the organism in the intestine may heal or it may progress in the intestines or disseminate to other organs [14]. Cervical lymphadenopathy, intestinal lesions, chronic skin tuberculosis, and other non-pulmonary forms are particularly common [15]. Infection due to *M bovis* in humans usually has a prolonged course and symptoms generally take months or years to appear. Sometimes, the bacteria remain dormant in the host without causing diseases [16]. Loss of appetite, diarrhoea, weight loss, intermittent fever, intermittent hacking cough, big conspicuous lymph nodes, weakness, and other symptoms are typical in ZTB. Young kids infected with *M. bovis* frequently develop stomach infections, whereas elderly individuals have swollen and occasionally ulcerated lymph nodes in the neck [17]. Pulmonary illness is more likely in persons who have had their infections reactivated, and this would only happen if any of the animals had active TB. Fever, cough, chest discomfort, cavitation, and haemoptysis are some of the symptoms. The pulmonary form of tuberculosis occurs less frequently and is usually occupationally related [12].

14.3 Clinical Diagnosis

TB is often a chronic debilitating illness in cattle, although it can also be acute and fast progressing. In nations with eradication programmes, most affected cattle are recognized early, and clinical infections are uncommon. In the latter stages, frequent symptoms include gradual emaciation, a low-grade fluctuating temperature, weakness, and loss of appetite. Animals with involvement of pulmonary system typically have a wet cough that worsens in the morning, during cold weather, or during activity, as well as dyspnoea or drowsiness [18]. The retropharyngeal or other lymph nodes swell in certain animals, burst, and drain. Lymph nodes that are greatly swollen can potentially impede blood arteries, airways, or the digestive tract. If the digestive tract is involved, intermittent diarrhoea and constipation may be seen [19]. Bovine TB symptoms generally appear months after infection. Infections can sometimes lie latent for years before reactivating during times of stress or old age. As a result, BTB can be difficult to identify merely on clinical indications, particularly in industrialized nations where the number of severe instances of animals with clinical evidence may be limited or missing, and the majority are diagnosed by regular testing or discovered at the abattoir [20].

14.4 Diagnosis of Tuberculosis

The diagnosis of BTB is crucial for illness prevention and therapy, but it is also crucial for pathogenesis, epidemiology, and transmission investigations, as well as for evaluating the success of vaccine trials. However, BTB diagnosis is difficult due to the wide taxonomic diversity, difficulties in collecting samples, a lack of gold standard diagnostic tests, a lack of knowledge about the true infection status, difficulty in interpreting and conducting experimental studies, and limited financial resources [4, 19].

The disease remains the leading cause of death and a major public health issue due to the management of disease in resource-limited settings, hampered due to the lack of a rapid, simple, sensitive, and cost-effective diagnostic test. Although several molecular and non-molecular diagnostic tests have been developed, conventional microbiological methods are still considered "gold standards" for the diagnosis of BTB. The details of the available tests are listed in Table 14.1 and Fig. 14.1. It has been believed that no single method or test is sufficient for detecting TB-infected cattle. Therefore, a multidisciplinary approach based on currently available methods must be used (Fig. 14.2).

14.5 Direct Methods: Evidential of the Agent

14.5.1 Smear Microscopy

Smear microscopy is the earliest, rapid, and simple procedure used to identify the presence of acid-fast bacilli (AFB). Although it is faster and less expensive than other methods but failed to distinguish among members of the Mycobacteriaceae family as well as between members of the genus *Mycobacterium* and other organisms that are positive results with acid-fast staining characteristics, such as certain *Nocardia*, *Legionella*, *Rhodococcus*, *Tsulunnurella*, *Cyclospora*, and *Cryptosporidium*. Additionally, it is still less sensitive than culture for detecting mycobacteria and requires not less than 5000 to 10,000 bacilli/mL in samples. Despite this quantitative inconsistency of smears, microscopy may be helpful in several ways [21].

14.5.1.1 Ziehl-Neelsen (ZN)

ZN smear microscopy is the simplest, rapid, and low-cost procedure. ZN staining method requires heat application during the carbol-fuchsin staining for the uniform penetration of the dye into the cell wall. To eliminate the heating step during the staining, a cold method has also been developed by Kinyoun [22]. The ZN staining method has been used widely for more than 100 years and is available in almost all

Table 14.1 Principal methods for diagnosis of bovine tuberculosis

Technology/tests	Stage of development	Developers/ supplier(s)	Detection level	DST utility
A. Active TB				
Direct visualization (Microscopy)				
Conventional microscopy with acid-fast staining	Routine diagnosis	Multiple	Primary	No
Fluorescent microscopy with non-specific cell-wall staining	Routine diagnosis	Multiple	Primary	No
Fluorescent microscopy with LED light source	Routine diagnosis	Various	Primary	No
Fluorescent microscopy with molecular in-situ hybridization (FISH)	In development	ID-FISH technology	Primary	No
Automated microscopy	In development	Various	Primary	No
Computer-assisted microscopy	In development	Various	Microscopy	No
Growth-based detection (Culture)				
Conventional solid media LJ, Middlebrook 7H10/7H11 agar, 7H9/7H12/Dubos medium	Commercialized reagents and prepared media	Multiple	Primary	Yes
Automated liquid culture systems MGIT 960	Commercialized	Becton, Dickinson and Company (BD)	Primary	Yes
Molecular detection				
Automated, non-integrated Nucleic Acid Amplification Test (NAAT)	Commercialized	GenProbe, Roche, BD, others	Primary	No
Automated, integrated NAAT	Commercialized	Cepheid	Primary/ Referral	Yes
VetMAX™ MTBC (M. tuberculosis complex) PCR kit	Commercialized	Thermo	Referral	No
Species identification				
Luminescent probe of culture isolate	Commercially available	GenProbe	Referral	No
Fluorescent probe of smear-positive sputum	In development	ID-FISH	Referral	No
Spoligotyping	Commercialized		Referral	No
Variable number tandem repeat (VNTR) typing	In development		Referral	No
Next generation sequencing	Commercially available		Referral	Yes
B. Immuno assay tuberculosis infection detection				
Skin test with purified protein derivative (PPD)	Commercialized	Multiple	Primary	No
Whole-blood IFN-γ" release assay	Commercialized	Cellestis	Primary	No
ELISPOT IFN-g release assay	Commercialized	Oxford Immunotech	Primary	No

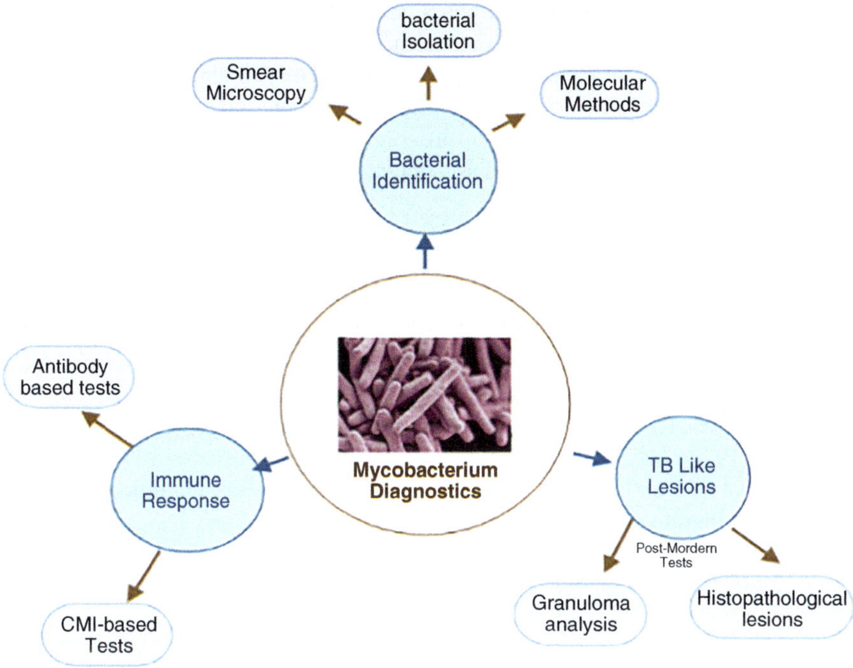

Fig. 14.1 Diagnostic methods available for zoonotic TB

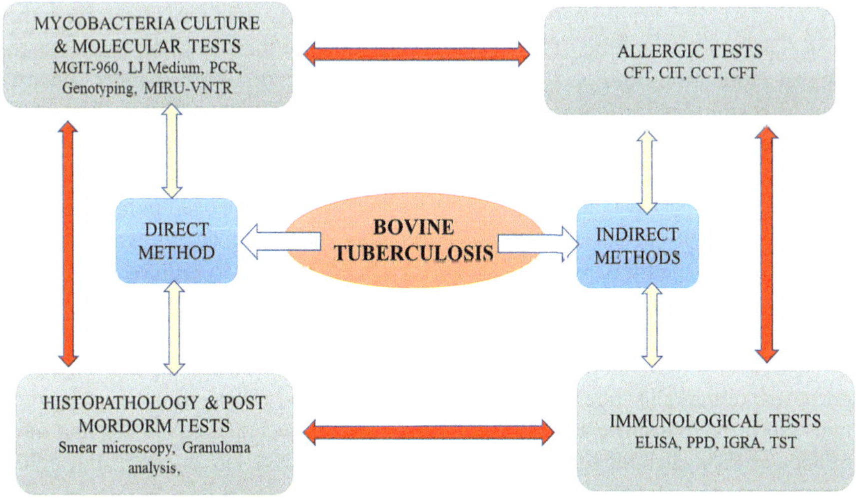

Fig. 14.2 The relationships between the various methods currently employed to diagnose bovine tuberculosis

primary healthcare laboratories. Although specificity is high, the sensitivity of the ZN staining has been reported to vary from 20 to 60%, and its utility is doubtful, especially in the case of paucibacillary load or difficulty in obtaining sputum samples in cattle. The sensitivity of smears is also dependent on staff expatriation with a well-functioning quality assurance programme in place [23]. Therefore, direct smear examination is not employed as a diagnostic method.

14.5.1.2 Fluorescence Microscopy

A powerful light source (halogen or high-pressure mercury vapour lamp) were used in fluorescence microscopy together with an acid-fast fluorochrome dye, such as auramine-O or auramine-rhodamine (AR). The International Union Against TB and Lung Disease (IUATLD) recommended a standardized fluorescence staining method in 1978 [24]. Fluorescence microscopy is more sensitive (10% more) than conventional ZN and slide smears are examined at lower magnification which allows examination of a larger area/ unit of time and consumes lesser time to read the smear than ZN methods [23]. The sensitivity of fluorescence microscopy ranged from 52 to 97% in diagnosing clinical samples [25].

14.5.2 Mycobacterial Culture

Mycobacteria are strictly aerobic and grow more slowly than most bacteria pathogens. The generation time of the mycobacteria is more than 12 h. *M. bovis* has the longest replication time of 16–20 h [26]. Mycobacterial culture is more sensitive than AFB smear microscopy and can detect 10 to 100 viable bacilli/mL. Therefore, smear AFB-negative clinical specimens should be further tested by culture to confirm the absence of *Mycobacterium*. The techniques used in the isolation of *M. bovis* especially in veterinary laboratories differ slightly from those used in medical laboratories, primarily because *M. bovis* strains grow poorly or not grown at all on a glycerol-based medium, which has historically been used to culture *M. tb*. As a result, sodium pyruvate-containing media are utilized instead of glycerol for *M. bovis* isolation.

Apart from higher sensitivity, cultural methods are used for further diagnostic purposes, remarkably phenotypic Drug Sensitivity Test (DST) to notify drug regimens, to provide a sufficient specimen for deoxyribonucleic acid (DNA) and other molecular tests such as genotyping, rapid molecular DST, or for molecular epidemiologic studies [4, 26, 27].

M. bovis isolation has become the "gold standard" for BTB diagnosis. However, one of its crucial characteristics is the lengthy duration necessary for isolation and biochemical identification, which may take more than 12 weeks to accomplish the final diagnosis and also exhibits limited sensitivity [28]. The major limitation of systematic cultivation of *M. bovis* is to obtain samples from living animals and unfortunately, it is usually collected during autopsies and slaughterhouses. Therefore, efforts should be made to ensure that the laboratory collects good quality of samples, to enable the correct diagnosis of BTB [10, 26].

14.5.2.1 Decontamination Methods

Due to the highly contagious nature of *Mycobacterium*, all the culture methods require minimum biosafety level 2 capacity infrastructure and skilled personnel for sample processing, inoculation, and monitoring of grown cultures. However, if there is manipulation with the *M. tuberculosis* TB culture suspension such as biochemical and molecular characterizations, then the use of a biosafety level 3 laboratory is recommended [19, 29]. To eradicate such competitive bacteria, collected samples must be handled using decontamination processes that include the addition of NaOH, oxalic acid, or quaternary ammonium compounds. Unfortunately, the toxic effects of these decontaminating chemicals may impair mycobacterial viability, preventing the organism from being cultured (Table 14.2). The N-acetyl

Table 14.2 Decontamination methods of specimen processing used for diagnosis of TB

Decontamination methods	Use	Advantage	Disadvantage
Sodium hydroxide	Laboratories using concentration by centrifugation	Digestion/decontamination at the same time when used at a final conc. of 2% Low cost	Precise timing needed to avoid killing mycobacteria; may kill some mycobacteria at 2% conc.
NALC-NaOH	Mostly used in developed countries Used in combination with centrifugation	Good mucolytic action Use of NaCl as mucolytic reduces NaOH concentration and its potential deleterious action on mycobacteria	Short shelf-life of prepared reagents (24 h) Higher cost
Oxalic acid	Recommended to eliminate *P. aeruginosa* contamination (e.g., in urine)	Effective in inhibiting overgrowth by *Pseudomonas*	Use restricted to inhibit *Pseudomonas*
Ogawa-Kudoh	Ideal method for low-resource settings	Centrifugation or concentration is not necessary Low cost and can be used in the field	May have higher contamination rates
Cetyl pyridinium-sodium chloride	For preservation and digestion/decontamination while in transport to the laboratory	Avoids overgrowth of contaminants for up to 8 days	Egg-based media is required since the compound remains active in agar and may be deleterious to mycobacteria

L-cysteine and sodium hydroxide (NALC-NaOH) are most widely used globally in routine service [10].

Using the culture method (solid and liquid) the *M. bovis* has been isolated from livestock and wildlife samples. The details of the solid and liquid method used for isolations are as follows:

14.5.2.2 Löwenstein-Jensen (LJ) Medium

LJ medium is an egg-based solid media that is highly rich in phospholipids and proteins that bind and/or neutralize hazardous chemicals in clinical specimens. They have been utilized for the primary isolation of mycobacteria from clinical samples, and they also produce a larger percentage of positive findings when compared to agar-based isolation media because of its luxuriant growth on egg media [30]. LJ is routinely used for mycobacterial culture and DST in low- and middle-income countries, due to its low cost and stability for several weeks. Although LJ culture is a more sensitive method than smear AFB, its suitability in diagnostic uses is hampered by the long time it takes to give a positive result or rule out infection. A recent study reported an average time of 21 days is taken to give culture-positive results by LJ [31] and another group observed that the time was longer (63 days) when it was used for the DST purpose [32].

14.5.2.3 BACTEC Mycobacteria Growth Indicator Tube-960 (MGIT-960)

MGIT-960 system is developed by BD, USA [33]. The system is applied for the early detection (7–12 days) of mycobacteria [34]. The MGIT-960 instrument detects bacterial growth by embedding an oxygen-sensitive fluorescent compound, tris 4,7-diphenyl-1,10-phenonthroline ruthenium chloride pentahydrate derivative, at the bottom of 16 100 mm round bottom screw cap tubes. The tube has 7 ml of modified Middlebrook 7H9 broth medium, an enrichment mixture of oleic acid, albumin, dextrose, and catalase (OADC) and to avoid contamination a mixture of antibiotics PANTA (polymyxin B, amphotericin B, nalidixic acid, trimethoprim, and azlocillin) is added to the medium at the moment of use [35]. The principle of the detection is based on a large amount of dissolved oxygen normally present in medium quenches the natural fluorescence of the ruthenium derivative compound. During the growth of bacteria in the tube, free oxygen is utilized due to their metabolism and is replaced with carbon dioxide. With the depletion of free oxygen in the medium, the quenching effect lowers accordingly and allows the fluorescence to be detected when exposed to UV light [36]. The growth index is automatically calculated by the instrument based on the intensity of fluorescence detected at every 60 min which is directly proportional to the amount of oxygen-depleted in the tube. When a certain level of fluorescence is reached, the instrument indicates the vial as positive which is equivalent to 10^5–10^6 CFU/mL. For negative results, it takes a minimum of 42 days [34].

In 2010, TiKa-MGIT (TiKa Diagnostics, UK), a novel specialized culture medium with the unique ability to stimulate MTBC growth when used in conjunction with the BACTECTM MGITTM system, has been used to enhance the sensitivity of the mycobacterial culture, even from samples with low mycobacterial numbers [37]. This technique should be used on a larger scale to increase the detection rate of MTBC infection in cattle that have a paucibacillary load or in cattle that are heavily contaminated with other environmental bacteria.

14.6 Post-Mortem Diagnosis

The diagnosis of pathologic BTB during autopsies or sanitary inspections of carcasses in refrigerated slaughterhouses was significantly difficult, because many pathogens, including *Actynomices bovis*, *Trueperella pyogenas*, and others, have granulomatous inflammation and morphologic characteristics similar to BTB [19]. Through conventional post-mortem examination, approximately 47% of presumptive BTB lesions (granuloma) were seen in slaughtered cattle carcasses. Despite this, anatomic-pathology analyses have been critical for BTB diagnosis in control programmes [38]. This disease is distinguished by the formation of granulomas in which bacteria reside. At first, epitheloid and large cells are present in the tubercle's center, but as the illness progresses, they become surrounded by lymphocytes, plasma cells, and monocytes, which causes peripheral fibroplasia and central caseous necrosis. The typical well-differentiated granulomas usually look yellowish, caseous, casino-calcareous, or calcified, and frequently encapsulated in structure. Some tubercles are so small that they cannot be seen with the naked eye unless the tissue is sectioned. Tubercles are found in the lymph nodes of cattle, particularly those in the head and thorax. They are also found on the surfaces of the lungs, spleen, liver, and body cavities [19].

The detection of these lesions exhibits a major lack of sensitivity (28.5%) as well as specificity. There may be limited opportunities to improve the sensitivity of post-mortem detection of BTB. Continuous education and training for slaughter inspectors are unquestionably critical.

14.6.1 Histopathological Diagnosis

To make the presumptive diagnosis of BTB, histopathology and/or microscopic demonstration of acid-fast bacilli can be used. Mycobacteria can be detected in a sample using Ziehl Neelsen staining followed by light microscopy or auramine-O staining followed by fluorescence microscopy [39]. Mycobacteriosis is presumed if the tissue exhibits characteristic histological lesions such as caseous necrosis, mineralization, epithelioid cells, multinucleated giant cells, and macrophages known as granulomas. Although *M. bovis* can be isolated in culture, lesions are frequently paucibacillary, making it difficult to detect the presence of organisms that grow quickly in histological sections. However, lesions in primates, felids, mustelids (badgers), and marsupials (brush-tailed possums) are found to contain large numbers of acid-fast organisms [40, 41]. Recent MTC research reveals that auramine O staining is more sensitive and selective than Ziehl-Neelsen staining. Despite these benefits, the demand for post-mortem samples restricts the diagnostic procedure, and the majority of lesions might be paucibacillary, resulting in false-negative findings [38, 39].

14.7 Molecular Tests

The ability to diagnose bovine tuberculosis using molecular assays has significantly increased over the past few years. Additionally, improvements in molecular characterization have given rise to new tools that have improved our understanding of *M. bovis* epidemiology and tuberculosis control. Pal et al. [42] provided an in-depth examination of the various molecular methods that have been created for the direct detection of mycobacteria from clinical samples. These approaches rely on polymerase chain reaction (PCR) amplification of particular mycobacterial DNA or RNA target pieces. Tests can be performed on sputum, blood, nasal swabs, and other tissues, with the benefit of finding non-viable bacilli quickly. However, its sensitivity is constrained when used for paucibacillary samples.

14.7.1 Polymerase Chain Reaction

To specifically identify MTBC organisms, a variety of PCR-based techniques have been created and modified. These techniques are based on detecting the presence of mycobacterial DNA, either directly from ante- or post-mortem samples or from cultured isolates [43]. Two emerging PCR-based molecular tools, i.e., VetMAXTM *M. tuberculosis* complex PCR kit developed and marketed by Thermo Fischer Scientific, Waltham, MA, USA, and GeneXpertR (Cepheid, Sunnyvale, CA, USA) technology are commercially available for the diagnosis of BTB. The VetMAXTM MTBC PCR kit detects IS6110 insertion element found in MTC using a variety of clinical samples including lymph nodes and other tissues. The assay requires DNA extraction of samples and an infrastructure capable of performing PCR, restricting its use to diagnostic or research laboratories [43].

14.7.2 Xpert MTB/RIF Assay

The Xpert MTB/RIF assay is a novel automated diagnostic test that performs sample processing followed by hemi-nested real-time PCR analysis in a single hands-free step for rapid and simultaneous detection of MTC resistance to RIF in clinical specimens in less than 2 h. This technique detects MTBC deoxyribonucleic acid (DNA) in a simpler and standardized manner, making it useful in high-throughput diagnostic or research contexts [44]. The Xpert MTB/RIF test identifies *M. tuberculosis* complex and RIF resistance using PCR amplification of the *M. tuberculosis* 81-bp rpoB gene segment and further probing of this region for mutations linked to RIF resistance [7].

In addition, it is readily accessible in locations with a high human TB prevalence and where mycobacterial culture labs might not exist. This rapid and simple approach is now being developed and optimized to identify *M. bovis* in elephant post-mortem tissue homogenates and ante-mortem bronchoalveolar lavage samples, trunk wash fluids, and mouth swabs [45].

14.8 Genotyping and Strain Identification

In order to genetically distinguish *M. bovis*, the following molecular typing techniques are frequently used: (1) IS6110 analyses (RD); (2) spacer oligonucleotide typing (spoligotyping); (3) the variable number of tandem repeats (VNTR) typing of mycobacterial interspersed repetitive units (MIRU); and (4) next-generation sequencing.

14.8.1 Insertion Sequence (IS) 6110

Insertion element found within species of the *M. tuberculosis* complex. These were related to the IS3 family of insertion sequences. These were discovered in members of the family Enterobacteriaceae [46]. IS6110 is 1361 bp long and contains 28 bp, defective inverted repeats at its extremities with three mismatches and 3-bp direct repeats that possibly is the outcome from the recurrence of the target sequence [47]. These are present in diverse copy numbers and are integrated at the different chromosomal sites. The number of IS6110 copies present in the genome is dependent on species and strain. In *M. bovis*, low copy number of IS6110 element positively influences the results, with good discriminatory power among MTC members. The polymorphism of restriction fragments produced by digesting the IS6110 fragment with the *Pvu*II restriction enzyme has been used as a method for genotyping of *M. tuberculosis* complex species.

14.8.2 MIRU-VNTR Typing

Mycobacterial interspersed repetitive units (MIRU)-variable number tandem repeat (VNTR) typing of *M. bovis* is also used to assess the transmission by measuring the genetic homology of dissimilar isolates. The technique utilized DNA fragments having tandem repeated sequences in which the number of copies of the repeated sequence varies among strains. VNTR sequences have come out as valuable markers for genotyping.

A total of 41 MIRU loci have been identified in MTC. However, locus 21 is absent in *M. bovis*. Each locus contains repetitive DNA sequences that can differ in the number of repeats between strains [48]. By using PCR and fragment sizing, the number of repetitions in each VNTR locus is determined. A VNTR profile (i.e., 42,235) is then created by concatenating each allele, which may subsequently be compared to other profiles to find matches. The tool can accurately describe samples, explain patterns seen in various herds or geographical regions, and provide an analysis of organism distribution within a given area. Additionally, it can monitor the molecular epidemiology of the disease, control the trade in animals, and assess the effectiveness of eradication efforts [48–50].

14.8.3 Spacer Oligotyping (Spoligotyping)

Spoligotyping was the initial genome-wide approach that allowed for the identification and classification of *M. tuberculosis* complex in clinical samples without requiring for culture [51]. It can determine the phylogenetic relationship between organisms from certain geographical regions and track sources of infection. The technique relies on a DNA polymorphism discovered in the mycobacterial direct repeat (DR) locus. This region comprises a number of conserved 36-bp DRs with distinct functions.,

The individual spacer sequences ranging from 34–41 bp in length scattered among each DR. MTB strains vary in the number and presence or absence of DRs [52, 53], which can be observed in hybridization patterns of spacers.

The approach is known as spacer oligotyping or spoligotyping because each spacer area has a corresponding oligonucleotide probe on the nylon membrane. *M. tuberculosis* and *M. bovis* have distinct spoligo patterns, allowing spoligotyping to distinguish between the two members of the M. TB complex. *M. bovis* strains characteristically lack the final four-spacer sequencers 39–43 [54, 55].

Spoligotyping is predominantly useful for subtyping isolates with low copy numbers of IS6110 (<5) especially *M. bovis* [56]. Spoligotyping has been used in several investigations of *M. tuberculosis* complex strains all over the world since it is a quick and reliable genotyping approach. Spoligotyping data from many studies have been consolidated into the SITVITWEB database for epidemiology, population genetics, and classification [57] (Fig. 14.3).

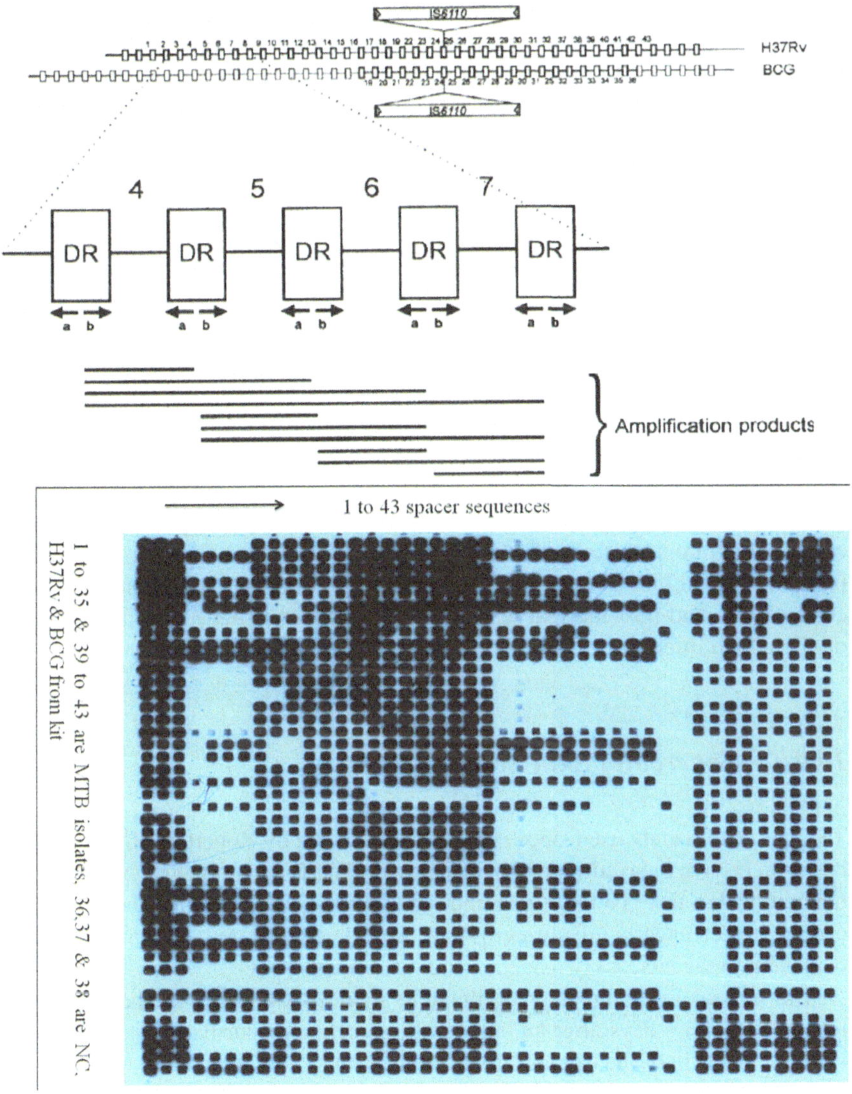

Fig. 14.3 Representative diagram of spoligotyping

14.8.3.1 Whole Genome Analysis

In comparison to other molecular-based genotyping techniques, next-generation sequencing (NGS) of MTBC has higher resolution and discriminatory power. Whole genome sequencing (WGS) has made it possible to compare different genetic profiles at the nucleotide level and to more precisely study the molecular epidemiology and genetic diversity of MTBC [58]. Currently, high-quality WGS is dependent on mycobacterial culture because the preparation of WGS libraries requires high-quality DNA. Furthermore, a significant drawback of WGS is the inability to determine the direction of transmission between species without using large sample sizes [59]. The method for obtaining diagnostic-quality MTBC WGS from the specimens is currently being developed. In future, it will provide better resolution with clinical isolates of *M. bovis* and will provide a database for future epidemiological investigations, especially in bovine TB outbreaks.

14.9 Indirect Methods

The caudal fold test (CFT), the (mid) cervical intradermal test (CIT), or the comparative cervical test (CCT) are used for the initial examination for BTB in young animals. Furthermore, the interferon-gamma (IFN-γ) assay is used as an additional confirmation test of the CFT or CIT (serial testing) or in conjunction with tuberculin skin tests (TSTs) to boost diagnostic sensitivity (parallel testing). The majority of control programmes are restricted to passive surveillance through post-mortem inspection of all killed livestock. The section that follows provides an overview of current ante- and post-mortem testing technologies and suggests adjustments that might help to improve disease control and eradication [41].

14.9.1 Tuberculin Skin Test

TST has been widely used since it was recommended by Robert Koch in 1890. It is based on in-vivo delayed-type hypersensitivity (DTH) reaction characterized by the tuberculin skin test (TST). It is an indirect method of diagnosing tuberculosis and can able to detect early infections (3–8 weeks after contact with *M. bovis*) by the use of standard reagents, assay reagents, and instruments [41].

The dermal swelling is predominantly induced by a cell-mediated immune response (CMI) 3 days after an intradermal injection of purified protein derivative (PPD) in the skin of the caudal fold (CFT) or neck (CIT), which is a critical indicator of Mycobacterium infection PPD response is solely dependent on the administrative sites such as caudal fold, neck etc. The skin of the neck is thought to be more sensitive to tuberculin-related hypersensitivity (DTH response) than the skin of the caudal fold. Many international studies have documented the disparities in performance. PPD possesses an overall sensitivity of 48–96.8% and a specificity of 96–98.8%. Similarly, the comparative cervical tuberculin Test (CCT) test involves intradermal injections of tuberculin and purified protein derivatives (PPDs) from *M bovis* and *M. avium*, followed by 72-h monitoring for swelling and indurations at the injection site. It is commonly used to distinguish between animals infected with *M. bovis* and those who have become sensitized to PPD-B as a result of previous exposure to other mycobacteria [60, 61].

The TST's main advantages and reasons for widespread use are its low cost, high availability, and the lack of alternative methods for detecting BTB. However, there are numerous known limitations to the test, including difficulties in administration and interpretation of results, the need for a second-step visit, a low degree of standardization, and imperfect test accuracy [62].

14.9.2 Interferon-Gamma Assay

The test compares Interferon-Gamma (IFN) production following stimulation with avium and bovine PPD which depends on the expulsion of IFN from sensitized lymphocytes over a 16–24-h period of incubation with a particular antigen [63]. Aside from the high logistical demands (culture must begin within 24 h of blood sampling) and high costs, the TST with tuberculin demonstrated identical problems with standardization as previously discussed [41, 64]. ESAT-6 and CFP-10, *M. tuberculosis* complex-specific antigens, have also been utilized to increase IFN assay specificity, particularly in TST-positive populations. The use of these antigens may also allow for the differentiation of BCG-vaccinated from unvaccinated animals [65].

In comparison to the TST, the benefits of the IFN-assay include increased sensitivity, the possibility of more rapid repeat testing, the elimination of the need for a second visit to the farm, and more truthful testing methods and data interpretation. Limitations include modest test sensitivity to detect mycobacterium infection, then high logistical demands (culture must begin within 24 h of blood sampling), a higher probability of non-specific response in young animals (due to natural killer (NK) cell activity), and its high, as well as the difficulties in tuberculin standardization previously discussed in relation to the TST. The sensitivity and specificity of the IFN-c assay have been estimated in a great number of international studies. Estimates of test sensitivity range from 73.0 to 100%, with a median value of 87.6%, and specificity from 85.0 to 99.6%, with a median of 96.6% [63, 66].

Therefore, this assay may be modified to provide a highly specific and sensitive screening test for use either alone or together with other screening tests such as TST. Furthermore, a multispecies IFN-c assay for non-bovine species such as camelids, cervids, dogs, and cats would be a useful tool for BTB screening and control in those species as well as overall BTB control.

14.9.3 Enzyme-Linked Immunosorbent Assays (ELISA)

The indirect ELISA technique assesses the affinity of specific antibodies for an antigen. The ELISA has the advantage of being simple, but its sensitivity is limited due to the late and irregular development of humoral immune responses in cattle during the disease [67].

The PPD and single or connected purified antigenic materials isolated from *M. bovis*, such as the antigens of the Ag85 complex that are responsible for large portion of the secreted proteins, as well as MPB70 and its highly homologous protein MPB83, secreted mycobacterial proteins with restricted species distribution, are typically used to diagnose cattle infected with *M. bovis* [68, 69]. The majority of these antigens have attained sensitivity and specificity of about 90%, despite the presence of anergic animals and higher antibody titres in more severe illness stages [70]. In recent years, the invention of a test known as the lateral flow that relies on the detection of numerous antigens in certain kinds of animals (e.g., elephants) has provided promising outcomes for TB diagnosis, but it may not be appropriate for others, such as buffaloes [71].

14.10 Conclusion

Despite all efforts to control BTB, the disease still exists with serious complications. This zoonotic disease has serious consequences for both human health and the agricultural industries. Many developed countries have successfully established TST-based eradication programmes before the slaughter of animals. It had a positive impact on the reduction of zoonotic cases in humans. However, the specificity and sensitivity of a tuberculin test are constrained, so the culture should be used to confirm the presence of *M. bovis*. Though molecular tests, i.e., PCR, detection of *M. bovis* directly from clinical samples is also possible. Additionally, *M. bovis* strains can be distinguished using genetic fingerprinting methods such as spoligotyping, and MIRU-VNTR. BTB is still present today due to a variety of factors, such as the sensitivity and specificity limitations of diagnostic tests, larger herd sizes, increased animal movement and trade, and the lack of effective control measures, such as restrictions on whole herd depopulation. However, additional tests may be necessary to enhance disease control, particularly in the latter stages of eradication programmes when the incidence of falsely reactive animals to skin tests is higher.

In conclusion, an integrated strategy including currently available tests such as bacteriological, molecular, histopathological, and immune-mediated tests must be used in a modern strategy for the diagnosis and control of BTB.

References

1. Rahman MT, Sobur MA, Islam MS, Ievy S, Hossain MJ, El Zowalaty ME, et al. Zoonotic diseases: etiology, impact, and control. Microorganisms. 2020;8(9):1405.
2. Chaber AL. The era of human-induced diseases. Ecohealth. 2018;15(1):8–11.
3. Luciano SA, Roess A. Human zoonotic tuberculosis and livestock exposure in low- and middle-income countries: a systematic review identifying challenges in laboratory diagnosis. Zoonoses Public Health. 2020;67(2):97–111.

4. Thomas J, Balseiro A, Gortázar C, Risalde MA. Diagnosis of tuberculosis in wildlife: a systematic review. Vet Res. 2021;52(1):31.
5. Radostits OM, Blood DC, Gay CC, Arundel JH. Veterinary medicine: a textbook of the diseases of cattle, sheep, pigs, goats and horses. Paris: Baillière Tindall; 1994. p. 1763.
6. Shitaye JE, Tsegaye W, Pavlik I. Bovine tuberculosis infection in animal and human populations in Ethiopia: a review. Vet Med. 2007;52(8):317–32.
7. Quadri NS, Brihn A, Shah JA, Kirsch JD. Bovine tuberculosis: a re-emerging zoonotic infection. J Agromedicine. 2021;26(3):334–9.
8. Kruse H, Kirkemo AM, Handeland K. Wildlife as source of zoonotic infections. Emerg Infect Dis. 2004;10(12):2067–72.
9. Backues KA. Chapter 36. Mycobacterial diseases in carnivores. In: Fowler ME, Miller RE, editors. Zoo and wild animal medicine. 6th ed. Saint Louis: W.B. Saunders; 2008. p. 294–8. https://www.sciencedirect.com/science/article/pii/B9781416040477500397. Accessed 20 Apr 2023.
10. Collins ÁB, Floyd S, Gordon SV, More SJ. Prevalence of Mycobacterium bovis in milk on dairy cattle farms: an international systematic literature review and meta-analysis. Tuberculosis (Edinb). 2022;132:102166.
11. Cuervo-Soto LI, López-Pazos SA, Batista-García RA, Cuervo-Soto LI, López-Pazos SA, Batista-García RA. Metagenomics and diagnosis of zoonotic diseases. Farm animals diseases, recent Omic trends and new strategies of treatment. London: IntechOpen; 2018. https://www.intechopen.com/chapters/58730. Accessed 20 Apr 2023.
12. Teppawar RN, Chaudhari SP, Moon SL, Shinde SV, Khan WA, Patil AR, et al. Zoonotic tuberculosis: a concern and strategies to combat. In: Basic biology and applications of actinobacteria. London: IntechOpen; 2018. https://www.intechopen.com/chapters/61347. Accessed 20 Apr 2023.
13. Une Y, Mori T. Tuberculosis as a zoonosis from a veterinary perspective. Comp Immunol Microbiol Infect Dis. 2007;30(5):415–25.
14. Grange JM, Collins CH. Bovine tubercle bacilli and disease in animals and man. Epidemiol Infect. 1987;99(2):221–34.
15. Thakkar K, Ghaisas SM, Singh M. Lymphadenopathy: differentiation between tuberculosis and other non-tuberculosis causes like follicular lymphoma. Front Public Health. 2016;4. https://doi.org/10.3389/fpubh.2016.00031.
16. Thoen C, Lobue P, de Kantor I. The importance of Mycobacterium bovis as a zoonosis. Vet Microbiol. 2006;112(2–4):339–45.
17. Smith RMM, Drobniewski F, Gibson A, Montague JDE, Logan MN, Hunt D, et al. Mycobacterium bovis infection, United Kingdom. Emerg Infect Dis. 2004;10(3):539–41.
18. Ali M. Bovine tuberculosis; 2013. https://doi.org/10.13140/RG.2.2.10600.55045.
19. Ramos DF, Silva PEA, Dellagostin OA. Diagnosis of bovine tuberculosis: review of main techniques. Braz J Biol. 2015;75(4):830–7.
20. Cousins DV. Mycobacterium bovis infection and control in domestic livestock. Rev Sci Tech. 2001;20(1):71–85.
21. Rimal R, Shrestha D, Pyakurel S, Poudel R, Shrestha P, Rai KR, et al. Diagnostic performance of GeneXpert MTB/RIF in detecting MTB in smear-negative presumptive TB patients. BMC Infect Dis. 2022;22(1):321.
22. Chopra S, Mahajan S, Singh Y, Chopra M. Comparative evaluation of Ziehl- Neelsen staining and kinyoun's staining in the diagnosis of clinically suspected cases of tuberculosis. IP Int J Med Microbiol Trop Dis. 2022;8(2):149–53.
23. Bansal R, Sharma PK, Jaryal SC, Gupta PK, Kumar D. Comparison of sensitivity and specificity of ZN and fluorescent stain microscopy with culture as gold standard. J Tuberc Res. 2017;5(2):118–28.
24. Balakrishna J, Shahapur P, Chakradhar P, Hussain S. Comparative study of different staining techniques-Ziehlneelsen stain, Gabbet's stain, Fluorochrome stain for detecting of mycobacterium tuberculosis in the sputum. J Pharm Sci Res. 2013;5(4):89–92.

25. Cattamanchi A, Davis JL, Worodria W, den Boon S, Yoo S, Matovu J, et al. Sensitivity and specificity of fluorescence microscopy for diagnosing pulmonary tuberculosis in a high HIV prevalence setting. Int J Tuberc Lung Dis. 2009;13(9):1130–6.
26. Rodriguez-Morales AJ, Castañeda-Hernández DM. Bacteria: Mycobacterium bovis. In: Reference module in food science. Amsterdam: Elsevier; 2019. p. B9780081005965226000. https://linkinghub.elsevier.com/retrieve/pii/B9780081005965226401. Accessed 20 Apr 2023.
27. Abdelsadek HA, Sobhy HM, Mohamed KF, Hekal SHA, Dapgh AN, Hakim AS. Multidrug-resistant strains of Mycobacterium complex species in Egyptian farm animals, veterinarians, and farm and abattoir workers. Vet World. 2020;13(10):2150–5.
28. Niu J, Wang D, Yan M, Chang Z, Xu Y, Sizhu S, et al. Isolation, identification and biological characteristics of mycoplasma bovis in yaks. Microb Pathog. 2021;150:104691.
29. Sorlozano A, Soria I, Roman J, Huertas P, Soto MJ, Piedrola G, et al. Comparative evaluation of three culture methods for the isolation of mycobacteria from clinical samples. J Microbiol Biotechnol. 2009;19(10):1259–64.
30. Kassaza K, Orikiriza P, Llosa A, Bazira J, Nyehangane D, Page AL, et al. Lowenstein-Jensen selective medium for reducing contamination in mycobacterium tuberculosis culture. J Clin Microbiol. 2014;52(7):2671–3.
31. Essawy TS, Saeed AM, Fouad NA. Comparative study between using Lowenstein Jensen, bio-FM media and mycobacteria growth indicator tube (MGIT) system in identification of mycobacterium tuberculosis. Egypt J Chest Dis Tuberc. 2014;63(2):377–84.
32. Sharp SE, Lemes M, Sierra SG, Poniecka A, Poppiti RJ, Löwenstein-Jensen media. No longer necessary for mycobacterial isolation. Am J Clin Pathol. 2000;113(6):770–3.
33. Siddiqi SH, Rusch-Gerdes S. Procedure manual for BACTEC MGIT 960 TB system; 2006.
34. Diriba G, Kebede A, Yaregal Z, Getahun M, Tadesse M, Meaza A, et al. Performance of Mycobacterium Growth Indicator Tube BACTEC 960 with Lowenstein–Jensen method for diagnosis of Mycobacterium tuberculosis at Ethiopian National Tuberculosis Reference Laboratory, Addis Ababa, Ethiopia. BMC Res Notes. 2017;10(1):181.
35. Robbe-Austerman S, Bravo DM, Harris B. Comparison of the MGIT 960, BACTEC 460 TB and solid media for isolation of Mycobacterium bovis in United States veterinary specimens. BMC Vet Res. 2013;9(1):74.
36. Kumari P, Thakur J, Kumar P, Kumar R, Parekh D. Comparison of LJ medium and BACTEC MGIT 960 culture system for the diagnosis of tuberculosis. J Clin Diagn Res. 2020;14:DC09.
37. Medeiros L, Marassi CD, Figueiredo EES, Lilenbaum W. Potential application of new diagnostic methods for controlling bovine tuberculosis in Brazil. Braz J Microbiol. 2010;41(3):531–41.
38. Biet F, Boschiroli ML, Thorel MF, Guilloteau LA. Zoonotic aspects of Mycobacterium bovis and Mycobacterium avium-intracellulare complex (MAC). Vet Res. 2005;36(3):411–36.
39. Marais BJ, Brittle W, Painczyk K, Hesseling AC, Beyers N, Wasserman E, et al. Use of light-emitting diode fluorescence microscopy to detect acid-fast bacilli in sputum. Clin Infect Dis. 2008;47(2):203–7.
40. Corner LA. Post mortem diagnosis of Mycobacterium bovis infection in cattle. Vet Microbiol. 1994;40(1–2):53–63.
41. Schiller I, Oesch B, Vordermeier HM, Palmer MV, Harris BN, Orloski KA, et al. Bovine tuberculosis: a review of current and emerging diagnostic techniques in view of their relevance for disease control and eradication. Transbound Emerg Dis. 2010;57(4):205–20.
42. Pal M, Berhanu G, Feyisa D, Mideksa B, Kandi V. Bovine tuberculosis: a review of molecular diagnostic methods and impact on public health. Am J Microbiol Res. 2021;9(1):1–8.
43. Goosen W, Kerr T, Kleynhans L, Buss P, Cooper D, Warren R, et al. The VetMAX™ M. tuberculosis complex PCR kit detects MTBC DNA in antemortem and postmortem samples from white rhinoceros (Ceratotherium simum), African elephants (Loxodonta africana) and African buffaloes (Syncerus caffer). BMC Vet Res. 2020;16:220.
44. Chakravorty S, Simmons AM, Rowneki M, Parmar H, Cao Y, Ryan J, et al. The new Xpert MTB/RIF ultra: improving detection of mycobacterium tuberculosis and resistance to rifampin in an assay suitable for point-of-care testing. MBio. 2017;8(4):e00812–7.

45. Bernitz N, Kerr TJ, Goosen WJ, Chileshe J, Higgitt RL, Roos EO, et al. Review of diagnostic tests for detection of Mycobacterium bovis infection in south African wildlife. Front Vet Sci. 2021;8:588697. https://doi.org/10.3389/fvets.2021.588697.
46. van Soolingen D, Hermans PW, de Haas PE, van Embden JD. Insertion element IS1081-associated restriction fragment length polymorphisms in mycobacterium tuberculosis complex species: a reliable tool for recognizing Mycobacterium bovis BCG. J Clin Microbiol. 1992;30(7):1772–7.
47. van Soolingen D, de Haas PE, Hermans PW, Groenen PM, van Embden JD. Comparison of various repetitive DNA elements as genetic markers for strain differentiation and epidemiology of mycobacterium tuberculosis. J Clin Microbiol. 1993;31(8):1987–95.
48. Supply P, Allix C, Lesjean S, Cardoso-Oelemann M, Rüsch-Gerdes S, Willery E, et al. Proposal for standardization of optimized mycobacterial interspersed repetitive unit-variable-number tandem repeat typing of mycobacterium tuberculosis. J Clin Microbiol. 2006;44(12):4498–510.
49. Darsanaki R, Azizzadeh A, Nourbakhsh M, Raeisi G, Aliabadi M. Biosensors: functions and applications. J Biol Today World. 2013;2:53–61.
50. Swift BMC, Barron ES, Christley R, Corbetta D, Grau-Roma L, Jewell C, et al. Tuberculosis in badgers where the bovine tuberculosis epidemic is expanding in cattle in England. Sci Rep. 2021;11(1):20995.
51. Driscoll JR. Spoligotyping for molecular epidemiology of the mycobacterium tuberculosis complex. In: Caugant DA, editor. Molecular epidemiology of microorganisms: methods and protocols. Totowa, NJ: Humana Press; 2009. p. 117–28. (Methods in molecular biology™). https://doi.org/10.1007/978-1-60327-999-4_10.
52. Hermans PW, van Soolingen D, Bik EM, de Haas PE, Dale JW, van Embden JD. Insertion element IS987 from Mycobacterium bovis BCG IS located in a hot-spot integration region for insertion elements in mycobacterium tuberculosis complex strains. Infect Immun. 1991;59(8):2695–705.
53. van der Zanden AGM, Kremer K, Schouls LM, Caimi K, Cataldi A, Hulleman A, et al. Improvement of differentiation and interpretability of spoligotyping for mycobacterium tuberculosis complex isolates by introduction of new spacer oligonucleotides. J Clin Microbiol. 2002;40(12):4628–39.
54. Dale J, Brittain D, Cataldi A, Cousins D, Crawford JT, Driscoll J, et al. Spacer oligonucleotide typing of bacteria of the mycobacterium tuberculosis complex: recommendations for standardised nomenclature. Int J Tuberc Lung Dis. 2001;5:216–9.
55. Kamerbeek J, Schouls L, Kolk A, van Agterveld M, van Soolingen D, Kuijper S, et al. Simultaneous detection and strain differentiation of mycobacterium tuberculosis for diagnosis and epidemiology. J Clin Microbiol. 1997;35(4):907–14.
56. Cronin WA, Golub JE, Magder LS, Baruch NG, Lathan MJ, Mukasa LN, et al. Epidemiologic usefulness of spoligotyping for secondary typing of mycobacterium tuberculosis isolates with low copy numbers of IS6110. J Clin Microbiol. 2001;39(10):3709–11.
57. Demay C, Liens B, Burguière T, Hill V, Couvin D, Millet J, et al. SITVITWEB—a publicly available international multimarker database for studying mycobacterium tuberculosis genetic diversity and molecular epidemiology. Infect Genet Evol. 2012;12(4):755–66.
58. Slatko BE, Gardner AF, Ausubel FM. Overview of next-generation sequencing technologies. Curr Protoc Mol Biol. 2018;122(1):e59.
59. Ekblom R, Galindo J. Applications of next generation sequencing in molecular ecology of non-model organisms. Heredity. 2011;107(1):1–15.
60. Doherty ML, Monaghan ML, Bassett HF, Quinn PJ. Effect of a recent injection of purified protein derivative on diagnostic tests for tuberculosis in cattle infected with Mycobacterium bovis. Res Vet Sci. 1995;58(3):217–21.
61. Monaghan ML, Doherty ML, Collins JD, Kazda JF, Quinn PJ. The tuberculin test. Vet Microbiol. 1994;40(1):111–24.
62. Bakker D, Eger A, McNair J, Riepema KH, Willemsen PTJ. Comparison of commercially available PPD s: practical considerations for diagnosis and control of bovine tuberculosis.

2008. https://research.wur.nl/en/publications/comparison-of-commercially-available-ppd-s-practical-consideratio. Accessed 20 Apr 2023.
63. Matulis G, Jüni P, Villiger PM, Gadola SD. Detection of latent tuberculosis in immunosuppressed patients with autoimmune diseases: performance of a mycobacterium tuberculosis antigen-specific interferon γ assay. Ann Rheum Dis. 2008;67(1):84–90.
64. de la Rua-Domenech R, Goodchild AT, Vordermeier HM, Hewinson RG, Christiansen KH, Clifton-Hadley RS. Ante mortem diagnosis of tuberculosis in cattle: a review of the tuberculin tests, γ-interferon assay and other ancillary diagnostic techniques. Res Vet Sci. 2006;81(2):190–210.
65. Wood PR, Corner LA, Plackett P. Development of a simple, rapid in vitro cellular assay for bovine tuberculosis based on the production of γ interferon. Res Vet Sci. 1990;49(1):46–9.
66. Li X, Xia A, Xu Z, Liu J, Fu S, Cao Z, et al. Development and evaluation of a Mycobacterium bovis interferon-γ enzyme-linked immunospot (ELISpot) assay for detection of bovine tuberculosis. J Dairy Sci. 2022;105(7):6021–9.
67. Sparks K, Ballow M. The indirect ELISA for quantitation of specific antibody: analysis of antibody dilution curves. Diagn Immunol. 1983;1(4):269–75.
68. Wiker HG. MPB70 and MPB83—major antigens of Mycobacterium bovis. Scand J Immunol. 2009;69(6):492–9.
69. Villafañe L, Forrellad MA, López MG, Garbaccio S, Garro C, Rocha RV, et al. Production of Mycobacterium bovis antigens included in recombinant occlusion bodies of Baculovirus. J Mol Microbiol Biotechnol. 2019;29(1–6):83–90.
70. McNair J, Corbett D, Girvin R, Mackie D, Pollock J. Characterization of the early antibody response in bovine tuberculosis: MPB83 is an early target with diagnostic potential. Scand J Immunol. 2001;53:365–71.
71. Songthammanuphap S, Puthong S, Pongma C, Buakeaw A, Prammananan T, Warit S, et al. Detection of mycobacterium tuberculosis complex infection in Asian elephants (Elephas maximus) using an interferon gamma release assay in a captive elephant herd. Sci Rep. 2020;10(1):14551.

Correction to: Diagnosis of Mycobacterium

Amit Singh and Divakar Sharma

Correction to: A. Singh, D. Sharma (eds.), *Diagnosis of Mycobacterium*, https://doi.org/10.1007/978-981-99-5624-1

The original version of this book was inadvertently published with errors.

1. Chapter 6 was published before incorporation of the final corrections. The chapter has been updated with author's corrections. The basic facts have not been changed.

2. Chapters 10 and 14 were published with incorrect affiliation of the author "Anil Kumar Gupta", which has been corrected to "IRCH, All India Institute of Medical Sciences, New Delhi, India".

The above errors have been corrected with this erratum.

The updated version of the chapters can be found at
https://doi.org/10.1007/978-981-99-5624-1_6
https://doi.org/10.1007/978-981-99-5624-1_10
https://doi.org/10.1007/978-981-99-5624-1_14

A. Singh, D. Sharma (eds.), *Diagnosis of Mycobacterium*,
https://doi.org/10.1007/978-981-99-5624-1_15

SPRINGER NATURE

GPSR Compliance

The European Union's (EU) General Product Safety Regulation (GPSR) is a set of rules that requires consumer products to be safe and our obligations to ensure this.

If you have any concerns about our products, you can contact us on ProductSafety@springernature.com

In case Publisher is established outside the EU, the EU authorized representative is:

Springer Nature Customer Service Center GmbH
Europaplatz 3
69115 Heidelberg, Germany

The manufacturer's authorised representative in the EU is Springer Nature Customer Service Centre GmbH, Europaplatz 3, 69115 Heidelberg, Germany. If you have any concerns regarding our products, please contact ProductSafety@springernature.com

Printed and bound by CPI Group (UK) Ltd, Croydon, CR0 4YY
01/10/2026
02225867-0008